QUALITY ASSURANCE OF POSTHARVEST STORED PRODUCTS

QUALITY ASSURANCE OF POSTHARVEST STORED PRODUCTS

Shannon Chapelle
K C Bose
Nicole Vogle

Kruger Brentt
Publishers
2023

Kruger Brentt Publishers UK. LTD.
Company Number 9728962

Regd. Office: 68 St Margarets Road, Edgware, Middlesex HA8 9UU

ISBN: 9781787150546

For information on all our publications visit our website at http://krugerbrentt.com/

PREFACE

Quality can be defined as the "fitness for the purpose." The "optimal" quality of a product is related to a determined development or ripening degree, where the composition or combination of physical attributes and chemical components has maximum acceptance by consumers. The primary role of an effective post-harvest handling system is ensuring that the harvested product reaches the consumer, while fulfilling market/consumer expectations in terms of volume, quality, and other product and transaction attributes, including nutrition, food security, and product safety. The aim of this review was how to manage postharvest losses of crops for food security. Biological (internal) causes of deterioration include respiration rate, ethylene production and action, rates of compositional changes (associated with color, texture, flavor, and nutritive value), mechanical injuries, water stress, sprouting and rooting, physiological disorders, and pathological breakdown. The rate of biological deterioration depends on several environmental (external) factors, including temperature, relative humidity, air velocity, and atmospheric composition (concentrations of oxygen, carbon dioxide, and ethylene), and sanitation procedures. Postharvest losses of crops commodities were managed or controlled by doing proper harvesting, transportation, packing, storage, processing, sorting and cleaning. Thus, reduction of post-harvest food losses is a critical component of ensuring future global food security.

The degree of governmental controls, especially on wholesale and retail prices of fresh fruits and vegetables, varies from one country to another. In many cases, price controls are counter-productive. Although intended for consumer protection, such regulations encourage fraud and provide no incentive for producing high-quality produce or for postharvest quality maintenance. On the other hand, regulations covering proper handling procedures and public health aspects (food safety issues) during marketing are, if enforced properly, very important to the consumer. Systematic sorting or grading coupled with appropriate packaging and storage, will extend shelf life, maintain wholesomeness, freshness, and quality, and substantially reduce losses and marketing costs. Sorting is done to separate poor produce from good produce, and further classify the good produce based on other quality parameters like size. Only crops with high initial quality can be stored successfully; it is therefore essential to ensure that only crops of the highest quality (mature, undamaged) are stored. Shelf life can be extended by maintaining a commodity at its optimal temperature, relative humidity and environmental conditions.

The present book contains fourteen chapters covering all related disciplines. These chapters include Introduction, Factors affecting the post-harvest quality of fruits, Microbial spoilage of fruits and vegetables, Post-harvest physiology and biochemistry, Biotic and abiotic factors affecting maturity and quality, Nutritional composition and antioxidant properties of fresh produce, Vase life improvement of flowers and foliage, Post-harvest management of spice, Medicinal, and plantation crops, Maturity indices, Harvesting and post-harvest handling of fruits and vegetables, Pre-harvest factors affecting quality on post-harvest life of fruits and vegetables, Post-harvest diseases and physiological disorders, Post-harvest operations: grading, ripening, De-greening, curing etc., Advances in post-harvest technologies to enhance shelf life of crops, Quarantine and phytosanitory treatments for fresh produce. This book provides an overarching account of current technologies to enhance shelf life of fruits and vegetables and to maintain their quality during storage, packing and transportation. It is written in a lucid style with theoretical and practical discussions and being supplemented with illustrations, and tables for easy understanding of the subject. Provides appropriate approaches, technologies, and control parameters necessary to achieve shelf-life extension without compromising produce quality The book is also supported with a comprehensive bibliography and index in the end. This text is designed for professionals working in the postharvest industry, postharvest technologists, fresh produce management, food safety and quality professionals, postharvest physiologists, students and researchers.

We are grateful to all those persons as well as various books, manuals, periodicals, magazines, journals etc. that helped in the preparation of this book. In spite of the best efforts, it is possible that some errors may have occurred into the compilation and editing of the book. Further queries, constructive suggestions and criticisms for the improvement of the book are always welcome and shall be thankfully acknowledged.

Shannon Chapelle
K C Bose
Nicole Vogle

CONTENTS

	Preface	**v**
1.	**Introduction**	**1**
	⦿ Definition of the post-harvest system	1
	⦿ Factors affecting postharvest quality	2
	⦿ Genetic/variety	2
	⦿ Cultural operations	3
	⦿ Mineral nutrition	4
	⦿ Pest and diseases	5
	⦿ Post harvest factors affecting post harvest quality of fruits	5
	⦿ Sorting and grading	7
	⦿ Packaging and packaging materials	7
	⦿ Storage	7
	⦿ Transportation	8
	⦿ Challenges in post-harvest management	9
	⦿ Effects of post-harvest management	9
	⦿ Importance of post-harvest management	10
2.	**Factors affecting the Postharvest Quality of Fruits**	**1**
	⦿ Pre-harvest factors	2
	⦿ Post-harvest factors	6
	⦿ Sweet cherry (*prunus avium* l.)	7
	⦿ Sweet cherry quality factors	7
	⦿ Health promoting compounds	10
	⦿ New frontiers for sweet cherry quality improvements	16
	⦿ Preharvest factors affecting the postharvest quality of tomatoes	19
	⦿ Postharvest factors affecting the postharvest quality of tomatoes	22

- Postharvest calcium chloride application 23
- Conclusion 24

3. Microbial Spoilage of Fruits and Vegetables 25

- Introduction 25
- Microorganisms associated with spoilage in fruits and juices 26
- Microbial spoilage of vegetables 29
- Microbiological spoilage of fresh-cut fruits and vegetables 30
- Unusual characteristics of spoilage microorganisms 34

4 Postharvest Physiology and Biochemistry 39

- Physical treatments 42
- Chemical treatments 45
- Gaseous treatments 47
- The detrimental effects of CA include: 51

5. Biotic and Abiotic Factors Affecting Maturity and Quality 57

- Introduction 57
- Effects of these factors 57
- Both biotic and abiotic factors have their own effects on crop 58
- Drought–pathogen stress combination: a model for understanding combined abiotic–biotic stresses 59
- Potential traits for screening genotypes for tolerance to combined drought and pathogen infection 60
- Genetic engineering strategies for biotic and abiotic stress tolerance and quality enhancement in crops 67
- Biotic stress management through transgenic approach 68
- Disease resistance 71
- Virus resistance 76
- Abiotic stress management through transgenic approach 77
- Drought tolerance 78
- Heat tolerance 79
- Salinity tolerance 80
- Herbicide tolerance 82
- Enhancing fruit quality for increased shelf-life and reduced post-harvest losses 82
- Color enhancement and increasing vase life in various ornamental crops 84
- Modification of plant architecture 85
- Marker-free transgenic technology 86
- Genome editing technology in horticultural crop improvement 88
- Conclusion 89

6. Nutritional Composition and Antioxidant Properties of Fresh Produce **91**

- Types of losses 93
- Post harvest losses in horticultural crops 93
- Areas for research and development 102
- Maturity indices, harvesting and post harvest handling of fruits and vegetables 103
- Factors affecting maturity 105
- Maturity of fruits and vegetables 105

7. Vase Life Improvement of Flowers and Foliage **111**

- Introduction 111
- Factors affecting vase life 112
- Ethylene and flower longevity 114
- Metabolic changes in cut flowers 115
- Role of carbohydrates in vase solution 117
- Mechanism of action of compost tea 120
- Organic substances used to extend vase life 121
- Improving ornamental's vase life through molecular approaches 123

8. Post-harvest Management of Spice, Medicinal, and Plantation Crops **129**

- Importance of Post Harvest Technology 129
- Need for Post harvest technology 130
- Post harvest handling of spices 131
- Harvesting and post-harvest processing of medicinal plants 141
- Post Harvest Technology of Plantation Crop 147
- Post Harvest Technology of Nuts 169

9. Maturity Indices, Harvesting and Post-harvest Handling of Fruits and Vegetables **179**

- Introduction 179
- Fruits and vegetable nutritive values 179
- FV quality and safety aspects 180
- Maturity 180
- Computational methods 182
- Factors affecting maturity 183
- Maturity Index 184
- Technologies and practices to reduce postharvest losses 185

10. Pre-Harvest Factors affecting Quality on Post-harvest Life of Fruits and Vegetables **187**

- Pre-Harvest Factors 187
- Quality of post harvest product 192

11. Postharvest Diseases and Physiological Disorders **193**

- Causes of postharvest disease 193
- Economical and health impact of postharvest diseases 194
- Challenges of postharvest losses in crops 195
- Detection of postharvest pathogens 195
- Physiological Disorders 195

12. Post-harvest Operations: Grading, Ripening, De-greening, Curing etc. **215**

- Sorting, sizing and grading 215
- Grading 216
- Ripening 217
- Degreening 229
- Postharvest Handling Systems: Underground vegetables (roots, tubers and bulbs) 229
- Harvesting 230
- Postharvest Procedures 231

13. Advances in Post-harvest Technologies to Enhance Shelf Life of Crops **235**

- Tree fruits 237
- Vine fruits 239
- Small fruits 239
- Maturity and harvesting indices 240
- Fruit quality 241
- Challenges of postharvest technology 242
- Technologies to extend the shelf life of crops 245

14. Quarantine and Phytosanitory Treatments for Fresh Produce **249**

- Quarantine treatments 249
- Cold storage 249
- Heat treatments 250
- Atmosphere 250
- Fumigation 251
- Irradiation 252
- Generic phytosanitary irradiation treatments used commercially 253
- Effect of phytosanitary irradiation on fruits and vegetables 255

Bibliography **257**

Index **275**

1

INTRODUCTION

DEFINITION OF THE POST-HARVEST SYSTEM

"Properly considered," Spurgeon says, "the post-harvest system should be thought of as encompassing the delivery of a crop from the time and place of harvest to the time and place of consumption, with minimum loss, maximum efficiency and maximum return for all involved" (The Hidden Harvest, 1976).

The term "system" denotes a dynamic, complex aggregate of logically interconnected functions or operations within a particular sphere of activity. The term "chain" or "pipeline" highlights the functional succession of various operations but tends to ignore their complex interaction.

Fruits and vegetables are a substantial portion of the food supply chain and are depended on globally for good health and nutrition. The supply of good food, however, greatly depends on good postharvest handling practices. Although substantial research has been carried out to preserve the quality of fresh horticultural produce, further research, particularly on safety, is still required.

Temperature management and dehydration control are the two most important strategies to extend shelf life and retain the quality of those fruits and vegetables. Other supplements, such as a controlled and modified atmosphere, 1-methylcyclopropene, edible coatings, fumigation, and heat treatment can further enhance their storability. On the other hand, due to growing concerns about food safety and environmental issues, a major goal for postharvest handling of tropical or subtropical and temperate horticultural should emphasise both the effective preservation of the quality attributes, and the use of appropriate technologies considered safe with low or no adverse environmental impact.

Quality can be described as "fitness for purpose." A product's "ideal" quality is linked to a determined degree of growth or maturation, where the structure or mixture of physical characteristics and chemical components has highest consumer acceptance. A key concept in maintaining quality features and extending shelf life is the understanding that distinct pieces of fruit and vegetables are still alive, not only when still connected to the mother plant, but also after harvest. Quality assessment must be carried out in all sections as the sector must have understanding of the real value of the item and its conservation ability based on pre-established norms until commercialization.

Consumers may not accept a product unless it has the demands or required quality attributes that can have a significant effect on the marketing chain, particularly on exports. The quality of a product can be assessed, whether destructive or not, through a sensory panel or using tools. Quality specifications can be categorized into three categories: sensory, dietary, and safety, and should be regarded together in order to meet the needs of customers and public health. The relative significance provided to a particular quality attribute differs depending on the commodity involved and the person (producers, consumers and handlers) or market involved with quality assessment. High returns, excellent looks, easy harvesting and the capacity to sustain long-distance shipping to markets are significant quality characteristics for manufacturers.

From the point of perspective of wholesale and retail marketers, appearance, firmness, and shelf life are essential. Consumers, on the other side, judge on appearance (including ' freshness ') at the moment of original purchase the quality of products, such as new fruits, ornamentals, and vegetables.

FACTORS AFFECTING POSTHARVEST QUALITY

Postharvest management starts with pre-harvest managements. Once the fruits are harvested, the overall quality of fresh fruits can hardly be improved but it can be maintained. The final market value of the produce and acceptance by the consumers depends upon the grower's ability to apply best available pre-harvest technology followed by harvesting and then to apply best available postharvest handling practices.

The pre-harvest factors influencing postharvest quality are frequency of irrigation, use of fertilizers, pest control, growth regulators, climatic conditions like wet and windy weather, natural climates such as hailing, high wind velocity, heavy rainfall, and tree conditions (age, training pruning, light penetration, etc), which influences overall fruit quality and suitability for storage by modifying physiology, chemical composition, and morphology of fruits. One such pre-harvest factor is spray of Gibberellic acid (10 ppm), if applied at color break stage, results in delay in color development and maintains firmness. This is important because it in extending harvesting period.

Factors affecting post-harvest quality of fruits and vegetable The most important factors affecting the post-harvest quality of fruits and vegetables are discussed below:

GENETIC/VARIETY

Varieties with shorter shelf-lives are usually susceptible to increased losses from post-harvest. Varieties with dense peel, elevated firmness, low breathing rate and low manufacturing levels of ethylene would normally have longer storage life. Cultivars that can resist marketing and distribution rigors will have lower post-harvest losses.

Varieties with low heat resistance and/or decay-causing pathogens can be stored well with minimal storage losses for longer length. Therefore, while growing horticultural plants, one must choose such varieties that inherently have excellent quality and potential for storage in relation to the potential for elevated yield and pest resistance.

Light

Light controls several physiological processes such as synthesis of chlorophyll, phototropism, respiration and stomatal opening. Duration, intensity and quality of light have an impact on the quality of harvest fruits and vegetables. Most products require elevated light intensity (3000-8000 F.C.). The carnation and chrysanthemum vase life is longer than small under elevated light intensity. In tomatoes, fruit leaf shading generated a deeper red colour during maturation than those subjected to light during maturation. The fruit side that has been subjected to sun will usually be firmer than the side that has not been exposed. Generally speaking, the reduced the light intensity, the reduced the plant tissue ascorbic acid content.

Temperature

Temperature influences all types of physiological and biochemical processes linked to crop development and yield. During field circumstances, the greater temperature reduces product life and quality. At high temperature, stored carbohydrates of fruits, vegetables and flowers are quickly depleted during respiration and plant respires at the faster rate. For instance, elevated temperature on and off the plant during the tomato fruiting season leads to rapid fruit ripening.

Orange cultivated in tropics t ends to have greater sugars and TSS than the sub-tropics grown. Tropical oranges, however, tend to be green and peel less readily, owing to the reduced diurnal temperature in the tropics.

Humidity

In some horticultural products, elevated moisture during the growing season outcomes in thin rind and enhanced size, which is more susceptible to elevated disease incidence during the post-harvest period. Humid atmosphere can cause fungal and bacterial diseases to develop, causing damage during storage and transportation. Damaged products rapidly remove water and emit a higher ethylene concentration than healthy ones. In crops with thin leaves or leaflets, low humidity can trigger browning of the leaf surface. High moisture can retain water–in a situation, carried pollutants can be absorbed more readily through the cuticles or stomata.

CULTURAL OPERATIONS

The capacity of leaf photosynthesis depends on the incidence of light, whereby the shaded parts of the canopy assimilate less and need more leaves than the well illuminated part for optimal fruit development. Growers can rely on a number of methods which directly or indirectly influence photosynthesis and sink activity (fruit growth). Among these, the most important are tree height, distance, fruit thinning, pruning, fertilization, and application of growth regulators, irrigation and phytosanitary control. Ringing (5 mm wide) the base of productive main branches of Sweet Orange trees, three weeks after anthesis, increases fruit retention by 38%, as compared to non-ringed trees. Heavy pruning diminishes leaf area, whole tree photosynthesis and translocation of photosynthates to fruits and roots, increasing the root/shoot ratio and favoring vegetative growth.

Climatic factors

Environmental factors such as light, CO_2, relative humidity, temperature and water availability are major direct or indirect constraints for plant photosynthesis. Environmental factors affect the content of bioactive compounds indirectly by giving the prerequisites for photosynthesis, and thereby providing energy or precursors of the synthesis of the bioactive compounds. Further, the syntheses of these compounds are also affected directly by various environmental factors. Abiotic conditions, i.e. soil fertility and water availability, vary from year to year and site to site, and can affect the level and quality of fruit after harvest. Increased exposure to light increases fruit size, total soluble solids and flesh firmness

MINERAL NUTRITION

In order to maintain plant growth and development, balanced application of all nutrient components is essential. The use of fertilizers on plants affects their rate of respiration after harvest. Certain elements' excess or deficiency may influence crop quality and post-harvest life.

a. **Nitrogen:** High N fertilization reduces pH life and quality of anthurium, cut flowers and many horticultural products while moderate to high K increases. K in water melon application tends to reduce the post-harvest respiration. High levels on N tend to decrease flavour, TSS, firmness and colour of the fruit and in stone fruits it increases physiological disorders and decrease fruit colour. Generally speaking, plants with elevated nitrogen concentrations typically have reduced quality keeping than those with reduced concentrations as. High nitrogen improves fruit breathing, thereby decreasing their storage life by quicker tissue decay.

b. **Phosphorous:** Phosphorous application minimizes weight loss, sprouting and rotting relative to lower implementation in bulb plants. Phosphorous diet may change some product's post-harvest physiology by influencing membrane lipid chemistry, membrane integrity, and breathing metabolism. The respiration rate of lowphosphorous fruits will be higher than that of high phosphorous fruits during storage.

c. **Potassium:** Potassic fertilizers enhance quality; their deficiency can lead to abnormal fruit and vegetable maturation. Potassium helps to reduce certain physiological storage illnesses, e.g. orange pitting of superficial rind.

d. **Calcium:** The fruit's storage capacity depends mainly on the Calcium level and is linked with the texture of the product. The greater concentrations of N, P and Mg and low concentrations of K and Bo result in fruit deficiency in calcium and decrease their storage life. Reduction in calcium absorption leads to lateral stem breakage of poinsettia. Treatment with calcium delays maturation, senescence, decreases sensitivity to chilling injury, increase firmness and reduces decay subsequent to storage in avocados and improves the quality.

e. Zn is known to be a vehicle for the transmission of ions across the tissue and to boost the fruit's Calcium content. Adequate supply of Bo enhances Calcium's mobility in leaves and fruits and consequently increases fruit firmness, TSS, organic acids and reduces drought spot incidence, bitter pit and cracking illnesses. And it gives resistance to illnesses. The incorporation of 4% Calcium into the centre lamella proto pectin forms a bond with the cell wall cellulose and thus delayed fruit softening.

Infused Calcium inhibits inner browning, delays breathing, and reduces endogenous substrates' metabolism. Apple's post-climacteric respiration reduced as peel calcium increased from 400 ppm to 1300 ppm. Calcium may reduce the endogenous substrate catabolism by limiting the diffusion of substrate from vacuole to the respiratory enzymes in the cytoplasm (limited membrane permeability). Flowers like carnation require 850 to 1200 g of water to create one gram of dry matter. By and large, <5% of water assimilated in the plant framework is used for the improvement of various plant segments. Dampness stress builds the rate of transpiration over the rate of ingestion and sporadic water system/dampness system leads natural products/vegetable splitting (potato and pomegranate breaking). More elevated amount of dampness stress influences both yield and quality by diminishing cell augmentation.

PEST AND DISEASES

Contamination by parasites, microscopic organisms, vermin and bugs decreases the life span just as customer acceptability. Tissue damage caused by them show wilting and produce ethylene leads to early senescence. Vascular diseases/stem rot /root rot of floral corps hinder the transport, affects the post-harvest life and quality. The potato tuber moth may infest tubers during growth if they are exposed above the soil and subsequently in the storage.

POST HARVEST FACTORS AFFECTING POST HARVEST QUALITY OF FRUITS

Maturity Stage

This is the starting point of postharvest quality management. Therefore, it must be ensured that properly matured fruits should be harvested. It must be harvested when it attains the appropriate stage of development based on physiological and horticultural maturity. Harvest maturity varies in accordance with the crop concerned. The fruit is harvested at different stages of maturity depending on how far the fruit will be transported, how long it will be kept in storage and the requirements for the specific market. Maturity always has a considerable influence on the quality of fresh produce as well as the storage potential and occurrence of many storage disorders. Maturity at harvest has a major impact on quality and postharvest life potential of fruits and vegetables. All fruits with a few exceptions avocados, bananas and pears reach their best quality stage when fully ripen on tree.

Methods of Harvesting

Selection of suitable method for harvesting of the produce is necessary otherwise bruises or injuries during harvesting May later manifest as black or brown patches making them unattractive. Latex coming out of stem in mango should not be allowed to fall on fruits as it creates a black spot. Injury to peel may become an entry point for microorganisms, causing rotting. Some harvesting gadgets have been developed, e.g. mango harvester (CISH).

There are basically three methods most commonly used for harvesting any fruits.

1. Harvesting individual fruits with hand by pulling or twisting the fruit pedicel
2. Harvesting individual fruits or fruit bunch with the help of fruit clippers/ secateurs/scissors
3. With harvester specially designed for harvesting

Time of Harvesting

Harvesting time also affects quality. Fruits harvested before 10 AM in the morning and transported to pack house for sorting, grading, and packing yield better quality and lasts longer. It is desirable that the fruits are harvested during the cooler parts of the day to reduce the risk of heat injury and sunburn. Therefore, morning harvesting and within 10 AM transportation to destination pack house or market is always preferred in order to control damage due to high temperature. In case of grapes harvesting in India, it starts at 6 o'clock in the morning and harvested produce reach pack house by 10 AM. It facilitates faster pre-cooling also and yield better quality.

Pre-cooling

The quality of fresh fruits largely depends on pre-cooling before storage and marketing. This is a compulsory postharvest treatment followed in developed countries for almost all perishable commodities. The rapid cooling of fresh produce from field temperature (pulp temperature at the time of harvesting) to its best storage temperature is called pre-cooling. It is an important postharvest operation recommended in almost all flowers, fruits, and few vegetables. Fruits and vegetables which require on farm precooling if transport time to reach them to cold storage is more than a few hours. It is desirable that fresh produce like grapes, mandarins, berries, cherries, leeches, melons, stone fruits, okra, tomatoes, capsicum, chili peppers, cucumbers, green beans, peas, spinach should be cooled as rapidly as possible. The main objective of any pre-cooling operation is to remove field temperature (field heat). This is important because it increases shell life of the produce. Removing field heat reduces rate of respiration and all biochemical reactions from newly harvested produce.

The act of cooling immediately after harvest is important to remove the field heat before the items are handled further. It is important that the cooling occurs as soon as possible after harvest. Delays in the pre-cooling will reduce the final quality and shorten the postharvest life. Pre-cooling to remove field heat as quickly as possible after harvest is essential for slowing down the rate of deterioration of highly perishable products. The method chosen is largely determined by the type of product in question and the cost to benefit ratio.

SORTING AND GRADING

This is one of the most important postharvest operations after harvesting. This is done primarily for quality packing and removal of diseased and defective produce from the lot. Proper sorting and grading gives assurance of quality produce. This is either done in the farmer's field or in the pack houses. Both manual and mechanical graders are used for grading. All round-shaped fruits and vegetables are easily graded by mechanical graders. Grading may be based on color, size, and extent of defects, while sorting is totally dependent on man power for removal of diseased, defected, and damaged fruits. Grading is done by simple to highly sophisticated graders. Today, many sophisticated graders are in use for fresh produce such as GREEFA. Both size and color grading simultaneously is possible and is being used on commercial scale in apples.

PACKAGING AND PACKAGING MATERIALS

Fruits are fragile products and therefore need packaging to protect them from mechanical damage. The packages should also be well ventilated. It is important to avoid compression damages on the fruit during storage and transportation. The packages should also hold a weight of maximum 20 kg, as the fruit can be damaged when a heavy box is dropped on top of another. When fruits are transported the main goal should be to have as low amount of losses as possible.

Packages used are big and heavy and result in significant fruit injury because of weight compression of upper fruits. Researchers also reported that most mangoes packed in wooden crates which apart from causing physical injuries and bruises during transit are being restricted in international markets on account of quarantine concerns and special disinfestations treatments necessary for international trade. Both packing and packaging materials play many important roles in quality maintenance of fresh produce. Packing starts with placing the produce in the box. While placing, care must be taken to place in line, pedicel end of all fruits should be in one direction, separation layers or trays must be used where it is necessary. The box should not be underfilled or overfilled.

STORAGE

These practices are supported by Liu who reported that in developing countries the common storage facilities are air-cooled common storage houses which rely on natural cold air.

Almost all fruits are seasonal in nature. Every year, harvesting season falls during a fixed period, say 2-3 months. This may be little early or late due to prevailing weather conditions during growing periods. The demand of any fruit beyond the harvesting season is called off-season demand. This demand can be fulfilled only if fruits are stored in the harvesting season and sold during off-season. The management of temperature, ventilation, and relative humidity are the three most important factors that affect postharvest quality and storage life of horticultural produce. Recommended storage temperature and relative humidity for cabbage, lettuce and carrots are 0 to 2 °C and

95 to 100% respectively. Recommended storage temperature and relative humidity for mangoes, avocados, papayas and potatoes are 13-15 °C and 85-90%, respectively.

TRANSPORTATION

According to the FAO transportation is a big and often the most important factor in the marketing of fresh produce. Ideally, transport would take produce from the grower directly to the consumer. Kader stated that in most developing countries, roads are not adequate for proper transport of horticultural crops. Also, transport vehicles and other modes, especially, those suited for fresh horticultural perishables are in short supply for local and export to other countries and the majority of producers have small holdings and cannot afford to own their own vehicles.

Transportation is done in two phases-namely, from the field to the homestead and from the home/company collection area to the market. Transportation for small scale farmers is relatively safe, because the product is either carried to the market or simply transported on carts or bicycles, rather than on trucks. However, for medium-scale farmers or groups of farmers, the transportation of produce is more complicated, and the produce is more susceptible to mechanical and heat damage. The fruits are either loaded onto trucks on wooden stacks, or simply piled onto the trucks.

Mechanical damage (fatigue) occurs during transportation because of vibrations that occur while traveling what are typically long distances, usually over untarred roads. High temperatures and the buildup of gases that accelerate enzyme activity (and thus cause over-ripening or softening) and microbial activity are factors that contribute to the deterioration of fruit harvests

Road Condition and Duration of Transportation

Both road condition and duration of transportation affect quality of fresh produce. In hilly tracks and rough road surface, more touching and bruising take place as compared to smooth surface. Longer duration during transportation also affects quality. Reefer van should not be hold unnecessary. It not only increases the cost of produce, but also affects quality.

Pattern of Loading

Pattern of loading also plays crucial role in maintaining quality of fresh produce. Here pattern of loading means number of packed boxes in one layer (stacking height). In case of fresh produce, stacking height depends on extent of perishable nature of packed commodities and strength of packing materials. If produce are more perishable or box strength is weak, stacking height is kept low and vice versa. For example, height of grape boxes is kept low or it is packed in five ply corrugated boxes or thermocol boxes.

This precaution must be taken to preserve postharvest quality of this highly perishable commodity. While loading, another important criteria is interlocking between the boxes. Loading and unloading fruits directly affect quality of fresh produce. It can be done either by hand or with the aid of a forklift. Forklift is used for palletized boxes and shipping containers only. Generally, fruits are stacked on pallets to ease the loading and unloading process and to prevent damage to the product and packages. Exposure to sun while awaiting loading at local mandis or transport can reduce quality drastically. The exposed portion turns black or brown and starts decaying. It is advised for non-reefer transport to move continuously while under sunlight and stop and park your vehicle under a tree shade, especially during sunny days.

CHALLENGES IN POST-HARVEST MANAGEMENT

Challenges for producers, traders, and handlers are always present. Here are some of the factors that can affect post-harvest management:

- Agricultural inputs and structures are hard to obtain. The lack of infrastructure, proper knowledge, and the right technology can hugely affect yield and, resulting from this, income. Investments on seeds, fertilizers, and other important pre-harvest inputs can go to waste if post-harvest handling is not done properly.
- For smallholder farmers, the lack of financial support can break their livelihood. A single year's poor harvest can amount to a loss of decades' worth of savings. This can be avoided by placing secure and reliable post-harvest systems. Financial support from both non-government organizations and government bodies can also give the difference needed by these farmers to have a successful, sustainable operation.
- Gender inequalities also play a role in agricultural challenges. Women farmers are often given less assistance and opportunities than what they actually need. Land ownership, as well as access to education, financial assistance, additional labor force, and infrastructure, are all challenges to the operation and livelihood of women farmers worldwide.
- Changing technlogy is a big factor for post-harvest management as well. International standards for fumigation, for example, has been drastically changed due to concerns such as overuse of toxic chemicals. These shifts in technology can be difficult for producers, handlers, and traders to keep up with, especially when tied with a lack of financial support.
- Another enormous challenge that the agricultural sector is facing is climate change. Adapting to the changes in weather and being prepared for natural disasters such as drought, flash floods, hurricanes, and many more can be difficult.

EFFECTS OF POST-HARVEST MANAGEMENT

Mismanagement during the post-harvest season can have serious effects. Post-harvest losses occur in large-scale, with almost a third of the world's food going to loss or waste. In Sub-saharan Africa alone, more than 333 million people were affected by food insecurity in 2016, according to World Hunger Organization.

At the same time, post-harvest loss greatly affects farmers' incomes. Losing a portion of their yield and the degrading quality of their commodities translates to a loss of their income. This further aggravates the impoverished status of these smallholder farmers, creating difficulties in funding their livelihood as well as in providing for their families' daily needs.

IMPORTANCE OF POST-HARVEST MANAGEMENT

Proper management of post-harvest systems can serve as a major help in resolving various social and economic issues. A significant decrease in post-harvest loss can alleviate food insecurity all over the world. Simultaneously, food safety can be ensured by protecting commodities from mold growth and contamination. This will result in better health and empower small communities, which can, in turn, alleviate poverty.

2

FACTORS AFFECTING THE POSTHARVEST QUALITY OF FRUITS

From many studies and field observations over the past 60 years, it has been reported that 40-50% of horticultural crops produced in developing countries are lost before they can be consumed, mainly because of high rates of bruising, water loss, and subsequent decay during postharvest handling. Nutritional loss (loss of vitamins, antioxidant, and health-promoting substances) or decreased market value is another important loss that occurs in fresh produce. Quality of fresh produce is governed by many factors. The combined effect of all decides the rate of deterioration and spoilage.

These factors, if not controlled properly, lead to postharvest losses on large scale. Approximately one third of all fresh fruits are lost before it reaches to the consumers. Another estimate suggests that about 30-40% of total fruits production is lost in between harvest and final consumption. Main causes of postharvest loss include lack of temperature management, rough handling, poor packaging material, and lack of education about the need to maintain quality.

Post harvest quality represents market quality, edible quality, transport quality, table quality, nutritional quality, internal quality and appearance quality. Quality means a combination of characteristics, attributes and properties that gives the values to human and enjoyments. Consumers consider good quality in relation to colour, flavour and nutrition. Quality of the produce is the final manifestation of inter-relation between the commodity and its environment.

The genetic characteristics and physiological status of the commodity determine the typical post-harvest behavior and quality of the produce and these two are the major bases for the interaction. Pre-harvest factors viz, environmental factors such as temperature, relative humidity, water potential, light, cultural practices and pest management techniques determined the inherent quality of the produce. However, the ultimate quality is the final manifestation of inter relation between the commodity and its environment. Several pre-harvest and post-harvest factors affect the quality of horticultural crops. Some of these factors are related to plant, others are related to environment or to cultural practices Several pre-harvest and post-harvest factors affect the quality and shelf life of horticultural crops. Some of these factors are related to plant, others are related to environment and cultural practices.

PRE-HARVEST FACTORS

Related to Plants

Type of fruit crop: Shelf life of the fruit are varies from crop to crop e.g. jackfruit, bael, apple, pumpkin, water melon having good shelf life while mango, cherry, strawberry, peach, plum, guava, tomato, etc having less shelf life.

Cultivars: The shelf life of fruits varies according to variety within a crop. Because several parameters of shelf life are controlled genetically. Varieties with shorter shelf-lives are generally prone to higher post harvest losses. Varieties with thick peel, high firmness, low respiration rate and low ethylene production rates would usually have longer storage life. The cultivars that have ability to withstand the rigors of marketing and distribution will have lesser losses after harvest. Varieties with resistance to low temperature disorders and/or decay-causing pathogens can be stored well for longer duration with minimum storage losses. Hence, while growing horticultural crops, one must choose such varieties that inherently have got good quality and storage potential in addition to the high yield and pest resistance potential. For example in Guava Lalit variety has less shelf life than L-49. Similarly in mango Mallika has more shelf life than Lucknow Safeda and Dashahari.

Planting density: It affects the quantity, quality and shelf life of the produce. High density planting increases competition between plants, reduces light availability, and thus may decrease quantity. Low density planting lead to large size, better colored fruit which may have shorter shelf life. Larger fruits are commonly more sensitive to physiological disorders.

Related to Intercultural Operations

Water relation and irrigation: Stress due to excessive or inadequate water in the medium reduces the longevity of the fruits. In general, <5 % of water absorbed in the plant system is utilized for the development of different plant components. Moisture stress increases the rate of transpiration over the rate of absorption and irregular irrigation/moisture regime leads fruit cracking (pomegranate cracking). Higher level of moisture stress affects both yield and quality by decreasing cell enlargement. Crops which have higher moisture content generally have poorer storage characteristics. An example of this is the hybrid onions, which tend to give high yield of bulbs with low dry matter content but which have only a very short storage life. If fully matured banana harvested soon after rainfall or irrigation the fruit can easily split during handling operations, allowing micro organism infection and post harvest rotting. If orange is too turgid at harvest (early morning) the flavdeo/oil gland in the skin can be ruptured during harvesting, releasing phenolic compounds and causes Oleocellosis or oil spotting (green spot on the yellow / orange coloured citrus fruit after degreening).

Mango hot water treatment is better if there is delay of 48 hr. between harvest and treatment and resulted better efficiency of hot water in disease control.

Generally, crops that have higher moisture content or low dry matter content have poorer storage characteristics. Keeping quality of bulb crops like onion and garlic will be poor if irrigation is not stopped before three weeks of harvesting.

Mineral nutrition: Balanced application of all nutrient elements is necessary for the maintaining growth and development of the plants. The application of fertilizers to crops influences their post harvest respiration rate. Excess or deficiency of certain elements can affect crop quality and its post harvest life. Numerous physiological disorders are also associated with mineral deficiencies which ultimately lead to post harvest losses.

Nitrogen: High nitrogen fertilization reduces while moderate to high potassium improves post harvest life and quality of pomegranate and many horticultural produce. Application of potassium in water melon tends to decrease the post harvest respiration. High levels of nitrogen tend to decrease flavor, TSS, firmness and color of the fruit and in stone fruits (peach, plum, apricot) it increases physiological disorders and decrease fruit colour.

Generally, crops that have high levels of nitrogen typically have poorer keeping qualities than those with lower levels. As high nitrogen increases fruit respiration, faster tissue deterioration thereby reducing their storage life.

Phosphorous: Phosphorous nutrition can alter the post harvest physiology of some produce by affecting membrane lipid chemistry, membrane integrity and respiratory metabolism. The respiration rate of low-phosphorous fruits will be higher than that of high phosphorous fruits during storage.

Potassium: Potassic fertilizers improves keeping quality, its deficiency can bring about abnormal ripening of fruits and vegetables. Potassium helps in reducing some physiological storage disorders, e.g. superficial rind pitting in oranges.

Calcium: the storage potential of the fruits is largely dependent on the level of calcium and it is associated with produce texture. The higher level of nitrogen, phosphorus and magnesium and low levels of potassium and boron lead to the Ca deficiency in fruits and reduce its storage life. Reduction in calcium uptake causes lateral stem breakage of poinsettia. Calcium treatment delays ripening, senescence, reduces susceptibility to chilling injury, increase firmness and reduces decay subsequent to storage in avocados and improves the quality.

Physiological disorders of storage organs related to low Ca content of the tissue are

- Bitter pit in apples
- Cork spot in pears
- Blossom end rot in tomato
- Red blotch of lemons

Zinc (Zn) is known to act as vehicle for carrying ions across tissue and increase Ca content of the fruit. Adequate supply of boron improves the mobility of Ca in the leaves and the fruits and subsequently increases fruit firmness, TSS, organic acids and reduce the incidence of the drought spot, bitter pit and cracking disorders and impart diseases resistance.

The incorporation of 4% Ca into proto pectin of middle lamella form bond with the cellulose of the cell wall and thus delayed softening in fruits.

Infused Ca inhibits the internal browning, retarded respiration, and reduced the metabolism of endogenous substrates. Post climacteric respiration of apple decreased as peel Ca level increased from 400 to 1300 ppm. Calcium may reduce the endogenous substrate catabolism by limiting the diffusion of substrate from vacuole to the respiratory enzymes in the cytoplasm (limited membrane permeability). Application of $CaCl_2$ delayed the accumulation of free sugars, decreased inorganic contents, mold development, softening and development of red colour in strawberry. In pears reduced cork spot, increased flesh firmness, total acidity and juiciness and in apple even after 90 day of storage at ambient condition shown acceptable quality.

Canopy Manipulation

- **Fruit thinning:** increases fruit size but reduces total yield. It helps in obtaining better quality produce.
- **Fruit position in the tree:** Fruits which are exposed to high light environment possesses higher TSS, acidity, fruit size, aroma, and shelf life compared to which lies inside the canopy. Hence better training system should be practiced to circulate optimum light and air. Eg.: Grapes, Mango, peaches, kiwifruits.
- **Girdling:** Increases the fruit size and advance and synchronized fruit maturity in peach, litchi and nectarines. Increases fruitfulness in many fruit tree species.
- **Use of Agrochemicals:** Use of chemicals on the plants to prevent the pathogen will have direct impact on extending the postharvest life. Generally, if produce has suffered an infection during development its storage or marketable life may be adversely affected. Banana which suffers a severe infection with diseases such as leaf spot may ripen pre maturely or abnormally after harvest and in mango it is rapid postharvest loss. Pre harvest application chemicals like MH on onion filed prevent them sprouting during storage.

Related to Rnvironment

Light: Light regulates several physiological processes like chlorophyll synthesis, phototropisum, respiration and stomatal opening. The duration, intensity and quality of light affect the quality of fruits and vegetables at harvest. Most of the produce needs high light intensity (3000-8000 f.c.). Absorption of red light (625-700 nm) through pigments, phytochrome, is essential for carbohydrates synthesis which determines the shelf life of the produce.

Citrus and mango fruits produced in full sun generally had a thinner skin, a lower weight, low juice content and lower acidity but a higher TSS. And citrus fruits grown in the shade may be less susceptible to chilling injury when subsequently stored in cold storage.

In tomatoes, leaf shading of fruits produced a deeper red colour during the ripening than in the case of those exposed to light. The side of the fruit that have been exposed to sun will generally firmer than the non exposed side. In general, the lower the light intensity the lower the ascorbic acid content of plant tissues. In leafy vegetables, leaves are larger and thinner under condition of low light intensity.

Humidity: High humidity during growing season results in thin rind and increased size in some horticultural produce and this produce is more prone to high incidence of disease during post harvest period. Humid atmosphere may cause the development of fungal and bacterial diseases, which damages produce during storage and transport. Damaged produce remove water very quickly and emit a larger concentration of ethylene than healthy ones. Low humidity may cause browning of leaf edge on plants with thin leaves or leaflets. High humidity can maintain the water – borne pollutants in a condition so that they can be more easily absorbed through the cuticles or stomata's. Reduced transpiration leads to calcium and other elemental deficiency.

Temperature: All type of physiological and biochemical process related to plant growth and yield are influenced by the temperature. The higher temperature during field conditions decreases life and quality of the produce. At high temperature, stored carbohydrates of fruits are quickly depleted during respiration and plant respires at the faster rate. The produce which is having higher amount of stored carbohydrates show longer storage. For Example: high temperature during fruiting season of tomato leads to quick ripening of fruits on and off the plant.

Orange grown in the tropics tend to have higher sugars and TSS than those grown sub tropics. However, tropical grown oranges tend to be green in colour and peel less easily and it is due to the lower diurnal temperature that occurs in the tropics.

Rainfall: Rainfall affects water supply to the plant and influences the composition of the harvested plant part. This affects its susceptibility to mechanical damage and decay during subsequent harvesting and handling operations. On the other hand, excess water supply to plants results in cracking of fruits such as cherries, plums, and tomatoes.

During Harvest Factors

Season: Quality of produce are greatly inf luenced by season e.g. Winter season harvest having more shelf life as compared to other season, while off season fruits and vegetables give more remunerative price. Harvesting during or immediately after rains should not be carried out since it creates most favourable conditions for multiplication of micro-organisms. Citrus fruits become susceptible to damage if harvested during rains as their rind becomes turgid and prone to easy bruising, sun-scald etc.

Time: Fruits and vegetables should always be harvested when temperature ismild. Because, higher temperature leads to faster respiration. Morning harvest of horticultural crop prefer for local market because they are fully fresh and turgid and having dew drop in this time. Evening harvesting is preferred for distant market due to higher accumulation of reserved carbohydrates and less amount of moisture which give the better quality of the produce to consumer.

Method of harvesting: Selection of suitable method for harvesting of the produce is necessary otherwise bruises or injuries during harvesting may later manifest as black or brown patches making them unattractive. Latex coming out of stem in mango should not be allowed to fall on fruits as it creates a black spot. Injury to peel may become an entry point for microorganisms, causing rotting. Some harvesting gadgets have been developed, e.g. mango harvester in Lucknow (CISH).

Stage of harvesting: Fruits must be harvested at right stage of maturity. A very common cause of poor product quality at harvest and rapid deterioration thereafter is harvesting immature fruits. Fruits harvested immature or over mature usually do not keep long. Fruits harvested too early lose water fast and are more susceptible to mechanical damage and microbial attack. An over mature fruit is more susceptible to decay, has passed its best eating quality, and deteriorates fast.

POST-HARVEST FACTORS

Pre-cooling: High temperatures are detrimental to keeping quality of fruits and vegetables, especially when harvesting is done during hot days. Pre-cooling is a means of removing the field heat. It slows down the rate of respiration, minimizes susceptibility to attack of micro-organisms, and reduces water loss.

Washing and drying: Most of the fruits and vegetables are washed after harvesting to improve their appearance, to prevent wilting and to remove primary inoculum load of microorganism. Hence, a fungicide/bactericide should be used in washing water. Washing, improves shelf life of bananas by delaying their ripening. After washing, excess of water should be removed which would otherwise encourage microbial spoilage.

Sorting and grading: Fruits and vegetables require sorting and grading for uniform packing at field level. Sorting is done on the basis of size and colour while grading practice is performed as per the defect or on the basis of marketable and unmarketable produce.

Disinfection: Papaya, mango, melon and other fruits are susceptible to fruit fly attack. Disinfection is done either by vapour heat treatment (VHT) at 43°C with saturated air with water vapour for 6-8 hr by Ethylene dibromide fumigation.

Waxing: Fruits and vegetables have a natural layer on their outer surface which is partly removed by washing. An extra discontinuous layer of wax applied artificially with sufficient thickness and consistency to prevent anaerobic condition within the fruits provides necessary protection against decay organism. Waxing also improves the appearance and glossiness, making them more acceptable.

Packing: It means more than carrying multiples of an object. Packing not only protects the horticultural produce but also makes a favourable impression on the buyers and May able to fetch higher income.

Delivery: Moving the harvest produce from the farm to the customer in good condition is important. All efforts up to delivery can be invalid if the fresh fruits and vegetables reach the destination in poor condition. Care should be taken to protect the produce and it becomes necessary when mixing load of fruits and vegetables to prevent violating the compatibility factors.

SWEET CHERRY (PRUNUS AVIUM L.)

Sweet cherry (*Prunus avium* L.) is one of the most popular table fruits. In 2014, the estimation of the world total cherry production was 2.294 thousand metric tons, with the top three producing countries (Turkey, USA, and Iran) producing 43% (FAO, 2017). Sweet cherry is greatly valued by consumers due to its taste, color, nutritional value, and beneficial health effects. Important quality characteristics of cherry fruits are weight, color, firmness, sweetness, sourness, flavor, and aroma. Sweet cherry is a highly perishable fruit containing significant levels of important nutrients such as potassium, dietary fiber, ascorbic acid, carotenoids, anthocyanins, and phenolic acids with only low caloric content.

The objective of this review is to provide understanding of the cherry pre-and postharvest changes, and packaging technologies with regard to critical quality indicators and health benefits. It also provides a peek on the road forwards with regard to the availability of new technologies. Part of these technologies are related to improving the circumstances in the chain as to gain access to long distance markets. Examples of these are the use of natural compounds and biodegradable films for modified atmosphere packaging (MAP) and edible coatings. The other part is related to improving the properties of the sweet cherry fruit itself; these are presented by showing the recent advances in marker assisted breeding preservation.

SWEET CHERRY QUALITY FACTORS

Maturity

Fruit maturity is one of the key factors determining overall fruit quality. Cherries should be harvested at the end of the maturation stage when they are fully ripe to ensure a good eating quality. Although, the harvest timing varies based on the sweet cherry cultivars, there is harvest window of <5 days for harvesting fruit of optimal quality. The harvest season for sweet cherries is short and labor intensive. Ripe fruit are prone to mechanical damage and cooling directly after harvest is important.

During early developmental stages, skin color changes, and softening starts, glucose, and fructose accumulate, together with a rapid increase in fruit size. In later development stages, ascorbic acid and anthocyanin accumulate, antioxidant activity increases, and fruit flesh darkens. A significant increase of total phenols was obtained in "Pico Negro" cherries during ripening and consequently, an increase in antioxidant activity, which may have beneficial health effects. Early-harvested cherries suffered from low acceptance due to the low sweetness while late-harvested cherries showed low acceptance due to low texture.

Nevertheless, higher sensory scores were obtained for late-harvested sweet cherry fruits, especially for skin color and flavor intensity. Harvested cherries show weight loss, changes in the sugar-acid balance, color, softening, and stem browning. During postharvest storage (4 days at 20°C) "Ambrunés" fruits showed a decrease in sugar levels, skin color, firmness, and TA, whereas SSC remained fairly stable, decreasing from 15.2 to 14.8°Brix. Malic acid levels decreased up to 20%, which may indicate that malic acid is an essential substrate for respiration in cherries.

Assessment of the optimal maturity in the orchard is not straightforward due to the variation in maturity within a tree, but especially within an orchard. Orchard mapping can support growers finding the correct harvest window by sampling°Brix and firmness during growth, just as now is being applied for dry matter assessment per plot for mango. This type of quality mapping may be helpful to scheduling future cherry harvest campaigns.

Color, Firmness, and Water Loss

Cherry fruit quality traits related to consumer purchase decisions are based on external quality attributes such as color, fruit size, stem freshness, absence of defects, and stem length. Aroma, flavor, sourness, sweetness, and texture are also essential attributes. Fruit color is, however, the main quality trait. Color development has been studied as function of cultivar, ripeness stage, and storage conditions. Large variation exists amongst sweet cherry cultivars grown in Italy with the skin color variation classified as dark—(e.g., "Black Star" and "Moreau") and light type cultivars (e.g., "Gabbaladri" and "Napoleona Verifica"). Dark red blush cultivars have a higher consumer preference when comparing with full bright red cherries. Color change during maturation is mostly due to an increase in anthocyanin content, specifically, cyanidin-3-O-rutinoside and cyanidin-3-O-glucoside.

Transcriptomic studies in cherry fruit maturation have been carried out to elucidate the role of abscisic acid (ABA) and ethylene and ripening. PacNCED1 transcription induced ABA synthesis and accumulates during cherry ripening, which should play a crucial factor in regulation on the sweet cherry fruit ripening. The ethylene production via regulation of PacACO1 expression (1-aminocyclopropane-1-carboxylic acid oxidase) might be stimulated with exogenous ABA, which indicates that the ethylene is synthesized by cherries (Ren et al., 2011) to start anthocyanin biosynthesis.

Positive effects of light on biosynthesis of phenolic compounds, mainly flavonoids are reviewed. Extensive studies have been carried out on the effect of supplemental UV-light treatments to improve anthocyanin levels in sweet cherry skin. Moreover, the application of UV-C light in postharvest has also been mentioned as a promise tool for extension of fruit shelf life, by delaying fruit senescence, increasing flavonoid content, and antioxidant activity in fruits. The application of blue light emitting diodes (LEDs) to increase anthocyanin levels in cherries could be a reality soon (Arakawa et al., 2017), increasing both color and health promoting potential.

In general, consumers prefer cherries fully ripe based on their dark skin color. However, consumers chose partially ripe over full ripe "Sweetheart" cherries which indicates the importance of a number of less well understood attribute interactions in consumer acceptance. In general, the chromatic values at harvest are lower at full ripeness stage than at the partially ripe stage. During postharvest storage, a further reduction in lightness (L*) and hue angle (h°) values were observed which indicates that the cherries became darker.

Firmness is related to storability, providing resistance to fruit deterioration and mechanical injury. Late season cultivars are frequently firmer than early season cultivars. Fruit firmness at harvest varies greatly, from 3.3 for "Moreau" to 27 N for "Minnulara" cherries.

Firmness at harvest of 17 sweet cherry cultivars varied between 2.56 and 4.71 N and was judged acceptable for 73–92% of the panelists. Softening is due to an increased peroxidase (POD), polyphenoloxidase (PPO), pectinmethylesterase (PME), polygalacturonase (PG), and b-galactosidase (b-Gal) activities. The level of these enzymes increased around 2 to 2.5-fold during a 5-day storage period causing a damage of cell wall components and subsequent softening.

The fruit cuticle likely plays an important role with respect to postharvest performance. The sweet cherry fruit cuticle is composed of a cutin polymer matrix, which consists essentially of esterified hydroxy- and epoxyhydroxy fatty acids with a chain length of 16 or 18 C-atoms and embedded in cuticular and surface-deposited epicuticular waxes. During cherry development, the levels of C16 and C18 monomers decreased, but increased during cold storage. During early sweet cherry fruit development, the cessation of cuticular membrane deposition is due to downregulation of genes involved in cuticular membrane deposition, such as PaATT1, PaCER1, PaGPAT4/8, PaLACS 1, PaLACS2, PaLCR, PaLipase, PaLTPG1, PaWINA, and PaWINB (Alkio et al., 2012). The main function attributed to the fruit cuticle is limiting water permeability, susceptibility to infections, and physiological disorders such as fruit cracking. The fruit cuticle composition is also associated with postharvest water loss from fruit respiration and firmness changes.

Cherries have a high respiration rate which leads to softening that is partly due to water loss. Dehydration during postharvest storage might induce PacNCED1 transcription and the accumulation of ABA resulting in ethylene production and subsequent fruit senescence. Water loss may lead to an increased susceptibility to infections and mechanical injuries. During postharvest storage, in particular for longer periods, fungal spoilage can lead to considerable economic losses. The development of off-flavors is promoted by bacteria and fungi due to ethanol and acetaldehyde synthesis.

Taste

Cherry fruit taste attributes, sweetness, and sourness, are important for consumer acceptance. Sweetness can be expressed as soluble solids content (SSC), sourness as titratable acidity (TA) with the ratio (SSC/TA) regarded as overall taste attribute). SSC values varied from 12.3 to 23.5°Brix and varied from 14.7 to 23.7°Brix for Spanish sweet cherry cultivars. SSC values ranged from 18.1 to 19.3 and from 17.8 to 19.8°Brix for

Brazilian and Chinese sweet cherries, respectively. TA values ranged from 0.20 in “Lapins” to 0.28 for the Canada Giant cultivars. High TA affects taste when the SSC value is below 16°Brix in “Brooks” or below 13°Brix in “Bing”. The optimal SSC/TA ratio is between 1.5 and 2.0, with SSC values ranging between 17 and 19°Brix).

Sweetness and flavor intensity are important parameters for sweet cherry liking. The most important sensory features are related to the ripening stage, parameters linked with accumulation of organic acids and aromatic alcohols. Several pathways in the biosynthesis of volatiles originate from amino acids, membrane lipids, and carbohydrates contributing for the ripe fruit flavor.

The aroma profile of sweet cherries is mainly the result of a complex mixture of alcohols, carbonyls, and organic acids. Hexanal and ethyl-2-hexenal are the most relevant compounds in the aroma profile of sweet cherries, that together with ethyl-2-hexen-1-ol, are associated with fresh green odors and green notes. Benzaldehyde, originates from the enzymatic hydrolysis of amygdalin, is likely the main contributor to the characteristic flavor of sweet cherry. Other important contributors are hexanoic acid, described as floral, and acetic acid, the main volatile acid found in sweet cherry fruits. Significant changes in volatile composition have been observed during ripening of sweet cherry.

The branched-chain alcohols, 3-methyl-3-buten-1-ol, and 3-methyl-2-buten-1-ol increase during ripening, whereas the organic acids, tetradecanoic and hexadecanoic acid, decrease during ripening, which may be due to reactions between alcohols and organic acids, catalyzed by alcohol acyltransferases. These enzymes seem to play a crucial role in ester biosynthesis. High variation was recorded in the composition of volatiles amongst four sweet cherry cultivars grown in Spain.

“Sweetheart” cherries were mainly characterized by higher levels of organic acids, such as 9-hexadecenoic acid, hexadecanoic acid, and aromatic alcohols and “Ambrunés” cherries by higher sweetness levels and presence of aliphatic alcohols. “Pico Colorado” cherries showed higher levels of hexanal and “Pico Negro” cherries were characterized by higher levels of branched alcohols, mainly 3-methyl-2-buten-1-ol and aromatic aldehydes, such as benzaldehyde. Similar research with four sweet cherry cultivars grown in Greece reported that “Ferrovia” and “Skeena” cherries showed higher levels of carbonyl compounds (2-propanone and 2-hexenal) and alcohols compared to “Lapins” and “Canada” cherries.

HEALTH PROMOTING COMPOUNDS

The role of several classes of health promoting compounds that are present in sweet cherry, lowering the risk of cancer, cardiovascular disease, diabetes, and other inflammatory diseases, has been recognized and is a field of many animal and now increasingly also epidemiological studies.

Phenolic Compounds

Sweet cherry phenolic compounds exhibit high antioxidant activity . These compounds consist of flavonoids, flavan-3-ols, and flavonols and non-flavonoid compounds like hydroxycinnamic and hydroxybenzoic acids. Phenolic compounds, concentrated in the fruit skin, increase during ripening together with polyphenolic compounds and anthocyanins that color the cherry skin from green to red.

Anthocyanins and flavonols contribute to sensory and organoleptic properties of fruits. The most abundant phenolic compounds are anthocyanins such as cyanidin-3-O-rutinoside and cyanidin-3-O-glucoside, peonidin-3-O-rutinoside and glucoside, as well as pelargonidin-3-O-rutinoside.

Anthocyanins have health promoting properties. For instance, cyanidin-3-O-rutinoside slows down absorption of carbohydrates which may help prevent or treat diabetes mellitus and cyanidin-3-O-glucoside showed cardio-protective effects by reducing blood lipid levels. Several studies reported higher phenolic content, SSC, TA and antioxidant activity in ripe cherries than in partially ripe. Sweet cherry contains phenolic acids such as hydroxycinnamic acid derivatives (neochlorogenic, p-coumaroylquinic, and chlorogenic acids), flavonols (quercetin-3-glucoside, quercetin-3-rutinoside, and kaempferol-3-rutinoside) and flavan-3-ols (catechin and epicatechin).

Consumption of cherries might induce health benefits such as inhibition of tumor growth, inhibition of inflammation and protection against neurodegenerative diseases. These fruits are also considered an excellent source of polyphenols, such as tannins. Tannins can be classified in hydrolysable tannins (gallic acid polymerization) and condensed tannins or proanthocyanidins (catechin polymerization).

Tannins are also secondary metabolites that confer astringency and have health-promoting properties. Tannin content ranged from 32 to 75 mg 100 g-1 of fresh weigh for Serbian sweet cherry cultivars. Sweet cherry extracts might be a serious candidate to prevent oxidative stress-induced disorders like intestinal inflammation disorders and neuronal cell death.

Vitamins and Carotenoids

Cherries are rich in C, A, E, and K vitamins and carotenoids, specially b-carotene, lutein, and zeaxantine. Ascorbic acid (vitamin C) levels vary between 4 and 7 g kg-1 of fresh weight among seven sweet cherry cultivars grown in Turkey, whereas the ascorbate levels for 22 sweet cherry cultivars grown in southern of Italy was considerably lower, varying from 0.034 to 0.260 g kg-1 of fresh weight.

Carotenoids, such as b-carotene, b-cryptoxanthin, and a-carotene have been reported in sweet cherry fruit.

These compounds are the main precursors of vitamin A and responsible for the red, yellow and orange hues in the cherry skin. b-carotene, b-cryptoxanthin and a-carotene, lycopene, and lutein are present in cherry at levels of 0.02 mg g-1 dry weight.

b-carotene, b-cryptoxanthin, and a-carotene, zeaxantine were identified in seven sweet cherry cultivars in small quantities varying from 0.2 to 16 ppm of fresh weight. Cherry carotenoids levels are generally quite low, and likely do not present a large contribution to human health.

Nowadays an increasing trend in society is health consumerism, for example functional drinks with cherry extracts rich in bioactive compounds. However, it is important to evaluate the impact of natural ingredients on sensory properties and consumer acceptance, as the optimal combination of taste and health promoting compounds is difficult to achieve.

How Do Growth Conditions Affect Fruit Quality and the Level of Health Promoting Compounds?

Scion × Rootstock Interaction

Several studies have shown the influence of the scion × rootstock combination on cherry fruit quality. Fruit firmness varies with scion x-rootstock combination. For instance, "Burlat" cherries are softer when grafted on CAB 11E (clone of Prunus cerasus L.; semi-vigorous) and firmer when grafted on Gisela 5 (P. cerasus × P runus canescens Bois. hybrid; dwarfing) or P. avium. Fruits grown on PiKu 1 and Weiroot 13 rootstocks had high levels of chlorogenic acid, neochlorogenic acid, p-coumaric acid and quercetin-3-rutinoside compared to F12/1, Gisela 5, Maxma 14, and Weiroot 158 rootstocks.

The lowest stone weight (7% of total fruit weight) and highest level of total sugars (250 g kg-1 fresh weight) was observed for "Lapins" on F12/1 rootstock, whereas PiKu 1 rootstock resulted in the highest yield (20 kg tree-1). The Weiroot 13 rootstock promoted the highest content of total organic acids (9 g kg-1 of fresh weight) and the Edabriz rootstock resulted in the highest SSC (15°Brix). "Lapins" grown on the Weiroot 72 rootstock presented higher fruit weight, firmness, SSC, and phenols content. Choosing the right rootstock continues to be widely studied due to its importance for productivity but the effects on fruit quality are also of increasing interest.

Orchard Management

Nowadays, the big challenge in cherry production is to reduce the costs and the labor requirements. This implies the use of the right rootstock and the best canopy architecture. Mechanization of pruning and a fully mechanical harvest system will facilitate achieving higher yield and fruit quality.

Summer pruning, which can provide adequate vigor and good light penetration throughout the canopy, and a high leaf area per fruit are important factors for the production of high quality fruits. Tree location in the orchard and fruit position in the tree are the most important factors that affect fruit weight. Fruit size and sweetness decrease with increasing plant density, which indicates competition between trees. High light intensity increases ascorbic acid content, and high temperature enhances anthocyanin and total phenolic contents.

Canopy architectures, such as Y-trellised and vertical, Tall Spindle Axe, Kim Green Bush, Super Slender Axe, and Upright Fruiting Offshoots, have recently been implemented in several orchards with the aim to improve the crop loading, the light input, as well as, the increase of health relating compounds content in cherries.

Especially in Mediterranean areas, due to water scarcity, improvement of cherry production under water stress is essential. Several studies evaluated irrigation systems with regard to productivity and fruit quality. Drip irrigation applied to "Lapins" cherry trees consumed only 21–29% of the irrigation water compared to micro-sprinkler irrigation, enhanced marketable fruit by 7 to 12% and did not impact fruit yield or firmness, color, and size. Low-frequency drip irrigation increased SSC in "Cristalina" and "Skeena" cherries.

However, irrigation levels did not affect SSC, pH, TA, and fruit weight of the cherries of the 900-Ziraat cultivar. Water management techniques, such as regulated deficit irrigation (RDI) and partial rootzone drying (PRD) may enhance fruit taste without reduction of yield.

Applying RDI, to "Summit" sweet cherry trees after the crop was harvested did not impact yield but could save water up to 45% without affecting firmness and limiting thinning costs. Limiting water supply during cherry tree development, either by drip irrigation or RDI, may help to produce healthier food a more sustainable manner, and would especially be important taking into account the effects of climate change.

Cherry production depends on the pollinators and utilization of pollinizer cultivar in the orchard. However, in a near future, these can be replaced by artificial precision pollination systems using liquid pollen suspensions applied with electrostatic sprayer technology.

Growth Regulators

Application of Ca(OH)2 0.7% reduced cracking and fruit size and increased SSC, firmness, and calcium levels in both fruit flesh and skin. Sweet cherry trees sprayed with CaCl2 0.5% once a week from petal fall until 2 weeks before harvest showed higher SSC and higher levels of phenolics, reduced decay, and cuticular fractures, but showed higher weight loss during storage. Preharvest application of ethephon reduced both fruit removal force and fruit flesh firmness at harvest.

Fruits treated with 20 ppm gibberellic acid (GA3) at the green to straw-yellow stage showed slower softening with lower PG and cellulase activity, but no effects on SSC and b-Gal activity. GA3 can induce, depending on the cultivar, increased fruit weight and SSC, including delayed ripening and reduced fruit cracking.

GA3 application increased fruit firmness of "Lapins" and "135-27-17"cherries, but no effect was observed in "Celeste" and "Merpet" fruits. Similar preharvest applications of GA3 (25 ppm) at the straw-yellow stage of the "0900 Ziraat" cherries showed increased fruit size and firmness.

Applications of GA3 (30 and 60 mg L-1) to cherries at the straw-yellow stage of the Regina, Sweetheart and 0900 Ziraat cultivars retarded fruit ripening and preserved flesh firmness. However, these fruits showed a decrease in total phenolics, anthocyanins, and antioxidant activity. Preharvest application of 25, 50, and 100 mg L-1 GA3 and GA3 combined with 0.05, 0.1, and 0.5 mg L-1 22S, 23S-homobrassinolide (HBR) to "Regina" and "Summit" cherries at full bloom and at the beginning of fruit development increased fruit size and decreased SSC.

Fruits treated with only HBR showed higher fruit firmness of the flesh and stem resistance. At the straw-yellow stage applications of "Noir de Guben" cherries with 50, 100, and 150 ppm benzyladenine (BA) and BA combined with 12.5, 25, and 50 ppm gibberellins (GA4+7) improved size and delayed skin color change. BA alone or in combination with GA4+7 improved storability by maintaining firmness during 30 days in cold storage (4°C). GA3 applied at stage I (mesocarp growth consisting of both cell division and elongation) and stage II (endocarp hardening and embryo development) improved final "Bing" fruit weight, but higher fruit weight was also obtained when combined with GA4+7.

Application of salicylic acid (SA) and acetylsalicylic acid (ASA) at 0.5 and 1 mM at three key points of fruit development improved fruit weight up to 59%, increased firmness and increased anthocyanin and total carotenoid content. Application of 2 mM oxalic acid (OA) increased fruit size up to 23%, increased levels of anthocyanins, flavonols, neochlorogenic, and chlorogenic acids. The growth regulators application improves final fruit size due to the stimulating cell division. However, the successful application of these compounds is highly dependent on climatic conditions and time of application. Methyl salicylate (MeSA) application increased fruit size, red coloring, firmness, and SSC.

MeSA also increased total phenolics, anthocyanins, and induced higher activity of antioxidant enzymes such as ascorbate peroxidase, catalase, peroxidase, and superoxide dismutase at harvest. Interestingly, MeSA coatings affect postharvest quality of cherry fruits while still attached on the tree. Total phenolic and anthocyanin contents, total antioxidant activity and antioxidant enzymes were higher in MeSA-treated fruits during storage, while also delaying postharvest ripening due to less coloring, lower acidity, and less softening in treated fruits after storage.

Application of preharvest coatings may be an exciting new trend in research that could be instrumental to enable further access to long distance markets as both storability and the levels of health promoting compounds are improved. In Portugal, the cherry growers are currently applying biostimulants based on seaweed and tertiary ammonium compounds (such as glycine betaine), and it is likely that, observing the positive (unpublished) results that have been obtained, that application of these compounds might be common in the near future.

Storage Conditions

Optimization of storage and transport conditions are important tools to extend and preserve shelf life and thus allow long-distance transport to destination markets. Temperature and relative humidity (RH) are critical factors that influence cherry quality during postharvest storage. According to researchers, the optimal temperature to store sweet cherries is 0.5°C at 90% RH.

During cold storage, the level of phenolic compounds such as neochlorogenic acid, p-coumaroylquinic acid, chlorogenic acid, rutin, catechin and epicatechin, and the content of anthocyanins such as cyanidin-3-O-rutinoside and peonidin-3-O-rutinoside increased in several sweet cherry cultivars.

The phenolics content in "Canada Giant" and "Ferrovia" cherries increased during at storage at 20°C during 8 days, probably due to the condensation of the phenolic compounds by water loss during ripening or by postharvest synthesis (Kalt et al., 1999). However, researchers observed a decrease of 41–52% of total anthocyanins content after 15 days at 1°C and 95% RH in two sweet cherry cultivars, which may indicate no net anthocyanin biosynthesis.

Hydro-cooling reduced sweet cherry stem browning and surface shriveling, while decay, external color, and SSC were not affected in "Tragana Edessis" and "Mpakirtzeika" cherries after 1-week of cold storage (0°C, 95% RH). Adding CaCl2 during hydro-cooling reduced the respiration rate, ascorbic acid degradation, and membrane lipid peroxidation. Moreover, total phenolic content, antioxidant capacity, and fruit firmness increased and TA slightly decreased in "Sweetheart" and "Lapins" cherries.

Hydro-cooling can effectively increase quality of cherries by the reduction in temperature of the cherries after harvest, the sooner the better. Hydro-cooling portable equipment that can be installed in the orchard might be a good option.

In order to extend the self-life of sweet cherries, MAP, and controlled atmosphere (CA) storage have been successfully implemented. Optimum CO2 and O2 levels for long time storage have been established at 2–10% for O2 and 5–20% for CO2 Lower O2 levels may result in a decrease of volatile aromatic compounds and development of off-flavors, while CO2 levels higher than 30% have been related to a brown discoloration of the skin.

High O2 levels (70% O2 + 0% CO2) can inhibited ethanol production, but also induced a rapid decrease in ascorbic acid content and often resulted in browning due to high-O2 injury. Super atmospheric O2 packaging (100 kPa O2), in polypropylene film during 10 days at 4°C and 70% RH delayed respiration rate of cherries, inhibited ethylene production and PPO and POD activities and maintained firmness, soluble proteins and SSC for the first 4 days of storage.

However, ascorbic acid content was lower compared to control cherries. MAP combined with low temperature storage reduces the incidence of fruit decay while preserving SSC, TA, and stem color. Fruits also displayed higher antioxidant capacity, firmness and, a brighter colored skin.

High CO_2 levels and cold storage both favor net anthocyanin synthesis as anthocyanin biosynthesis is induced during the postharvest cold storage of several red fruits. This increase may be due to inhibition of PPO activity promoted by CO_2-enriched atmospheres, since changes in fruit color occurred due to oxidation of phenols by PPO. MAP and CA stored sweet cherry fruits at 5% O2 combined with 10% CO_2 at 1°C reduced PPO and POD activity, prevented flesh browning, decreased fruit decay and promoted firmness retention.

Weight loss, pitting, stem visual appearance, and PG and PPO activity, of "0900 Ziraat" cherries were lower when applying modified atmosphere-modified humidity packaging (MA/MH) using 0.20 μm low-density polyethylene during 8 days at 0°C and a 2-day shelf-life. Certainly, MAP can be useful for sweet cherry shelf-life extension, but large scale implementation in practice is limited as it is expensive and somewhat risky due to possible gas mixture mismanagement and fungal growth.

NEW FRONTIERS FOR SWEET CHERRY QUALITY IMPROVEMENTS

Natural Compounds

Natural compounds have been applied to improve the shelf life of cherries. Application of 1 mM SA or ASA or OA on cherries stored for 20 days at 2°C showed lower acidity, less color changes and firmness loss and maintained higher levels of health promoting compounds and antioxidant activity. Immersion in 30 mM b-aminobutyric acid (BABA) for 10 min reduced respiration rate, weight loss and increased TA levels, total phenolics, sugars and ascorbic acid of "Hongdeng" cherries during 5 days at 20°C.

Moreover, BABA-treated cherries showed higher ascorbate peroxidase, catalase, glutathione reductase, and superoxide dismutase activities (Wang et al., 2016). The use of these compounds is considered safe and environmentally friendly, resulting in extending storability of sweet cherry with increased health-promoting properties.

Currently, these postharvest treatments have only been applied at the research level, but the future is promising for application in practice.

Biodegradable Films for MAP and Edible Coatings

Preservation technologies have emerged using biodegradable films for MAP. Biodegradable films have a reduced environmental footprint and delay color changes, softening and loss of acidity. Micro-perforated polylactic acid (PLA) film inhibited fungal growth, maintained brightness, skin color, firmness, SSC, and decreased fruit weight loss in sweet cherry when compared with macro-perforated bags.

Cherries stored in PLA packages showed better appearance, texture, flavor, and overall acceptability. This packaging technology may become practice, but depends on the balance between (less) waste production and (higher) price for these type of films.

Edible coating is a promising technology in the preservation of fruits and vegetables quality. Edible fruit coatings can also provide an alternative to modified atmosphere storage and act as carriers of health promoting compounds. Bioactive compounds, such as antioxidants, antimicrobials, probiotics, flavors, and nutraceutical substances, can be transported from the coating onto the fruit skin by diffusion, enhancing shelf life, and nutritional quality. Edible coatings may reduce respiration and transpiration rates, reduce firmness loss, and limit decay.

Locust bean gum, shellac, polysorbate 80, glycerol, and beeswax coatings applied to "Burlat" cherries showed lower weight loss, less bruising injury, less surface pitting and lower flesh firmness loss resulting in a prolonged shelf-life. Chitosan is an antimicrobial compound that induces fungal defense-related enzymes and enhances the level of phenolic compounds. Chitosan-coated sweet cherries showed lower water loss, delayed color changes, higher total phenolics, flavonoid, and anthocyanin levels.

Researchers reviewed the application of aloe vera gels as an edible coating for fresh fruits to enhance the postharvest life and quality attributes of, amongst others, sweet cherries. Aloe vera gel reduced weight loss and lowered the respiration rate, delayed color changes, softening, and TA losses in cherry fruit compared to control fruits.

According to the European Directive (ED European and Parliament Council Directive N 95/2/EC, 1995; ED European Parliament and Council Directive N 98/72/EC, 1998) and USA regulations (FDA, 2016), edible coatings can be classified as food products, food ingredients or food additives. This allows for application of edible coatings in practice, although more studies are needed in order to improve functionality and consumer acceptance.

Marker Assisted Breeding

The recent publication of the huge and complex sweet cherry genome will allow for identification of candidate genes, signal transduction-, and metabolic pathways, which can have a relevant role in fruit quality. Sweet cherry linkage maps have been constructed to facilitate identification and characterization of quantitative trait loci (QTLs) associated with quality traits. QTLs have been identified for fruit weight and firmness (Campoy et al., 2015), fruit size, skin and flesh fruit color.

About 133 genes are putatively involved in fruit texture, color, flavor, and chilling injury resistance. Moreover, in sweet cherry, exocarp-specific transcripts associated with epidermis development (from flowering to maturity) and stress responses were established. In stone fruit breeding programs, marker-assisted selection (MAS) is an important biotechnological tool, although its use is currently limited due to the polygenic nature of traits linked to fruit quality.

Genotyping by sequencing (GBS) generates more saturated genetic maps, thus facilitating the identification of (more) accurate QTLs and the discovery of single nucleotide polymorphisms (SNPs), with focus mainly on finding sweet cherry quality QTLs and not on finding QTLs that affect the level of health promoting compounds. The use of DNA information, as clustered regularly interspaced short palindromic repeats (CRISPR) along with CRISPR-associated protein 9 (Cas9) system (CRISPR/Cas9), appears to be relevant in understanding genetic variation, inheritance, genomic organization, and phenotypic performance, contributing to the development of new cultivars of Rosaceous crop breeding.

Although, the CRISPR/Cas9 gene-editing technique has not yet been used in cherry, it has been effectively used to confer resistance against citrus canker, regulating ethylene synthesis (Ripening inhibitor gene) in tomato and in the efficient knockout of the L-idonate dehydrogenase gene (IdnDH), involved in the tartaric acid pathway in grape.

Taking this into account, it is expected that CRISPR/CAS gene editing will be instrumental with regard to improving fruit quality and the level of health-related compounds.

Future Perspectives

We are at the beginning of acquiring knowledge to understand the variation in health-related compounds throughout the fruit maturation and the postharvest phase. How environmental factors, dose and time of preharvest compounds application affects tree performance and consequently fruit quality is at the moment largely unknown. Further research should be aimed to develop cultivar-specific strategies that improves both fruit quality and fruit nutritional characteristics, without significantly affecting yield.

Optimizing cherry quality attributes needs a better understanding of the interaction between genetics, environmental factors, and plant growth regulators to be able to gain full benefit of new preservation technologies. The incorporation of bioactive compounds may provide advantages in food preservation and contributes to the development of functional foods.

Application of bioactive compounds and encapsulation with edible coatings are promising techniques that also need further investigation, also with regard to diminishing off flavors.

New cultivars with improved characteristics such as fruit size and color, anthocyanin biosynthesis, and cracking resistance are needed in order to produce cultivars with exceptional quality, high consumer appeal, well adapted to local growing and future climatic conditions. With the recent advancements in marked assisted breeding, especially now that the draft sweet cherry genome has been published, exciting times await.

Tomatoes

Tomato is one of the most popular produced and extensively consumed vegetable crops in the world. Tomatoes can be consumed in many ways. The fresh fruits are eaten in salads and sandwiches and as salsa whilst the processed ones are consumed dried or as pastes, preserves, sauces, soups, juices, and drinks. Tomatoes and tomato-based foods provide a wide variety of nutrients and many health-related benefits to the body. Tomato contains higher amounts of lycopene, a type of carotenoid with antioxidant properties which is beneficial in reducing the incidence of some chronic diseases like cancer and many other cardiovascular disorders.

In regions where it is being cultivated and consumed, it constitutes a very essential part of people's diet. Tomatoes production accounts for about 4.8 million hectares of harvested land area globally with an estimated production of 162 million tonnes. China leads world tomato production with about 50 million tonnes followed by India with 17.5 million tonnes. Tomato production can serve as a source of income for most rural and periurban producers in most developing countries of the world. Despite the numerous benefits that can be derived from the crop, postharvest losses make its production in most parts of the world unprofitable. Postharvest losses in tomatoes can be as high as 25–42% globally. These losses bring low returns to growers, processors, and traders as well as the whole country which suffers in terms of foreign exchange earnings.

Postharvest losses in tomatoes can be either quantitative or qualitative. Even though emphasis in crop research nowadays is increasing shifting from quantity to quality of produce there is still little improvement in the quality of commercially produced tomato varieties, hence resulting in high amount of qualitative losses. However, qualitative loss in tomato production can have a negative impact on many parameters like consumer acceptability, nutrient status of fruits, and financial income to producers.

An investigation into the possible factors that can affect the postharvest quality of tomatoes is therefore necessary. The postharvest qualities of tomatoes are dependent not only on postharvest handling and treatment methods but also on many preharvest factors such as genetic and environmental conditions. Many cultural practices such as types of nutrient, water supply, and harvesting methods are also believed to be factors influencing both pre- and postharvest quality of tomato.

Many postharvest quality losses are as a result of many preharvest factors. Tomato fruits that are diseased and infected by pest, inappropriately irrigated, and fertilised or generally of poor quality before harvesting can never be improved in quality by any postharvest treatment methods.

This indicates that the postharvest quality of the fruit cannot be improved after harvest but can only be maintained. It is therefore important to know the preharvest factors that can produce superior qualities in fruits during harvest whilst using appropriate postharvest handling and treatment methods to maintain the quality after harvest. The purpose of this paper therefore is to look at some preharvest and postharvest factors and how they can affect the postharvest qualities and shelf life of harvested tomatoes.

PREHARVEST FACTORS AFFECTING THE POSTHARVEST QUALITY OF TOMATOES

Below are some preharvest factors or activities that can affect the postharvest shelf life and qualities of harvested tomato fruits.

Fertiliser Application

Consumers in the past few decades have become increasingly concerned about the quality of the food products they are consuming. Researchers have therefore investigated the impact of plant nutrition on the quality of fruits produced. The quality traits required or purpose for which the crop is grown will help in selecting not only the type of fertiliser but the quantity used during production. For instance, an adequate supply of potassium fertilizer in tomato production improves fruit colour and reduces the incidence of yellow shoulder, whilst enhancing the titratable acidity of the fruit.

Yellow shoulder is a physiological disorder of tomatoes that is characterized by discoloured regions that border the stem scar. Insufficient supply of potassium in soilless tomato production can also result in ripening disorders. Unlike potassium, an increase in nitrogen supply to greenhouse-grown tomatoes, beyond a certain threshold level, may reduce fruit quality by decreasing the sugar content of the fruits. High nitrogen supply of about 250 kg/ha can impair some important quality traits of fruits, such as total soluble solids, glucose, fructose, and pH. A supply of reduced forms of nitrogen, such as ammonium, can result in improved fruit flavours. However, the variation of phosphorus supply in soils for growing tomato crops does not significantly influence quality traits such as the total soluble solids, pH, acidity of the tomato juice, or the fruit colour characteristics.

For trace elements use, the quality of tomato fruit is affected predominantly by the amount of boron used, although other micronutrients may affect fruit quality only when the plants show severe deficiency symptoms. Lower amounts of boron supply reduce fruit firmness which is of major concern during storage. Calcium application in tomato production has recently been studied and found to have a positive effect on the prevention of some diseases, whilst slowing the reduction in fruit firmness during ripening.

Spraying tomato leaves with different combinations of calcium salts is effective in controlling powdery mildew on the crop. Also an enhanced supply of calcium significantly increases the resistance to bacterial wilt caused by Ralstonia solanacearum in tomato. Calcium can be applied to the root zone in the soil or sprayed on the leaves. The type of application is determined by the growth stage of the crop and type of disease prevalent. Postharvest calcium application can also have a positive storage effect in calcium deficient harvested fruits.

Pruning

Controlling the number of flowers, fruits, or fruit trusses in tomatoes is an effective way of reducing the competition between fruits. Pruning therefore ensures nutrients are channelled to fewer fruits sinks which can lead to increased fruit size whilst increasing sugar content of fruits in some cases. Pruning clusters to three fruits increased total marketable yield and fruit weight and reduced cull yield of all cultivars under investigation.

Meanwhile, the effect of pruning on other quality traits of the fruit produced depends on many factors including the sink developmental stage, fruit to leaf ratio, truss position, and genetic background. Pruning can result in increased fruit size in most cultivars, under the right growing conditions, and thus can be used to improve the marketability of high total soluble solids (TSS) fruits which in most cases tend to be smaller in size. Larger fruits which are within a certain size range are usually considered to be of better value by most consumers.

Maturity Stage

The maturity stage of tomato fruit at harvest is an important determinant of many quality traits. Tomato, being a climacteric fruit, can be harvested at different stages during maturity, like mature green, half ripen, or red ripen stage. Each stage at harvest has its own postharvest attribute that the fruit will exhibit. Moneruzzaman et al. reported that the shelf life of all tomato cultivars under investigation is longest when harvested at green mature stage. Although shelf life has been the most important aspect in loss reduction biotechnology of fruit and vegetables, other aspects may be of interest rather than shelf life.

Fruit nutritional values and appearance may be affected when harvested green. For instance, sugar transport to fruits in a vine-ripened tomato appears to increase during the latter part of maturity and, therefore, when fruits are harvested immature or in a green state sugar import to fruits will be cut off making postharvest degradation of starch, the main source of carbohydrates, which is both undesirable and inadequate. Meanwhile, harvesting later also promotes higher sugar accumulation in riper fruits which are susceptible to mechanical injuries with a shorter shelf life. The pH of tomatoes is an important parameter in the tomato processing industry. Tomatoes are processed as high-acid foods and therefore the higher the acidity the better for processing.

Cultivars with high pH therefore may not be suitable for processing. A pH of 4.4 has been suggested to be the maximum and the optimum of a target of 4.25. The acidity of tomatoes is highest at the pink stage of maturity with a rapid decrease as the fruit ripens. Moneruzzaman et al. and Cliff et al. suggested tomato fruits can be harvested at mature green to give producers enough time for long distance marketing but for local marketing harvesting at the fully ripe stage is preferred to maximise nutritional value.

Cultivar Type

The potential quality of fruit is dependent on the cultivar type. Different cultivars are characterised by different quality parameters making some more desirable to the producers and consumers than others. The choice of an adequate-yielding tomato cultivar with desired fruit qualities and longer shelf life is therefore a vital decision a producer must take. Failure to select an appropriate cultivar may lead to lower yield, low quality fruits, or less market acceptability. Fruits of different cultivars differ in size, colour, texture, and flavour as well as storage potential. Getinet et al. report the influence of tomato cultivar on some postharvest qualities of tomatoes stored under different conditions. Getinet et al. established that tomato cultivar Roma VF has higher sugar content whilst maintaining lower weight loss as compared to cultivar Marglobe. Cultivar selection is therefore critical to the postharvest storage life and eating qualities of tomatoes.

Irrigation

Tomato is not a drought resistant crop and therefore yields decrease considerably after short periods of water deficiency during production. Proper irrigation scheduling in tomato production is therefore crucial to the crop development. However, with water being a scarce resource in most production areas, growers in recent years have therefore had to develop a more efficient water management scheme that maintains crop yield but has a moderate and controlled level of moisture stress on their crops.

In a study conducted by Mitchell et al., it was revealed that deficit irrigation reduced fruit water accumulation and fresh fruit yield but increased fruit total soluble solids levels. They also discovered that irrigating with saline water had no significant effect on total fruit yield but moisture content of fruits was slightly reduced. Ismail et al. also established that early morning irrigation every three days resulted in higher yields than daily irrigation.

The use of trace elements or the practice of soilless tomato production can be made possible during irrigation where the fertilisers (trace elements) are added to the irrigation water in a form of solution and administered. These trace elements are selected depending on the specific postharvest quality traits needed in the fruits.

POSTHARVEST FACTORS AFFECTING THE POSTHARVEST QUALITY OF TOMATOES

After harvesting, the fruit still remains alive and performs all functions of a living tissue. The climacteric burst of ethylene which makes the fruit palatable also triggers senescence and subsequent ripening in the fruits. The goal of any postharvest handling practice or treatment is to manage the concentration and timing of ethylene synthesis so that the fruit reaches the consumer at optimal eating quality. The following factors discussed below can affect the quality of tomatoes after harvest when not properly managed.

Temperature

Proper temperature management between the period of harvesting and consumption has been found to be the most effective way to maintain quality. Keeping harvested fruits cool at low temperatures of about 20°C will slow down many metabolic activities which lead to ripening, hence allowing more time for all the postharvest handling of the produce. Generally, one hour of delay between harvesting the crop and cooling it will lead to one day loss of shelf life.

Respiration and metabolic activities within harvested climacteric fruits like tomatoes are directly related to the temperatures of the ambient environment. High temperatures can hasten the rate of respiration (CO_2 production) in harvested or stored fruits products. CO_2 production in stored climacteric products like tomatoes can trigger ethylene production although this depends on other factors like O_2 or CO_2 levels, exposure time, and ripening stage.

Minute amounts of ethylene can cause ripening in fruits even at levels of tens of nL per L. Heat stored in field-harvested fruits is a major source of high temperatures in fruits. The time of the day at which harvesting is done must therefore be considered to avoid excessive field heat which can cause more rapid deterioration in the harvested fruits.

Low temperature storage can protect nonappearance quality attributes like texture, nutrition, aroma, and flavour [43]. Meanwhile, tomato being a tropical fruit is also adversely affected by exposure to extremely low temperatures. Chilling injury can occur in tomato fruits stored at temperatures below 10°C. The effect of chilling injury includes premature softening, irregular colour development, surface pitting, browning of seeds, water-soaked lesions, off-flavour development, and increased postharvest decay. It is therefore important to determine the optimum temperature needed when handling tomato fruits during storage.

Relative Humidity

Water loss from harvested fruit produce is predominantly caused by the amount of moisture present in the ambient air expressed as relative humidity. At very high relative humidity, harvested fruits maintain their nutritional quality, appearance, weight, and flavour, whilst reducing the rate at which wilting, softening, and juiciness occur.

Tomato fruits are very high in water content and susceptible to shrinkage after harvest. Fruit shrivel may become evident with any small percentage of moisture loss. The optimal values of relative humidity for mature green tomatoes are within the range of 85–95% (v/v) but 90–95% (v/v) for firmer ripe fruits.

Below the optimal range, evapotranspiration increases resulting in shrivelled fruits. Storage of tomato fruit at a lower relative humidity can result in shrivelling. Addition of moisture (wetting fruits) in lower relative humidity storage can reduce weight loss and prevent fruit from shrivelling. Meanwhile, completely saturated atmospheres of 100% relative humidity should be avoided, as moisture condensation on the fruit surfaces may encourage mould and fungal development.

Combination Gases

The combination of different gases in a storage environment is very important in extending the storage life of tomato fruits. The optimal atmosphere needed to inhibit senescence in mature green and ripe fruit of tomatoes is 3–5% (v/v) of oxygen but for carbon dioxide it is 1–3% (v/v) and 1–5% (v/v) in mature green and ripe fruit, respectively, whilst 94–96% (v/v) of nitrogen gas is required. A very low supply of oxygen can have a detrimental effect on fruits by causing anaerobic respiration. Carbon monoxide (CO) has been investigated as a gas for treating fruits and has been found to speed up ripening. It is therefore necessary to balance the carbon monoxide with low oxygen to delay senescence in the fruits.

Carbon monoxide slows down postharvest pathogenic infestations whilst improving some quality traits of tomatoes. For instance, tomatoes stored in 5–10% (v/v) carbon monoxide with 4% (v/v) oxygen were found to have superior total soluble solids (TSS) and titratable acid (TA) profiles as compared to control samples stored in air. The use of carbon monoxide in the food industry is however complicated because of its health hazards to humans and must be used with great caution.

POSTHARVEST CALCIUM CHLORIDE APPLICATION

Higher plants generally contain appreciable amounts of calcium, usually in the range of 1–50 mg Ca g-1 dry matter. Calcium deficiency in plants can be minimised using calcium based fertilisers. Calcium deficiency in tomato during fertilisation can result in susceptible plants which are exposed to many calcium related disorders like blossom end rot. Calcium has been found to increase the yield of tomato when used as a fertilizer. Postharvest calcium application has also been shown to have a positive effect on many storage parameters of various fruits and vegetables. For instance, the use of calcium chloride in controlling rapid ageing in harvested fruits in general has been investigated by many authors. Postharvest calcium chloride application reduces respiration, decreases ethylene production, and delays senescence in fresh produce such as tomatoes. Exogenous application of calcium maintains cell-wall integrity and protects it from degrading enzymes, enhancing better linkages between pectic substances within the cell-wall whilst increasing the cohesion of cell-walls. This therefore gives an indication that the rate of senescence in fruits is closely related to the amount of calcium in the plant tissue and varying the calcium status affects the rate of senescence. It has been established that altering the levels of exogenous calcium application affects parameters for senescence such as protein and chlorophyll content, respiration rates, and cell membrane fluidity.

Physical Handling

Physical handling can have a drastic effect on the postharvest quality or life of harvested fruits. Rough handling during harvesting and after harvesting can result in mechanical injuries which affect quality. Typical industrial production systems associated with tomatoes may include mechanical harvesting, packing into crates, sorting, grading, washing, and transporting over long distances. At each of these stages there may be significant occurrence of mechanical injury which may be bruising, scarring, scuffing, cutting, or puncturing the fruits. In small-scale tomato production, mechanical injuries may result from the use of inappropriate harvesting containers and packaging materials.

The effects of mechanical injuries on fruit are cumulative. Injuries which are equivalent to or greater than the bioyield point lead to a total breakdown of the structure of the affected cells which is accompanied by unwanted metabolic activities which may include increased ethylene production, accelerated respiration rates, and ripening, which results in either reduced shelf life or poor quality. It is therefore important to handle tomato fruit with care during the harvest and postharvest activities to minimise mechanical injuries to avoid losses.

CONCLUSION

Postharvest quality management of tomatoes starts from the field and continues until it reaches the final consumer. The postharvest quality status of the fruits in part depends on some preharvest practices carried out during production. The quality of any fruit after harvest cannot be improved by the use of any postharvest treatment method or handling practices but can only be maintained. Understanding and managing the various roles that preharvest factors like fertiliser application, pruning, maturity stage, cultivar selection, and irrigation can play in the quality of fruits at harvest is very important in order to produce high quality fruits at harvest. Tomatoes are highly perishable and are subjected to rapid quality loss after harvest.

3

MICROBIAL SPOILAGE OF FRUITS AND VEGETABLES

INTRODUCTION

Consumption of fruit and vegetable products has dramatically increased in the United States by more than 30% during the past few decades. It is also estimated that about 20% of all fruits and vegetables produced is lost each year due to spoilage. The focus of this chapter is to provide a general background on microbiological spoilage of fruit and vegetable products that are organized in three categories: fresh whole fruits and vegetables, fresh-cut fruits and vegetables, and fermented or acidified vegetable products. This chapter will address characteristics of spoilage microorganisms associated with each of these fruit and vegetable categories including spoilage mechanisms, spoilage defects, prevention and control of spoilage, and methods for detecting spoilage microorganisms.

Fruits are natural sources of minerals, vitamins besides carbohydrates and other essential substances. Naturally fresh fruits and juices made out of them contain high amount of water thereby making them highly prone to attack by microorganisms. While most of the fruits are naturally provided with coatings and coverings in the form of skins, but these are fragile enough to be easily disturbed by various biological and mechanical factors. Like vegetables, fruits being produce of plants get contaminated through different sources by a variety of microorganisms which may play significant role in their spoilage. These are soil, water, diseased plant, harvesting and processing equipments, handlers, packaging and packing material and contact with spoiled fruits.

Most microorganisms that are initially observed on whole fruit or vegetable surfaces are soil inhabitants, members of a very large and diverse community of microbes that collectively are responsible for maintaining a dynamic ecological balance within most agricultural systems.

Vectors for disseminating these microbes include soil particles, airborne spores, and irrigation water. Most bacteria and fungi that arrive on the developing crop plant either are completely benign to the crop's health or, in many instances, provide a natural biological barrier to infestation by the subset of microorganisms responsible for crop damage. The even smaller subset of bacteria and fungi responsible for causing spoilage to the edible portion of the crop plant is the subject of this section.

Spoilage microorganisms can be introduced to the crop on the seed itself, during crop growth in the field, during harvesting and postharvest handling, or during storage and distribution. Those same types of soil-borne spoilage microbes that occur on produce are the same spoilage microorganisms that are present on harvesting equipment, on handling equipment in the packinghouse, in the storage facility, and on food contact surfaces throughout the distribution chain. Therefore, early intervention measures during crop development and harvesting through the use of good agricultural practices (GAP) will provide dramatic reductions in yield loss due to spoilage at all subsequent steps in the food-to-fork continuum.

Examples of GAPs include foliar fungicide application in the field, cross-contamination prevention measures (stringent sanitation standard operating procedures) in the packing house and storage facility, and use of postharvest fungicides. These practices also will enhance substantially the food safety and shelf life of fresh-cut produce.

In 1998, FDA published the Guide to Minimize Microbial Food Safety Hazards for Fresh Fruits and Vegetables, recommending GAPs that growers, packers, and shippers implement to address the common microbiological hazards that may be associated with their operations. These GAPs are organized in eight categories:

- Water
- Manure and municipal biosolids
- Worker health and hygiene
- Sanitary facilities
- Field sanitation
- Packing facilities sanitation
- Transportation
- Traceback

In addition, FDA worked with the produce industry to develop commodityspecific food safety guidelines for sprouts, lettuce and leafy greens, melons, and tomatoes that provided metrics for soil and water amendments as well as adjacent land usage. These should also improve substantially the food safety and shelf life of fresh-cut produce.

MICROORGANISMS ASSOCIATED WITH SPOILAGE IN FRUITS AND JUICES

The microorganisms associated with fruits depend on the structure of fruit. The fruits contain different organic acids in varying amounts.The types of acids which are predominately found are citric acid, malic acid and tartaric acid. The low pH of fruits restricts the proliferation of various types of organisms.

Due to the low pH, a large number of microorganisms are restricted to grow on fruits. Fungi are most dominating organisms to grow on fruits because of the ability of yeasts and molds to grow under acidic conditions. A small number of bacteria which are aciduric (ability to resist acidic conditions) also grow. Also the dry conditions prevailing on the skin and surface do not allow the growth of certain microorganisms. Besides these plants also produce certain antimicrobial components too.

Despite the high water activity of most fruits, the low pH leads to their spoilage being dominated by fungi, both yeasts and molds but especially the latter.

Yeasts

Yeasts are unicellular fungi which normally reproduce by budding. Of the 215 species important in foods, about 32 genera are associated with fruits and fruit products . Only a few species of yeasts are pathogenic for man and other animals. None of the pathogenic species are common contaminants of fruits and fruit products. Fruit that has been damaged by birds, insects, or pathogenic fungi usually contain very high yeast populations.

The yeasts are introduced into the exposed tissue, often via insects, and are able to use the sugars and other nutrients to support their growth. Types of yeasts growing in fruits depend upon the nature of the fruit and the strain of yeast. Growth of a strongly fermentative type such as certain strains of Saccharomyces cerevisiae may produce sufficient CO2 (90 lb/in. or more) to burst the container.

Growth of some species in a clear fruit juice may produce only slight haze and sediment. While carbon dioxide and ethanol are the predominant metabolic products of yeasts, other products such as glycerol, acetaldehyde, pyruvic acid, and a -ketoglutaric acid are also formed. Oxidative yeasts such as species of Brettanomyces produce acetic acid in wines and other fruit products. Although yeasts produce hydrolytic enzymes which degrade pectins, starch, and certain proteins, enzymatic activity is usually much less than that exhibited by other aciduric microorganisms, molds in particular.

Molds

These are filamentous fungi which are important group of microflora of fruit products due to following reasons

1. Some of the members are xerophilic, thereby having potential to spoil foods of low water activity such as dried fruits and fruit juice concentrates.
2. Some of the species have heat resistant spores such as ascospores which can survive the commercial pasteurization treatments that are given to most fruit products.
3. Growth of molds on processing equipment such as wooden tanks can result in the generation of off-flavors in wines, juices, and other fruit products.
4. Mold-infected raw fruit may become soft after processing because pectinases were not inactivated by the thermal treatment.

5. The metabolic products of many molds are toxic to humans. Of these toxins, mycotoxins are important components.

Molds are aerobic microorganisms, but many of them are very efficient scavengers of oxygen. Due to this property of molds, processed fruits, including those hermetically sealed in cans or glass, are susceptible to spoilage. In case of limited vegetative growth, evidence of spoilage may be the changes produced by fungal enzymes such as the breakdown of starch or pectins while in case of heavy growth, colonies develop in the headspace or as strands throughout a beverage or similar product.

- R. stolonifer cause soft and mushy food ,cottony growth of mold.
- Anthracnose -Colletotrichum lindemuthianum, cause spotting of leaves and fruits and seedpods
- Downy mildew:Initially, the lesions tend tobe small and confined to the upper surface of wrapper leaves. As the areas enlarge, they turn from light green or yellowish to brown and become soft. It is caused by Phytophthora
- Watery soft rot. This rot occurs on the lower part of heads. The tissue is water soaked and light or pinkish brown.A white, cottony mold spreads over the decayed tissue , and the head eventually becomes a watery mass.

Bacteria

Various groups of bacteria have ability to grow on fruits and its juices. These bacteria by virtue of their diversity in metabolism grow on fruits and produce different types of compounds. The major group of bacteria which are involved are:

- Lactic acid bacteria
- Acetic acid bacteria
- Spore formers

Lactic acid Bacteria

The lactic acid bacteria are Gram-positive, catalase negative organisms which can grow under anaerobic conditions. These are rod-shaped (lactobacilli), or cocci (pediococci and leuconostocs) . The homofermentative species produce mainly lactic acid from hexose sugars; the heterofermenters produce one molecule of lactic acid, one molecule of carbon dioxide, and a two-carbon compound, which is usually acetic acid or ethanol or a combination of the two.

Growth of lactic acid bacteria in juices and other fruit products cause the formation of haze, gas, acid, and a number of other changes. Certain heterofermentative lactobacilli lead to slime in cider. The lactobacilli and leuconostocs that are present in citrus juices generate acetylmethylcarbinol and diacetyl, compounds that give the juices an undesirable, buttermilk-like flavor. Some strains, being extremely tolerant to ethanol grow in wines. Lactobacillus fructivorans can grow in appetizer and dessert wines containing as much as 20% ethanol. Lactic acid bacteria have the ability to decarboxylate malic acid to lactic acid. This malolactic fermentation is often desirable in high-acid wines because the acidity is reduced and desirable flavors are produced. Oenococcus oenos is the most acid and alcohol-tolerant species and often is isolated from wines that are undergoing a malolactic fermentation.

Acetic acid Bacteria

These are Gram negative, aerobic rods having two genera, viz. Acetobacter and Gluconobacter. Both of these species oxidize ethanol to acetic acid under acidic condition, Acetobacter species can oxidize acetic acid to carbon dioxide thus, the genus is called as over oxidizer. Because the bacteria are obligate aerobes, juices, wines, and cider are most susceptible to spoilage while held in tanks prior to bottling. Some strains of Acetobacter pasteurianus and Gluconobacter oxydans produce microfibrils composed of cellulose, which leads to formation of flocs in different fruit juice beverages.

Spore formers

Spores are heat resistant, so role of organisms producing spores is important in heat treated juices and beverages. Variuos spore formers such as Bacillus coagulans, B. subtilis, B. macerans, B. pumilis, B. sphaericus , and B. pantothenticus have been found to grow in different types of wines. Some of these organisms have also been involved in canned fruits. Spore-forming bacilli that actually prefer a low pH have been responsible for spoilage of apple juice and a blend of fruit juices.

MICROBIAL SPOILAGE OF VEGETABLES

Introduction

Vegetables form an integral part of diet due to their role in providing various types of vital nutrients such as carbohydrates, minerals, vitamins, roughage etc. Vegetables being a part of fresh produce, contain high moisture which makes them highly perishable foods and hence more prone to spoilage. Microorganisms gain entry into vegetables from various sources. These sources include:

- Soil
- Water
- Diseased plant
- Harvesting and processing equipments
- Handlers
- Packaging and packing material
- Contact with spoiled vegetables

The conditions in which vegetables are stored and transported after harvesting also contribute to rate of spoilage. Other than microbial, sources, the spoilage of vegetables can also occur due to the activity of native enzymes.

Types of Spoilage in Vegetables

The microbial spoilage of vegetables is predominately of following types

Spoilage due to pathogens

The plant pathogens which infect stem, leaves, roots, flowers and other parts or the fruit itself.

Spoilage due to saprophytes

Vegetables have general microflora inhabiting them. These organisms under certain conditions can grow on these vegetables and spoil them. There are certain secondary invaders which may enter the healthy food or grow after growth of pathogens.

It is well known that plant diseases are mostly caused by fungi. Thus most of the spoilage causing pathogens in vegetables are fungi. Fungi have specific characteristics when spoiling food as it leads to mushy areas which may be water soaked.

The fungi produce characteristic spores which may be pigmented. The pigmentation helps in identification of the type of spoilage by fungi. The bacterial diseases too cause spoilage of vegetables but to a lesser extent.

Spoilage in vegetables is largely affected by composition of vegetable. The non acidic foods are thus spoiled by bacterial rot while acidic foods with dry surfaces are more prone to mold spoilage. The product on which organism grows and types of organisms growing largely determine the character of spoilage.

Bacterial Soft Rot

- Caused by *Erwinia carotovora and Pseudomonas* such as *P. marginalis. Bacillus* and *Clostridium* spp. are also implicated.
- Breaks down pectin, giving rise to a soft, mushy consistency, sometimes a bad odour and water-soaked apperance.
- Vegetables affected- onions, garlic, beans, carrot, beets, lettuce, spinach, potatoes, cabbage, cauliflower, radishes, tomatoes, cucumbers, watermelons
- Fungal spoilage of vegetables
- *Penicillium, Cladosporium, Rhizopus, Aspergillus* spp. are responsible for various defects in vegetables.
- *Gray mold rot* – caused by Botrytis cinera in vegetables. Favoured by high humidity and warm temperature.

MICROBIOLOGICAL SPOILAGE OF FRESH-CUT FRUITS AND VEGETABLES

Fresh-cut fruits and vegetables are "any fresh fruit or vegetable or any combination thereof that has been physically altered from its original form, but remains in a fresh state". Fresh-cut fruits and vegetables offer consumers ready-to-eat produce that is one of several convenient, nutritious and fresh-like tasting, and are a rapidly growing category of value-added produce products that are minimally or lightly processed. Processing of fresh-cut products involves sorting, cleaning, washing, heating/pasteurization (such as BiosteamR), trimming/peeling, coring, slicing, shredding, packaging and/or other related steps, depending on the product.

The processing can be as simple as fresh-cut grape tomatoes for which raw tomato fruit is only sorted and washed with sanitized water or as complicated as cut cantaloupe, for which cantaloupe is sorted, cleaned with brush and spray water, heat treated with hot water or steam, peeled, deseeded, chunked, and rinsed with sanitized water before packaging in rigid containers. The fresh-cut fruits include melon chunks and slices; cored and sliced pineapple; apple wedges treated with antibrowning preservatives; peeled citrus fruits and segments; de-capped strawberry; de-stemmed and washed grapes; sliced kiwifruits, and fruit salads.

Examples of fresh-cut vegetables are shredded lettuce, shredded and diced cabbage, washed and trimmed spinach, peeled "baby" carrots, cauliflower and broccoli florets, sliced or diced tomatoes, peeled and sliced potatoes, snapped green beans, trimmed green onions, cleaned and diced onions, and mixed salads. Compared with whole fresh produce, fresh-cut produce is ready-to-use (ready-to-eat), contains 100% usable product, and always requires processing, refrigeration (including chilling-sensitive fruits and vegetables that can be injured after a period of exposure to chilling temperatures below 10–15°C (50–59°F) but above their freezing points), and packaging, specifically modified atmosphere packaging (MAP, a packaging technique that utilizes package atmosphere other than air, <0.1% CO2, 20.9% O2, 78% N2, in a sealed package to extend shelf life of foods). Fresh-cut products are in a raw state or fresh-like, nutritious, and contain live tissues without freezing, canning (heat sterilization), dehydrating, fermentation, acidification, or treatments with additives or preservatives to prevent spoilage.

Fresh-cut is also described as precut, minimally processed, minimally processed refrigerated, lightly processed, partially processed, fresh processed, and pre-prepared in literature and application communications. Fresh-cut fruit and vegetable sales are approximately $12 billion per year in the North American foodservice and retail market and account for nearly 15% of all produce sales. Fresh-cut products offer produce growers/shippers an opportunity to increase sales by adding value to raw agricultural commodities.

Shelf Life of Fresh-Cut Fruits and Vegetables

The shelf life (*i.e.*, the length of time that corresponds to a tolerable loss in quality of a processed food and other perishable items) of fresh-cut fruits and vegetables ranges from 1 to 35 days depending on types of shelf life (such as marketing shelf life, food safety shelf life, sensory shelf life, or microbiological shelf life), food safety concerns, marketing strategies of fresh-cut processors, produce commodities, raw materials, refrigerated storage temperatures, preparation methods, and packaging methods.

Marketing shelf life of fresh-cut products, indicated by sell-by-date, best if use-by-date, use-by-date, or best use-by-day, is not necessarily the same as sensory shelf life or microbiological shelf life, depending on marketing concerns and competition. The average shelf lives of fresh-cut fruits and vegetables are typically 10–14 days.

However, the shelf life for kitchen- or store-prepared fresh-cut fruits for the catering industry and restaurants is only 1–2 days. FDA (2000) has suggested a 7-day shelf life at 5°C for fresh-cut melons prepared at home for microbiological safety reasons. Sensory shelf life and microbiological shelf life are usually the same; however, they can differ significantly.

For example, white surface formation on "baby" carrots can occur weeks before microbial spoilage is observed even though the peeled carrots are packaged in polyethylene pouches and stored at 2°C. Poubol and Izumi (2005) reported that fresh-cut mango made from cultivar "Nam Dorkmai" (50–60% yellow) had a shelf life as short as 2 days at 5°C due to browning discoloration and a water-soaked appearance.

However, fresh-cut celery of 10-cm sticks or 1.9-cm crescents retained its fresh appearance and flavor up to 28 days when stored at 2°C. Researchers reported that cut pineapple had a shelf life of more than 5 weeks at 1°C. Fresh-cut melons (such as cantaloupe and watermelon) prepared by retail stores (store-cut) and displayed on retail shelves typically have a 2-day shelf life. The shelf life of cut cantaloupe that is prepared by fresh-cut processors could be as short as 5 days at 7°C. Cantaloupe washed with hydrogen peroxide, cut and packed at research laboratories could last for more than 18 days at 4°C. Sterile diced cantaloupe stored in controlled atmospheres at 4.5°C had a shelf life of up to 28 days. Researchers reported a shelf life of 12 weeks at 1°C for sliced melon, using modified atmosphere packaging (MAP) and gas-impermeable containers.

Impact of Microbiological Spoilage

As processing and packaging technologies have improved during the last decade, microbiological spoilage or microbiological shelf life has become a major reason for sensory quality shelf life failure for most packaged fresh-cut fruits and vegetables, followed by surface discoloration (e.g., pinking of cut lettuce, browning of cut potato, graying and browning with processed pineapple, and gray discoloration with cabbage), water-soaked appearance or translucency (e.g., cut watermelon, papaya, honeydew, and tomatoes), moisture loss (e.g., "baby" carrots and celery sticks), off-aroma (e.g., broccoli florets and diced cabbage in low percent O2 and high CO2 packages), flavor changes (e.g., cut kiwifruit), and texture changes (e.g., processed strawberry, grated celery, kiwifruit, and papaya).

Microbial spoilage including off-flavor (e.g., fermented aroma with cut lettuce, sour taste with cantaloupe and bell pepper) formation, slimy surface (e.g., "baby" carrots), wetness and soft rot (e.g., cut bell pepper), discoloration (e.g., apple wedges), and visual microbial growth/colonies (such as apple wedges, cantaloupe chunks, and cored pineapple) has been used as a main or exclusive objective criterion to determine shelf life of fresh-cut products. Researchers concluded that microbial decay can be a major source of spoilage of fresh-cut produce. Researchers reported that microbial spoilage is a limiting factor for shelf life of fruit pieces stored under controlled atmosphere conditions.

Shelf life, including microbial spoilage, results in 30–50% shrinkage of fresh-cut fruits. Microbial spoilage has been used by quality assurance departments in the fresh-cut industry as the objective indicator for quality failure for more than 50% of fresh-cut vegetable commodities and almost 100% of fresh-cut fruit products that have been treated with preservatives (such as antibrowning reagents) and/or packaged properly using MAP technologies.

Under equilibrium modified atmosphere (MA) conditions, mixed fresh-cut bell pepper (including green, yellow, and red bell pepper) was unacceptable by day 6 of storage at 7°C due to acidic flavor, water loss, and texture change. Processed Lollo Rosso lettuce had a shelf life of shorter than 7 days at 5°C due to high microbial counts and off-odor formation under MAP. Grated carrots became wet and slimy, lost firmness, and produced off-odors during storage at 10°C under MAP.

The first indicator of changes in freshness for fresh-cut lettuce packed using active/passive MAP to prevent pinking or browning is fermented aroma formation (Table 1). Studies of cut cantaloupe revealed evidence of visual spoilage, including presence of microbial colonies, slime, and turbidity in juice, within 15 days of storage at 4°C.

Table 1 Changes in intensity of sensory attributes of fresh-cut lettuce during storage at refrigeration temperature (15-point universal scale was used in the test)

Shelf life (days)	Flavor	Hardness (texture)	Fermented aroma
0	6.3	5.3	3.8
4	5.8	5.2	4.3
5	5.6	5.8	6.1
6	4.6	7.3	5.9
7	6.0	6.7	4.6
10	4.3	7.6	6.3
11	6.2	6.3	6.8
12	5.9	6.6	6.9
P-value	0.671	0.131	0.047

Sources of Microbial Contamination

Contamination sources of fresh-cut fruits and vegetables include raw materials and contact with processing equipment. The microorganisms that exist on the surfaces of raw, whole produce appear to be the major source of microbial contamination and consequent spoilage of fresh-cut fruits and vegetables. Researchers reported that, compared with good surface sanitization practices, no decontamination treatment or an ineffective antimicrobial treatment on whole cantaloupe resulted in premature microbiological spoilage of freshcut cantaloupe.

Studies have also revealed over a 1-year period of sampling that there is a close relationship between the total mesophilic aerobic counts on lettuce raw material and those on finished shredded lettuce product. Researchers determined that the most common bacteria on raw celery plants, including fluorescent Pseudomonas spp. and Aeromonas spp., were also the most common bacteria on cut celery products. Researchers reported that the microbial decay of fresh-cut lettuce is largely due to the growth of microorganisms originating from preharvest environments.

Researchers determined that the types of microorganisms found on shredded lettuce were highly associated with the microorganisms detected on lettuce before shedding. Several studies have revealed that yeast species identified on fresh-cut produce can also be isolated from raw materials prior to processing. Researchers concluded that peel is the major source of microbial contaminants on carrot sticks. Several outbreaks of salmonellosis that were associated with cut cantaloupe and watermelon have resulted from Salmonella present on the rind contaminated in the field or packinghouse.

Inoculation of *Listeria monocytogenes* and Salmonella on the surface of entire cantaloupes resulted in the contamination to fresh-cut pieces during cutting. These results indicate that the bacteria on the surface of whole produce are the same as those on freshcut produce and can contaminate finished product through processing. Fresh-cut products can also be contaminated by spoilage microorganisms through contact by people or equipment during processing possibly by air during processing and packaging steps, especially in facilities that have been used for produce processing over an extended period of time.

Researchers found significantly higher bacterial counts during processing on automated cutters and package fillers of a lettuce processing line, indicating that clean product can become recontaminated after passing through operations where vegetable and fruit debris can accumulate, such as cutters and package-filling equipment. Shredding and slicing steps in fresh-cut processing resulted in increased microbial populations by 1–3 logs on cut cabbage, lettuce, and onions and at least a 1-log increase for lettuce and chicory salads.

Researchers found statistically significant differences for total plate counts and total and fecal coliforms in the last wash water step (rinsing water) compared to the water used during the previous disinfecting stage (3.52 log10 vs. 0.94 log10 for TPC, 1.86 log10 vs. 0.04 log10 for total coliforms). Researchers reported that the microbial population significantly increased on broccoli florets after the washing step. Researchers determined that shredding, rinsing, and centrifugation of red lettuce Lollo Rosso increased coliform, lactic acid bacteria, and psychrotrophic bacterial counts. Yeast populations on cut cantaloupes increased after packaging.

Several researchers have suggested that the large numbers of lactic acid bacteria and fungi present on fresh-cut products indicate likely contamination from processing, such as cutting machines. For example, Geotrichum candidum has been termed "machinery mold" because it can accumulate on fruit-processing equipment.

UNUSUAL CHARACTERISTICS OF SPOILAGE MICROORGANISMS

Many fruits and vegetables present nearly ideal conditions for the survival and growth of many types of microorganisms. The internal tissues are nutrient rich and many, especially vegetables, have a pH near neutrality. Their structure is comprised mainly of the polysaccharides cellulose, hemicellulose, and pectin. The principal storage polymer is starch. Spoilage microorganisms exploit the host using extracellular lytic enzymes that degrade these polymers to release water and the plant's other intracellular constituents for use as nutrients for their growth. Fungi in particular produce an abundance of extracellular pectinases and hemicellulases that are important factors for fungal spoilage.

Some spoilage microbes are capable of colonizing and creating lesions on healthy, undamaged plant tissue. Spoilage microorganisms also can enter plant tissues during fruit development, either through the calyx (flower end) or along the stem, or through various specialized water and gas exchange structures of leafy matter. Successful establishment, however, requires the spoilage microbe to overcome multiple natural protective barriers. Fruits and vegetables possess an outer protective epidermis, typically covered by a natural waxy cuticle layer containing the polymer cutin.

A diverse community of epiphytic microorganisms that present a further competitive barrier to the spoilage organism also typically colonizes the outermost fruit surface. Overcoming these barriers requires an exquisite set of biochemical tools that allow the spoilage microorganism to (1) identify and recognize the plant surface; (2) employ one or more strategies to achieve irreversible attachment to the plant surface; and (3) initiate steps leading to internalization of the tissue. On plant structures other than the fruit, internalization can be achieved through a number of specialized vessels and surface structures employed by the plant to absorb and release water and to provide $CO2$ and $O2$ exchange.

However, the fruit of the plant lacks many of these structures, requiring the spoilage microbe to employ other methods to become internalized. This may partially explain the rather limited success of bacteria to spoil fruits and an improved ability to spoil vegetables that are not the fruit of the plant.

The natural acidity of most fruits also serves as a barrier to many spoilage microbes, especially bacteria. By contrast, spoilage fungi that typically produce more diverse and greater amounts of extracellular depolymerases successfully attack and spoil both fruits and vegetables.

Colonization and lesion development more typically and more rapidly occurs within damaged or otherwise compromised plant tissue. External damage such as bruising, cracks, and punctures creates sites for establishment and outgrowth of the spoilage microbes. Lesion development can be relatively rapid, occurring within days or weeks. This presents the risk that rapidly reproducing spoilage microorganisms will arrive within open wound sites at the packing facility, and thereby, through shedding from the asymptomatic wound, present the potential for cross-contamination within the facility during handling, culling, washing, sorting, and packing before storage.

Such cross-contamination to some degree is inevitable and, if not carefully managed with a robust facility sanitation program, could lead to the establishment of a population of spoilage microbes endemic to the facility that may be difficult to eradicate.

A further and potentially more serious complication is the introduction into the cold storage facility of spoilage microorganisms already established in wound sites on product, whether the product is in bins or boxed and palletized. Depending upon storage conditions and storage time (greater than 12 months for certain robust crops), and if not carefully managed, these "primed" spoilage microorganisms can have a devastating impact on the stored product. Apples, for example, are stored in very large, controlled atmosphere storage rooms, either in wooden bins or boxed and ready for distribution.

Two wound pathogens, *Penicillium expansum* and *Botrytis cinerea*, if not scrupulously cleaned from fruits prior to storage or if fruits with infected wounds have not thoroughly been culled from the lot, can cause significant crop loss as these spoilage fungi eventually degrade the wound sites, create lesions, and cross-contaminate adjacent fruits. If fruits receive improper preharvest fungicide application, poor washing, and/or inadequate culling, an expanding infestation of spoilage microorganisms can destroy a substantial portion of a stored lot of fruits. *P. expansum* and *B. cinerea* are pathogens of apples, pears, and a number of other pectin-rich fruits. *B. cinerea* is an especially sophisticated and selective plant pathogen that possesses multiple cutinases and lipases that are capable of degrading plants rich in pectin.

The bacterium Erwinia carotovora subsp. carotovora is a highly effective spoilage microbe that causes soft rot across a broad host range of vegetables and some fruits. One of six known genera of soft-rot bacteria (including *Xanthomonas, Pseudomonas, Clostridium, Cytophaga,* and *Bacillus*), *E. carotovora* subsp. carotovora is one of several species of *Erwinia* that infect and destroy plant tissues both pre- and post-harvest and is the species that causes the greatest damage to harvested vegetables. Soft rot is a form of decay characterized by a watery transparency in infected leafy plant parts and watery disintegration of nonleafy plant materials.

"Soft-rot erwinia" tend to initiate infection and decay at wound sites and, once established, can quickly advance to total destruction of the product. Soft-rot erwinia express four pectin-degrading extracellular enzymes: pectin lyase, polygalacturonase, pectin methylesterase, and pectate lyase. Of these enzymes, pectate lyase is primarily responsible for extensive decay. E. carotovora has built-in redundancy for this apparently critical pathogenicity factor, expressing four distinct extracellular pectate lyase isozymes.

Soft-rot erwinia are active only at temperatures of 20°C and above, which reinforces the need to maintain a continuous cold chain from immediately postharvest to retail to successfully manage this ubiquitous spoilage bacterium. Another group of soft-rotting bacteria, the fluorescent pseudomonads (i.e., Pseudomonas fluorescens and Pseudomonas viridiflava), can decay plant tissue at temperatures at or below 4°C. This is one explanation for the high prevalence of these bacteria on decayed vegetables at wholesale and retail markets. Researchers revealed through several P. viridiflava mutants defective in pectate lyase expression that these mutants completely lost the ability to induce soft rot on potato tuber slices. The soft-rotting fluorescent pseudomonads, when considered together with soft-rot erwinia, present a formidable challenge to commercial fresh product operations, and fresh vegetables in particular, from the farm to retail and wholesale outlets.

Pseudomonas tolaasii, another fluorescent pseudomonad and fresh produce spoilage bacterium, has a much narrower range of host-specificity than *P. fluorescens and P. viridiflava. P. tolaasii* causes spoilage of the white mushroom, Agaricus bisporus. Similar to *P. fluorescens and P. viridiflava, P. tolaasii* produces siderophores that fluoresce under ultraviolet light. However, unlike the soft-rot pseudomonads, *P. tolaasii* does not cause soft rot on plants (*i.e.*, it does not produce pectin depolymerases) but instead creates unsightly blemishes on the caps and stems of the Agaricus fruiting body as a result of localized infection and decay of those parts of the mushroom. Researchers identified three pathotypes of mushrooms based on pathology and fatty acid analysis: *P. tolaasii* and *P. gingeri* which cause severe and yellowed lesions and *P. reactans* which causes a mild discoloration of the infected area.

4

POSTHARVEST PHYSIOLOGY AND BIOCHEMISTRY

Postharvest physiology is about the plant response to technologies and other applications that extend shelf life and quality and delay senescence (plant death). Quality itself includes aspects that are appealing to consumers, such as appearance, aroma, taste, color (sensory), reduced risk of foodborne pathogens or pesticide residues (food safety), or more dense nutrient or phytonutrient content. Growers and shippers realized that keeping produce or flowers cold greatly extended the sales life and this stimulated much interest in exploring the causes of plant senescence at physiological, cellular, and molecular levels. As consumers continue to increase their daily intake of fruits and vegetables, pressure to deliver high-quality products with a longer shelf life is critical.

The postharvest loss might be defined as the degradation in both quality and quantity of a produce occurred after harvest till the consumption. The term "quality" refers external, internal, and hidden attributes, including weight loss, changes in color, visual quality "the acceptability", occurrence of decay, changes in nutrient content, flavor, etc. And "quantity" refers to the loss of the amount of a product. Sometimes, loss in quality does not change the quantity but highly affect marketability and price of the product. Losses arise due to the fact that freshly harvested fruits, vegetables, and flowers are living things which breathe and undergo changes during postharvest handling. There are some biological factors which cause deterioration on harvested produce, these are:

- Respiration: Respiration is a basic process of all plant materials. It might simply be defined as the reverse of photosynthesis. It is a set of metabolic reactions take place in the cells of organisms to convert biochemical energy from stored food (produced during photosynthesis, i.e. starch and sugar) into adenosine triphosphate (ATP) for growth, ripening etc. During respiration, plants take oxygen from the air and give off carbon dioxide, moisture, and heat. Respiration continues until the stored starch and sugar reserves are depleted and crops would age and die. Temperature dictates the speed of respiration and is the most important factor influencing the postharvest life of the given produce. The oxygen in the surrounding environment is at utmost important, while it is the primary gas used during respiration. Lowering the oxygen in the air lowers the respiration rate, but if there is not enough oxygen, and then, product goes in anaerobic fermentation and produces alcohols with abnormal flavors.

- Transpiration (water loss): Most fruits and vegetables contain between 80 and 95% water by weight. The loss of water in a vapor state from living tissues is known as transpiration. It causes shriveling, wilting, softening, poorer texture, loss in weight, and lower quality. It can be reduced in storage by (1) raising the relative humidity, (2) reducing the air movements, (3) lowering the air temperature, (4) using protective coverage, i.e. waxing, and (5) protective packaging, i.e., polyethylene film, modified atmosphere packaging etc. Furthermore, many studies conducted about the protective and anti fungal effects of natural extracts and/or plant and animal derived products on the postharvest quality of fruits and vegetables. For example, propolis extract, eucalyptus oil, and *Aloe vera* jell are reported to be successful in slowing down transpiration and having anti fungal effects.
- Ethylene: Ethylene gas (C_2H_4) is a colorless, odorless, natural hormone produced by some fruits (climacteric) and vegetables as they ripen and promotes additional ripening of produce exposed it. This can lead to the premature ripening of produce in storage facilities. Ethylene is capable of stimulating many reactions within plants. It is involved in the normal ripening process in many fruits, such as apples and bananas. Ethylene can also have undesirable effects on fruits, i.e. premature ripening, skin damage etc. The presence of CO_2, lack of O_2, and low temperatures can inhibit ethylene production on fruits, while on the other hand ethylene production is higher at injured produce. Ethylene can also be produced artificially and used as an environmental factor to stimulate ripening. An important point here is that the plants which produce ethylene, i.e. apple, should not be stored with fruits, vegetables or flowers known to be sensitive to it, i.e., cabbage. It may cause injuries on produce, loss of quality, and reduce shelf life.
- Postharvest diseases: Stored products are subject to a variety of rots and decay caused by fungi or bacteria. Most known fungus are *Penicilliumexpansum*, *Botrytis cinerea*, *Alternaria alternata*, *Rhizopus stolonifer*, *Phytophthora infestans* and *Fusarium* spp and the bacteria are *Ervinia carotovora* and *Pseudomona* spp. These diseases might cause light brown and soft spots on fruits and vegetables. Infection of diseases may start before or after harvest. When products transferred to storage, infections continue to develop. Mechanical damages, wounds or bruises are known to be the common entry points for bacteria and fungi. To prevent postharvest diseases, careful monitoring and management of diseases need to be started during growing period and continue in the storage. Preventing mechanical damage and harvesting the products during the cool times of the day are crucial points. Preharvest and postharvest application of suitable fungicides, bactericides might be helpful in managing disease problems. However, it must

be kept in mind that environmental conditions are highly important for the development of diseases. They usually require warm temperatures and high moisture. On the other hand, sanitation is of utmost importance for handlers not only to protect products from postharvest diseases but also to protect consumers from foodborne illnesses, i.e., *Escherichia coli* 0157: H7 and *Salmonella*.

On the other hand, there are some environmental factors (temperature, relative humidity, atmospheric composition, and light) which accelerate or retard deterioration by directly or indirectly influencing biological factors:

- Temperature: Temperature is the most important environmental factor which influences the postharvest life of a produce. The optimum storage temperature of commodities is differing among not only commodities but for species. Typically, for every increase of 10°C, the rate of deterioration increases between 2 and 3 fold. Temperature greatly affects water loss. Lowering the temperature also slows the pathogen development. Temperatures outside the optimal range can cause chilling injury, freezing or heat injuries. The severity of chilling, freezing and heat injuries depend on the storage duration and temperature. Among these three injuries, chilling injury is the most known and threat for storage. The critical temperature for chilling injury is mainly below 5–13°C depending on produce and maturity stage. Chilling injury symptoms are changing among commodities but generally include brown discoloration of the skin, necrotic pitting, and increased susceptibility to decay. Intermitted warming, application of some essential oils, salicylic acid, jasmonic acid, calcium chloride, etc. are reported to be helpful in delaying or preventing chilling injury.
- Relative humidity: While fruits and vegetables contain between 80 and 95% water by weight, it is utmost important to provide favorable environmental conditions to reduce transpiration of the produce. Higher relative humidity (85–95%) slows water loss from the commodity. However, at the same time, high relative humidity (free moisture) might stimulate pathogen development and might weaken the packaging materials i.e. cartoon boxes.
- Atmospheric composition: Respiration is the basic process causing deterioration on the harvested produce and is mainly depend on the atmospheric composition (level of O_2 and CO_2) as well as on the temperature, ethylene and water vapor. Therefore, regulating the gas concentrations in the surrounding atmosphere of the produce is highly important for reducing respiration and increasing preservation time. Reduction of O_2 and elevation of CO_2can delay deterioration of fresh horticultural crops. However, it is highly depended on the type of commodity, cultivar, maturity, and temperature. Modified atmosphere packaging (MAP) is a useful system which makes it possible to regulate the composition of the atmosphere in the packaging headspace. During respiration, O_2 is consumed, and CO_2, ethylene, and water vapor are generated, thus the packaging material allows the transfer of all of these gasses through the packaging material by regulating the inner composition at favorable levels to preserve the produce. MAP slows

down respiration and other metabolic processes, reduces sensitivity to ethylene, reduces the development of some physiological disorders i.e. chilling injury and may inhibit pathogen development.

- Light: Light is also a cause of some abnormal changes in product quality. It might affect some biological process. For example, exposure of potatoes to light would results in formation of chlorophyll, which appears as greening and formation of solanine which is known to be toxic to humans.

The big question here is "how to handle produce and the surrounding environment to retard this aging process and prevent quality?" For example, it is important to reduce temperatures to prevent respiration and transpiration, but decreasing temperatures may cause chilling or freezing injuries. On the other hand, increasing relative humidity is important for reducing transpiration and weight loss, but free moisture stimulates pathogen developments. Moreover, factors affecting postharvest losses vary widely from produce to produce. In conclusion, correct management of the environmental factors is crucial for the prolongation of the postharvest life of produce. Pre harvest applications, correct maturity at harvest, harvesting gently, pre cooling, transportation conditions, sorting, sanitation, fungicide treatment, protective coverage, grading, sizing, packaging, and storage are the basic steps in postharvest handling. In the light of this information, the present book is intended to provide useful and scientific information about the postharvest handling of different produces.

PHYSICAL TREATMENTS

Heat treatment

Heat treatment has been studied as an alternative to chemical treatments for harvested FFV. Treatments include hot water dip (HWD), saturated water vapour heat, hot dry air and hot water rinse (HWR) with brushing. Beneficial effects of these heat treatments are linked (i) through changes in physiological processes such as a reduction in chilling injury and delay of ripening processes by heat inactivation of degradative enzymes, (ii) by killing of critical insect contaminations, and (iii) by controlling the onset of fungal decay. Heat treatments can be of short- (up to 1 h) or long-term duration (up to 4 days). Heat treatments have been applied to firm potatoes, tomatoes, carrots and strawberries; to preserve the colour of asparagus, broccoli, green beans, kiwi fruits, celery and lettuce; to prevent development of overripe flavours in cantaloupe and other melons; and to generally add to the longevity of grapes, plums, bean sprouts and peaches, among others.

It has been demonstrated that heat shock by using hot water washing at temperatures ranging from 37 to 55°C for a duration of 30 s to 3 min can improve the postharvest quality of spinach, rocket leaves, apples and mandarin fruit. A clear mode of action of any water treatment is to wash-off the spores from the fruit surface.

Hot water is a better vector of energy than air and has provided comparable reductions in fungal decay. Blue mould on grapefruit caused by *Penicillium* sp. has been controlled by dipping fruit in hot water for 2 min at 50°C. Improvements in the quality of bell pepper, apples, melons, sweet corn, kumquat and grapefruit have been reported with cold water cleaning in combination with brushing and a short HWR. Hot water treatments also influence the structure and composition of epicuticular waxes. Covering of cracks and wounds and the formation of anti-fungal substances in the wax after heating are thought to be possible modes of action.

Researchers suggested that the combination of *Bacillus amyloliquefaciens* HF-01, sodium bicarbonate and hot water could be a promising method for the control of postharvest decay on citrus while maintaining fruit quality after harvest. To date, commercial applications of heat treatments are limited. Heat treatment provides an alternative to fungicide applications, and in Germany HWD has been used in the storage of organic apples. Treatment of fruit after a few days of cold storage or immediately after the opening of a long-term controlled atmosphere (CA) storage room provides new options for prolonging their subsequent storage life, although acceptance of this technology by fruit growers has been hampered by high-energy costs and also the need for added labour at the peak work period during harvest time.

Edible coating

Edible coatings are thin layers of external coatings applied to the surface of fresh produce to enhance the waxy cuticle or as replacements for natural barriers where the produce cuticle has been removed. The application of edible coatings on fresh produce provides a partial barrier to the movement of moisture on the surface of fresh produce, thereby minimizing moisture loss during postharvest storage; a gas barrier, thereby establishing a modified atmosphere around the product, which slows down respiration, senescence and enzymatic oxidation and preserves colour and texture; helps to retain volatile compounds contributing to produce a natural aroma and restrict foreign odours; maintains fresh produce structural integrity, and protects against mechanical damages; and serves as carriers of functional or active compounds, such as nutraceuticals, flavouring and colouring agents, antioxidants and antimicrobials, that will maintain/improve product quality and safety. Edible coatings are composed of hydrophobic groups, such as lipid-based waxes; hydrocolloid/hydrophilic groups, such as polysaccharide or protein-based materials; or an integration of both groups in order to improve the functionality of the coating. Within the last decade, there has been a considerable amount of research and innovations focused on the development of edible coatings from natural or synthetic sources in order to control physiological and pathological challenges of fresh produce (Table 1).

TABLE 1. SUMMARY OF EDIBLE COATING(S) USED ON FRESH/FRESH-CUT FRUIT AND VEGETABLES.

Coating material	Purpose of coating
guar gum; pea/potato starch ± potassium sorbate	antimicrobial
candelilla wax-based	antimicrobial; antioxidant; quality
soya bean gum; jojoba wax; glycerol and arabic gum	overall quality

Coating material	Purpose of coating
Shellac ± Aloe vera gel	keeping quality
soy protein; carboxymethyl cellulose	antioxidant; H_2O barrier
chitosan; zein	antioxidant; H_2O barrier
beeswax; coconut and sunflower oil	antimicrobial; antioxidant; quality
pectin base; alginate; carboxymethyl cellulose	antioxidant; H_2O barrier
chitosan; methyl cellulose	antimicrobial; antioxidant; $O_2/CO_2/H_2O$ barrier
soy protein; carboxymethyl cellulose	antioxidant; H_2O barrier
pectin base	overall quality
Aloe vera gel	overall quality
agar; chitosan; acetic acid (combined)	antimicrobial; O_2/CO_2 barrier
whey protein; rice bran oil	H_2O barrier; overall quality
chitosan	overall quality
sucrose-polyester based	H_2O barrier; antioxidant activity
alginate and gellan based	$O_2/CO_2/H_2O$ barrier

Several edible coatings including chitosan, *Aloe vera*, polyvinyl acetate, mineral oils, cellulose and protein based have shown desirable attributes on fresh produce with good barrier properties, without residual odour or taste and efficient antimicrobial activity.

However, more research is required to enhance moisture barrier properties of hydrophilic edible coatings, improve coating adhesion and durability during storage. To maximize the benefits of edible coatings for fresh produce, it is important to understand the effect of storage conditions on the desired functions and the adverse effect on fresh produce quality. The main limitation for the application of edible coating at the industrial level is the cost of scaling up research concepts or investment for new installation of film production and coating equipment, the lack of edible materials with desired physical and functional properties as well as the challenges of regulatory status for the different coating materials. Furthermore, process parameters, such as the method of coating and the amount of additives, can affect the film barrier properties and overall quality of the food product.

One of the commercial coating products is Natureseal, which maintains colour, texture and shelf life of a number of fresh-cut fruits e.g. apples, pears, carrots, celery, etc., has recorded a good success. However, further research development is required to investigate the influence of edible coatings on individual cultivars of fresh-cuts in order to understand the variation in shelf life.

Irradiation

Irradiation exposes food to radiant energy from *ã* rays and e-beam (high-energy electrons) that penetrate objects and break molecular bonds, including the DNA of living organisms. Ionizing radiation from cobalt-60 or caesium-137, or machine generated electron beams are used as a source of irradiation for extending shelf life of fresh produce. By inhibiting cellular reproduction, irradiation can neutralize pest and food safety problems. The effect depends on the doses, measured in kilograys (kGy). Low doses of irradiation (less than 1 kGy) only disrupt cellular activity enough to inhibit sprouting of tubers, bulbs and roots and delay senescence. Medium doses (1–10 kGy) reduce microbial loads while high doses (more than 10 kGy) kill a broad spectrum of fungi and bacteria spp. and pests.

Most medium- and high-level doses are not appropriate for fresh produce because they can cause sensory defects (visual, texture and flavour) and/or accelerated senescence due to the irreparable damage to DNA and proteins. Irradiation presents an effective postharvest treatment for destroying bacteria, moulds and yeasts, which cause food spoilage, and also control insect and parasite infestation resulting in reduced storage losses, extended shelf life and improved parasitological and microbiological safety of foods. Irradiation has been commercialized for control of potato and onion sprouting, and strawberry decay. Low-dose γ-irradiation on mango (0.3–0.7 kGy) resulted in delay in ripening and extension of shelf life by a minimum of 3–4 days.

Recently, researchers reported an irradiation dose of 1 kGy to be the only effective dose in which enhanced shelf life was achieved without any deterioration of various quality attributes of litchi fruit. While much of the focus of irradiation use on FFV has been for extending shelf life and reducing decay, it has been known for many decades that irradiation is effective at killing, sterilizing or preventing further development of a wide variety of insect pests of quarantine importance on perishable FFV.

Despite some misconceptions, exposing food to irradiation does not make the food itself radioactive. The irradiation process produces very little chemical change in food and does not change the nutritional value of food. Extensive research and testing has demonstrated that irradiated food is safe and wholesome.

CHEMICAL TREATMENTS

Antimicrobial and anti-browning agents

Over the past decade, the increasing number of reported outbreaks of foodborne illnesses has heightened the concern of regulatory agencies, producers and the consumers about the microbial safety of FFV. Outbreaks have been associated with vegetables such as cabbage, celery, cucumber, leeks, watercress, lettuce and sprouts. Antimicrobial and anti-browning agents offer the possibility to maintain safety and can be grouped into chemical- and natural/bio-based agents.

Chemical-based agents include chlorine-based solutions, peroxyacetic acid (PAA), organic acids, hydrogen peroxide (H_2O_2) and electrolysed water. A chlorine-based solution such as NaClO has been one of the commonly used disinfectants for fresh produce, owing to its very potent oxidizing properties and cost effectiveness.

However, its efficacy as an antimicrobial agent is dependent on the levels of chlorine and at high levels may cause taste and odour defects on treated products. Additionally, chlorine-based compounds have been reported to have limited effectiveness in the reductions of microbial load on fresh produce.

Surfactants, detergents and solvents, alone or coupled with physical manipulation such as brushing, may be used to reduce hydrophobic nature of the waxy cuticle or remove part of the wax to increase exposure of microorganisms to chlorine. However, chlorine has been associated with the possible formation of carcinogenic chlorinated compounds and this may lead to new regulatory restrictions in the EU.

PAA is a very strong oxidizing agent, with no harmful by-products. PAA has been reported to be effective in controlling *E. coli* O157:H7 and *L. monocytogenes* on apples, strawberries, lettuce and cantaloupe. A 5 log reduction in *Enterobacter sakazakii* was reported for lettuce when treated with PAA. Researchers reported that decontamination treatment of fresh-cut carrot with PAA reduced the initial load of aerobic mesophilic bacteria by about 4 log units and yeasts and moulds by 3.5 log units and no further microbial growth was observed during storage.

H_2O_2 possesses a bactericidal, sporicidal and inhibitory ability, owing to its property as an oxidant and being able to generate other cytotoxic oxidizing species, such as hydroxyl radicals. Treatment with H_2O_2 can extend the shelf life and reduce natural and pathogenic microbial populations in melons, oranges, apples, prunes, tomatoes, whole grapes and fresh-cut produce. However, H_2O_2 treatment requires a long duration of application and can cause injury on some produce. Also, it is accepted as a generally recognized as safe for some food applications but not yet approved as an antimicrobial agent. However, a recent study by Lopez-Galvez *et al.* found that the newly developed H_2O_2-based sanitizers provoked a significant increase in the respiration rate and the electrolyte leakage of fresh-cut iceberg lettuce compared with tap water washing.

Organic acid, ascorbic acid and calcium-based solutions have been applied largely to slow down enzymatic and non-enzymatic browning, deterioration of texture and microbial growth on fresh produce. Treatment of fresh-cut melon dipped in 0.52 mM citric acid for 30 s prior to modified atmosphere packaging (MAP) maintained microbial safety and prevented translucency and discoloration. Inhibitory effects of organic acids (acetic, lactic and malic acids) combined with MAP on foodborne pathogens, including *E. coli* O157:H7, *S. Typhimurium* and *L. monocytogenes*, on cabbage was reported by Bae *et al.* However, there are factors limiting the efficacy of antimicrobial and anti-browning agents such as internalization of bacteria and inaccessible sites within fresh produce such as the calyx. These limitations highlight the need for novel means of applying of antimicrobial and anti-browning agents.

Nitric Oxide

Nitric oxide (NO) is a highly reactive free radical gas and acts as a multifunctional signalling molecule in various plant physiological processes, such as fruit ripening and senescence of FFV. Endogenous NO concentrations decrease with maturation and senescence in FFV, thereby offering an opportunity for modulation of their levels with exogenous application to exert the opposite effects.

Optimum NO levels delay the climacteric phase of many tropical fruits, prolong the postharvest storage life by impeding ripening and senescence, suppress biosynthesis of ethylene, reduce ethylene production and, consequently, delay in fruit ripening. NO gas is applied as a fumigant or released from compounds such as sodium nitroprusside,

S-nitrosothiols and also diazeniumdiolates used as a dipping treatment. Reduced ethylene production during ripening in NO-fumigated fruit has been claimed owing to binding of NO with 1-aminocyclopropane-1-carboxylic acid (ACC) and ACC oxidase to form a stable ternary complex, thus limiting ethylene production. Other mechanism of NO action include the inhibition of ethylene biosynthesis, cross-communication with other phytohormones, regulation of gene expression and amelioration of oxidative postharvest stress.

Successful application of NO has been reported for apple, banana, kiwifruit, mango, peach, pear, plum, strawberry, tomato, papaya, loquat, Chinese winter jujube fruit and Chinese bayberry. NO treatments reduce the rate of respiration, water loss and inhibit browning, as well as reducing the incidence of postharvest diseases. NO fumigation in combination with cold storage has a synergistic effect in extending storage life of fruit such as plum and mango. Similarly, NO combined with modified atmosphere conditions extended the postharvest life of green beans, broccoli and bok choy. Commercial application of NO in FFV depends upon the development of a smart carrier/controlled release system for NO.

Sulfur Dioxide

Sulfur dioxide (SO_2) is widely used on table grapes to prevent decay during storage, by either initial fumigation of fruit from the field followed by weekly fumigation of storage rooms or slow release from in-package pads containing sodium metabisulfite. SO_2 technology has also been tested for control of postharvest decay on other fruits such as litchi, fig, banana, lemon or apple. Cantín *et al.* reported that SO_2 fumigation followed by CA storage (3%O_2+6 or 12%CO_2) is a promising postharvest strategy for fresh blueberries to reduce decay, extend market life and maintain high nutritional value. Rivera *et al.* demonstrated that SO_2 is an effective and practical technology for reducing the risk of blueberry grey mould decay during storage and it could be used for the export market. There are disadvantages to SO_2 use; the SO_2 concentration necessary to inhibit fungal growth may induce injuries in grape fruits and stems, and sulfite residues pose a health risk for some individuals, as well as firming of the texture of some fruit species (pomaces), incomplete de-sulphiting and incomplete re-colouring of red fruits. Nevertheless, SO_2 treatment is a widespread process because of its advantages of universal antiseptic action and economic application.

GASEOUS TREATMENTS

Ozone

Recent research and commercial applications have verified that ozone can replace traditional sanitizing agents. Ozone is a very pungent, naturally occurring gas with strong highly reactive oxidizing properties. Ozone is reported to have 1.5 times the oxidizing potential of chlorine and 3000 times the potential of hypochlorous acid. Contact times for antimicrobial action are typically four to five times less than that for chlorine. Ozone rapidly attacks bacterial cell walls and is more effective than chlorine against the thick-walled spores of plant pathogens and animal parasites, at practical and safe concentrations. Researchers reported that the fruit exposed to 2.5 ppm ozone had higher levels of total soluble solids, ascorbic acid content, β-carotene content, lycopene content, and antioxidant activity and also reduced weight loss at day 10 compared with untreated fruit. The sensory attributes of papaya of ozone-treated fruit was also superior in sweetness and overall acceptability endorsing ozone as a non-thermal and safe food preservation technique for FFV.

Ozone can be employed in cold storage, washing system or process water sterilization. Huyskens-Keil *et al.* reported that irradiation and washing with ozonated water slightly reduced respiration in white asparagus spears, but increased spear tissue toughness. However, neither washing the asparagus spears with ozonated water (3 or 4.5 ppm) nor treating them with radiation (1 kJ m^{-2}) systematically and significantly affected their microbial loads during storage. Some commercial use has occurred with commodities such as apples, cherries, carrots, garlic, kiwi, onions, peaches, plums, potatoes and table grapes. However, ozone does not penetrate natural openings or wounds efficiently. Additional research is needed to define the potential and limits of the effective use of ozone for postharvest treatments for the quality and safety of FFV.

Ethylene

Endogenous ethylene production and its exogenous application exhibit both beneficial and deleterious effects on horticultural fresh produce. Beneficial effects of exogenously applied ethylene includes triggering ripening, improving fruit colour and quality in some crops, such as bananas and avocados, kiwifruit, persimmon, tomato, mangoes, de-greening of citrus fruit. The deleterious effects of ethylene in postharvest phase horticultural commodities has also been documented, such as shorter storage life, promotion of senescence, fruit softening, discoloration (browning) and russet spotting in lettuce, yellowing of leafy vegetables and cucumbers and increased susceptibility of FFV to decay.

Therefore, ethylene management plays a pivotal role in maintaining postharvest life and quality of climacteric and non-climacteric horticultural produce. Most commercial strategies for maintaining horticultural commodities involve storing at low temperatures, blocking ethylene biosynthesis and its action, minimizing exposure produce to ethylene during ripening, harvest, storage and transport by controlling temperature and atmospheric gas composition. Newly developed ethylene measurement devices will enable to detect critical concentrations during storage and transportation.

Beneficial effects of ethylene biosynthesis inhibitors such as amino-ethoxyvinylglycine alone on postharvest quality have been demonstrated in apples and stone fruits and in combination with controlled atmosphere (CA) storage. Treating FFV with inhibitors of ethylene action, such as 1-methylcyclopropene (1-MCP) or NO alone or in combination with MAP or CA storage, also impedes ethylene production and action consequently extends storage life and maintains quality of FFV.

Commercialization and limitations of different ethylene inhibitors are discussed in detail by Martínez-Romero *et al.* A recent development has been palladium-promoted zeolite materials, which can be effective ethylene scavengers to prolong the shelf life of climacteric fresh produce, such as bananas and avocados. The material has the potential to be used commercially, as an alternative and/or supplemental treatment to 1-MCP. Martínez-Romero *et al.* developed a carbon-heat hybrid ethylene scrubber for fresh horticultural produce storage purposes. The device comprised a cartridge heater tightly joined to the activated carbon–1% palladium. Application of heat pulses leads to an increase in ethylene oxidation and to auto-regeneration of the activated carbon.

1-Methylcyclopropene

The discovery and patenting of cyclopropenes as inhibitors of ethylene perception represents a major breakthrough in controlling ethylene responses of horticultural products. The process of discovery of the effects of cyclopropenes, and their proposed method of action, has been described. Of the cyclopropenes, 1-MCP proved to be extremely active, but unstable in the liquid phase. However, 1-MCP can be complexed with *á*-cyclodextrin to maintain its stability; this development represented a major step towards its commercialization as it was then possible to release 1-MCP from the complex to expose to the horticultural products. Regulatory approval for use of 1-MCP has been obtained in more than 50 countries, and approval for use of the technology continues to occur around the world. 1-MCP is registered for use on a wide variety of FFV including apple, avocado, banana, broccoli, cucumber, date, kiwifruit, mango, melon, nectarine, papaya, peach, pear, pepper, persimmon, pineapple, plantain, plum, squash and tomato.

1-MCP affects many ripening and senescence processes, including pigment changes, softening and cell wall metabolism, flavour and aroma, and nutritional properties, but to varying degrees in both non-climacteric and climacteric products. While aqueous 1-MCP shows similar responses as those treated with gaseous 1-MCP, ripening actors such as activity of cell wall-associated enzymes, e.g. lycopene, antioxidant and volatiles of avocado, are delayed but recover to reach levels similar to those of untreated fruit. The range of responses reflects the enormous diversity of these crops in terms of both inherent diversity and morphological derivation. Several generalizations can be made about responses of crops to 1-MCP:

- Genotype, cultivar and maturity effects can be highly variable;
- Climacteric fruit are affected by 1-MCP treatment, but the capacity to interrupt ripening once initiated varies by fruit and attributes studied;
- Non-climacteric fruit also respond to 1-MCP, providing insights into ethylene-dependent and -independent events during ripening;
- Treated fruits are firmer, slower to soften, slower to change peel colour, and develop aroma and flavour. If 1-MCP concentrations and exposure periods for each fruit must be appropriate to allow ripening to occur, so the final quality of the treated fruit is similar to that of the untreated product; and

- The effect of 1-MCP on physiological disorders is dependent on the role of ethylene. 1-MCP decreases senescent-related and ethylene-induced disorders (senescent breakdown of apples and water soaking of watermelons). 1-MCP can increase susceptibility of apple fruit to CA storage-related disorders (CO_2 injury). Some chilling-related disorders (woolliness and internal breakdown of peaches and nectarines, chilling injury of citrus and bananas) can be increased by inhibition of ethylene production. Other chilling-related disorders (superficial scald of apples and pears and internal flesh browning of avocados and pineapples) are decreased by inhibition of ethylene production.

Registration of 1-MCP for FFV has focused on major or specialty products important to specific countries. The apple has been an excellent crop for use of 1-MCP, and the technology is used extensively around the world to maintain quality through the whole marketing chain from storage to consumer. Success of 1-MCP technology for apples is largely associated with a fruit where maintenance of 'at harvest' quality and only moderate softening to a crisp fracturable texture is desirable. Watkins *et al.* found that rapid treatment of fruit with 1-MCP after harvest can afford storage operators more freedom to delay CA storage application, but attention to cultivar, fruit maturity and susceptibility of fruit to storage disorders must be considered.

Challenges exist for the effective use of 1-MCP for fruit that ripen uniformly to a melting texture and/or have major colour, flavour and aroma changes that are expected by the consumer. Failure to ripen normally has been shown in avocado, banana, pear and tomato, where fruits were treated at an early ripening stage or where applied 1-MCP concentrations were too high. Despite the challenges, successful commercialization of SmartFresh for treatment of avocados, bananas, melons, persimmons and tomatoes has resulted from careful attenuation of 1-MCP concentrations and/or ripening stage at harvest.

Recently, 1-MCP formulations have been approved by the Environmental Protection Agency and other regulatory authorities for preharvest applications, and these are marketed as Harvista for FFV. Semi-commercial trials have been carried out in several US locations, Argentina, Brazil, Canada, Chile, New Zealand and South Africa. Harvista has useful effects on delaying fruit drop, slowing fruit maturation and ripening, and in maintaining postharvest quality, including improving the effects of SmartFresh.

Controlled atmosphere storage

CA storage refers to the monitoring and adjustment of the CO_2 and O_2 levels within gas tight stores at optimum storage temperature. Thus, the atmosphere is controlled rather than established passively as in the case of MAP outlined in §4e, though the effects of altered atmospheres on metabolism of FFV are essentially the same. In most cases, the concentrations of CO_2 are higher and those of O_2 are lower, optimum concentrations depending on the specific product and the purpose of the CA storage conditions. Reduced O_2 and elevated CO_2 levels affect both primary (glycolysis, fermentation and aerobic respiration) and secondary (e.g. processes involved in ethylene production and action, pigments, phenolics and volatiles) metabolism.

Each FFV has an optimal range of O_2 and CO_2 for maintaining quality and extending shelf life, and these can differ for whole and fresh-cut products of the same fruit or vegetable. Benefits of CA as well as detrimental effects resulting from exposure of FFV to atmospheres outside of the safe range were summarized by Kader. The beneficial effects of CA include:

- Retardation of senescence and associated biochemical and physiological changes, e.g. slowing down rates of respiration, ethylene production and softening; —
- Reduction in sensitivity to ethylene action at O_2 levels less than 8% and/or CO_2 levels more than 1%;
- Alleviation of certain physiological disorders, e.g. chilling injury of avocado and superficial scald of apples and pears;
- Direct or indirect effect on postharvest pathogens (bacteria and fungi) and consequently decay incidence and severity e.g. CO_2 at 10–15% inhibits development of *Botrytis* rot on strawberries and cherries; and
- Low O_2 (less than 1%) and/or elevated CO_2 (40–60%) can be a useful tool for insect control in some fresh and dried fruits, flowers and vegetables; and dried nuts and grains.

THE DETRIMENTAL EFFECTS OF CA INCLUDE:

- Initiation and/or aggravation of certain physiological disorders, such as internal browning in apples and pears, brown stain of lettuce and chilling injury of some commodities;
- Irregular ripening of fruits, such as banana, mango, pear and tomato, can result from exposure to O_2 levels below 2% and/or CO_2 levels above 5% for more than one month;
- Development of off-flavours and off-odours at very low O_2 concentrations (as a result of anaerobic respiration) and very high CO_2 levels (as a result of fermentative metabolism); and
- Increased susceptibility to decay when the fruit is physiologically injured by insufficient O_2 or too-high CO_2 concentrations.

Use of CA technology is limited to relatively few FFV, the major crop being apples. Other FFV include cabbages, sweet onions, kiwifruits, avocados, persimmons, pomegranates, nuts and dried fruits, and vegetables. Atmospheric modification during long-distance transport is used on apples, asparagus, avocados, bananas, broccoli, cranberries, cherries, figs, kiwifruits, mangos, melons, nectarines, peaches, pears, plums and strawberries. Limited use of CA for many FFV is related to the high levels of capital investment required for high-quality storage rooms and the maintenance and monitoring of atmospheres; therefore, fruit volumes must be high enough to fill rooms, and extended storage periods are needed to make investments economical. For shipping containers, limitations include maintaining container identity for those with equipment.

Specific CA storage technologies that are used are a function of the FFV, growing region and size of the industry. Where used, static systems are still used for the majority of FFV CA storage, ranging from simple controlled ventilation systems to conventional CA (more than 2 kPa O_2). The use of low O_2 (LO) and ultra low O_2 (ULO) CA storage, where O_2 levels are as low as 1.5–2.0 and 0.8-1.2 kPa, respectively, is increasingly becoming common.

Other developments include dynamic CA (DCA), initial low O_2 stress (ILOS) and hypobaric storage. The principle of DCA is that responses of fruit to decreasing O_2 in the storage atmosphere are monitored. As O_2 levels reach the anaerobic compensation point, stress responses by the fruit can be detected and the O_2 increased to slightly higher levels. The most common commercial DCA system is based on chlorophyll fluorescence (DCA-DF), but others include measurement of ethanol in the fruit (dynamic controlled systems, DCS) and the respiration quotient (DCS-RQ).

Most research has been carried out with apples and pears, a major impetus for research being the loss of diphenylamine, the antioxidant used widely for control of the physiological disorder superficial scald in Europe. However, DCA has been tested with avocado and should work with any chlorophyll-containing product.

ILOS involves the use of low partial pressures of O_2, e.g. 0.4 kPa, that result in ethanol accumulation for short periods of time that result in delay of fruit ripening. Hypobaric or low-pressure storage results in very low O_2 around the product and delay ripening and senescence but has yet to find commercial acceptance because of factors such as cost and safety. These issues may be addressed in the future.

Modified Atmosphere Packaging

MAP generally involves the packaging of a whole or fresh-cut product in plastic film bags, and can be either passive or active. In passive MAP, the equilibrium concentrations of O_2 and CO_2 are a function of the product weight and its respiration rate, which is affected by temperature and the surface area, perforations, thickness and permeability to gases of films used in packaging. In active MAP, the desired atmosphere is introduced in the package headspace before heat sealing, but the final atmosphere will eventually be a function of the same factors that affect passive MAP.

Correct equilibrium atmosphere can delay respiration, senescence, and slow down the rate of deterioration, thereby extending product storage life. More recently, active MAP also includes technologies to adsorb substances, such as O_2, ethylene, moisture, CO_2, flavours/odours, and release substances such as CO_2, antimicrobial agents, antioxidants and flavours.

An extensive number of models have been developed to predict respiration rates of FFV under MAP conditions. An example of prediction models where the Pack-in-MAP software for optimum packaging solutions for FFV was developed. Behind the software is an extensive database on product respiration rate, optimum temperature, optimum ranges of O_2 and CO_2 and gas permeability of packaging materials commonly used in MAP (Fig 1). The software is based on a series of mathematical algorithms to simulate the evolution of internal gas composition in the packaging as a result of food respiration and mass transfer through the packaging material and, when used in a reverse manner, to identify the window of gas permeability that satisfy food requirements, size and number of micro-perforations, if needed.

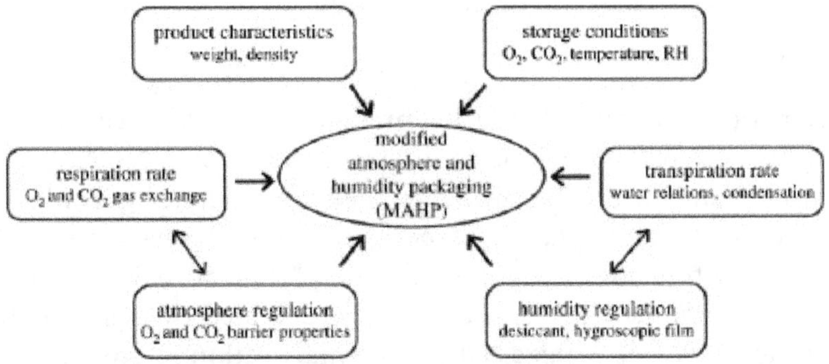

Fig. 1. Factors involved in designing MAHP for FFV

Current MAP design considers the respiration rate of product as the only important parameter for deciding target gas barrier properties required to achieve an equilibrium-modified atmosphere. However, besides in-package gas composition it is also important to take into consideration the in-package level of humidity, in order to avoid condensation and/or mould and bacterial development in MAP systems. It is well known that the in-package humidity is influenced by respiration and transpiration of the fresh produce as well as the water vapour permeability of the packaging material.

However, most polymeric materials (polyethylene, polypropylene or polyvinyl chloride) used in MAP have lower water vapour permeability relative to transpiration rates of fresh produce; therefore, most water molecules evaporated from the produce do not escape through the film and remain within the package, enhancing the water vapour pressure in the package microenvironment. Under these near-saturation conditions, even minor temperature fluctuation may result in condensation inside the package resulting in produce sliminess and enhancement of microbial growth and decay of produce.

Therefore, the major challenge of modified atmosphere and humidity packaging (MAHP) is finding a solution for creating optimal atmosphere and reducing the risk of water condensation in the package while still maintaining produce weight loss as low as possible. Recent developments on high water vapour permeable films such as Xtend, NatureFlex or biodegradable films or use of hygroscopic additives located within the package headspace or directly integrated into the packaging material. However, the hygroscopic additives should not be used for fresh produce, which have high water activity, in order to avoid excessive weight loss of the packed food. Recently, Singh *et al.* reported that the humidity absorption of the trays with NaCl improved the quality and shelf life of fresh mushrooms. In this study, different percentage of NaCl was incorporated in the polymer matrix of a film from which trays were produced. The results indicate that the amount of water vapour absorbed by the tray is directly proportional to the percentage of salt incorporated in the trays, which enhanced the total appearance of the package. The maximal capacity and the rate of moisture absorption by the humidity regulating trays need to be studied further in order to confirm whether the trays have enough moisture absorption capacity to prevent condensation for the selected horticultural crops. Additionally, questions about sustainability and consumer expectations of such humidity regulating trays have to be taken into consideration.

Emerging Technologies

Plasma is an emerging technique for decontaminating FFV. Plasma is composed of ionized gas molecules, which have been dissociated via an energy input. Depending on the mode of particles activation and the excitation energy, they can generate high or low temperatures, referred to as thermal or cold plasma, respectively. Cold plasma at atmospheric pressure can be generated by transforming argon gas into plasma at radio frequency of 27 MHz or by electric discharge between two electrodes separated by dielectric barriers. Three basic mechanisms have been suggested for the inactivation microbial spores in plasma environments, including the erosion of microbial spore surface atom by atom through adsorption of reactive free radicals 'etching'; direct destruction of DNA via UV irradiation and volatilization of compounds from the spore surface by UV photons through intrinsic photo-desorption. Researchers revealed that at the optimal operating conditions of cold gas plasma treatment about 15 min treatment time was required to achieve 2.72, 1.76 and 0.94 log-reductions in viable cells of *S. enteric* sv. *Typhimurium* on lettuce, strawberry surfaces and potato tissue, respectively.

Recent study by Baier *et al.* on fresh corn salad leaves showed that the plasma treatment at 20 W for 1 min successfully inactivated *E. coli* by 4 log-cycles. However, more research is required for a complete understanding of the role of microbial cell structure, physiology and stress resistance mechanisms involved in plasma resistance. Also, the effect of plasma treatment on food enzymes and postharvest quality attributes of FFV requires more detailed study. Safety of gases, consumer perception and the translation of laboratory scale to large commercial scale, also requires further investigation.

Conclusion and Future Prospects

A wide range of physical and chemical treatments exist to maintain and extend shelf life FFV. Specific treatments may only be applicable to certain types of product and spoilage conditions and the effectiveness of existing treatments on emerging quality issues need to be assessed. Postharvest treatments, such as CA and MAP, in combination with appropriate temperature control are the basis for maintaining physical, nutritional and sensory attributes, and by reducing decay incidence. These can be supplemented by chlorine, SO_2, irradiation, hot water, hot air, antimicrobial agents and edible coatings as appropriate for the specific product. Newer technologies include ULO and postharvest technologies based on ethylene oxidation, inhibitors of ethylene action and modulators of ripening, such as NO. Research on these technologies is continuing on a range of FFV, including climacteric and non-climacteric types. Research with DCA represents a new era where the technology is applied in a dynamic fashion, recognizing that the metabolism of the product changes in response to the applied storage conditions. To date, use of this technology is limited but the introduction of nanotechnology to the postharvest arena may open new opportunities. For example, the development of nanocomposite packaging materials with tuneable architecture can act as a smart carrier/controlled release system for 1-MCP, NO or antimicrobial agents. Future research in development of delivery systems will not only improve efficacy of postharvest systems but may also address the safety issues.

5

BIOTIC AND ABIOTIC FACTORS AFFECTING MATURITY AND QUALITY

INTRODUCTION

Agriculture is a key activity of human being since it provides basic needs such as food, clothing and shelter. It has been demonstrated that every 1% increase in agricultural yield translates into a 0.6–1.2% decrease in the numbers of absolute poor households in the world. Meanwhile, population growth was predicted to be 9.7 billion by 2050 and this will require an increase of about 70% in food production to meet the demand. Rainfed agriculture is projected to produce one-third or more of the food increase in global food output for the coming decades. Unfortunately, agricultural productivity depends on increasingly extreme weather phenomena. Thus, water availability, air pollution, and temperature have a large impact in agriculture.

Biotic factors like insects, rodents, pests, and many more spread the disease and reduce crop production. Biotic factors like insects, parasites, diseases, etc.

Abiotic factors like humidity, temperature, moisture, wind, rain, flood and many more destroy the crop raise. Abiotic factors like hot, cold, drought, salinity, etc. affect crop production. Resistant variety can withstand these factors and give a high yield.

EFFECTS OF THESE FACTORS

Several factors pose significant risk to farms leading to yield reduction when they are not correctly monitored and well managed. These factors can be grouped into three categories which are technological, biological and environmental. The pressure to increase crop production in many countries, has resulted in the expansion of land area dedicated to agriculture and the intensification of cropland management through practices such as irrigation, use of large quantities of inputs like inorganic fertilizers and synthetic chemicals for pest and weed control. These practices have resulted in degradation of soil properties and water quality, acceleration of soil erosion, contamination of groundwater and decline of food quality. This has prompted sustainable intensification initiatives to increase yields on existing farmland while decreasing the environmental impact of agriculture.

BOTH BIOTIC AND ABIOTIC FACTORS HAVE THEIR OWN EFFECTS ON CROP

Biotic Factors

Biotic factors impact crop production in the subsequent ways :

- Earthworms improve the texture of the soil.
- Microbes such as nitrogen-fixing bacteria increase the fertility of the soil.
- Insects serve as predators and keep the number of crops under control.

Abiotic Factors

Abiotic factors impact crop production in the subsequent ways :

- Sunlight aids in photosynthesis.
- Water and minerals from the soil along with gases are the raw materials for the synthesis of food and energy.
- Suitable temperature is required for the proper growth of the plant and various processes such as transpiration.
- Rainfall is required for natural irrigation of plant.

Organic crop production is one of the alternative agricultural practices promoted for the reduction of environmental pollution. As a result, several countries have introduced organic farming practices to replace the chemical-dependent ones. To conserve and regenerate soil properties, the maintenance of soil organic matter (SOM) has received considerable attention.

Although SOM is considered key to soil health, its relationship with yield is contested because of local-scale differences in soils, climate, and farming systems. The relationship between these factors should be quantified and proper soil management strategies set up to ensure sustainable crop production.

The impact of climate change in our agricultural systems is undoubtable. For example, drought followed by intense rain can increase the flooding potential, thereby creating conditions that favor fungal infestations of leaves, roots and tuber crops. In addition, reduction of bees' density due to global climate change has led to local extinction of several plant species.

The production of enough food to match population growth while preserving the environment is a key challenge, especially in the face of climate change. This chapter will review factors affecting yields of crops and provide some strategies to overcome yield loss while preserving the environment.

DROUGHT–PATHOGEN STRESS COMBINATION: A MODEL FOR UNDERSTANDING COMBINED ABIOTIC–BIOTIC STRESSES

Drought stress interacts with pathogen infection both additively and antagonistically. On the basis of the number of reports of plant diseases being affected by drought stress and the frequency of occurrence of drought stress, this combination can be considered as one of the most important stress combinations affecting crop yields worldwide. Drought stress is reported to enhance the susceptibility of *S. bicolor, T. aestivum, Senecio vulgaris* (groundsel), *Hordeum vulgare* (barley), *Gossypium* spp. (cotton), and *C. arietinum* to *M. phaseolina, Puccinia* sp., *Erysiphe graminis* f. sp. *hordei, Fusarium oxysporum* f. sp. *vasinfectum*, and *Rhizoctonia bataticola*, respectively.

On the other hand, drought stress is reported to provide endurance to tomato, *Medicago sativa* and *Arabidopsis thaliana* against *Botrytis cinerea* (causal agent of gray mold), *Oidium neolycopersici* (causal agent of powdery mildew), *Verticillium albo-atrum* (causal agent of verticillium wilt), and *Pseudomonas syringae* (causal agent of bacterial speck disease), respectively.

In some cases, concurrent pathogen infection helps plants to endure drought stress, resulting in increased yield. For example, infection with *Cucumber mosaic virus* (CMV) led to improved drought tolerance of *Capsicum annum* (pepper), *S. lycopersicum*, and *Nicotiana tabacum* (tobacco).

One of the important diseases known to be aggravated by high temperature and water deficit conditions is dry root rot (DRR), caused by a necrotrophic fungus *R. bataticola*, have shown the interaction between *R. bataticola* and drought stress in laboratory conditions by infecting *C. arietinum* plants grown at different soil moisture contents with this fungi.

Long periods of drought accompanied with warm days and cool nights generally favor powdery mildew in *Beta vulgaris* (sugar beet) caused by the fungus *Erysiphe betae*. Increased occurrence of powdery mildew infection was observed in several parts of United States in the drought year of 1988.

Occurrence of powdery mildew infections also coincided with extended periods of drought in Germany. In contrast to the above report, drought stress delayed powdery mildew disease development in *Alliaria petiolata* (garlic mustard), which could have been due to osmotic stress mediated stomatal closure that typically reduces the pathogen's ability to enter through the leaf.

However, the exact reason for the same is not yet known showed that drought stress reduced plant growth, resulting in the powdery mildew fungi infecting all available leaf area by the end of the experiment, though there was a delay in disease development.

In cases such as this, although drought did not aggravate disease development, the net impact of the two stresses resulted in loss of plant performance.

Drought stress accompanied by high soil temperature has been correlated with increased charcoal stalk rot development, caused by *M. phaseolina*, in *S. bicolor*. This disease has also recently emerged as a threat in regions with warmer summers and low rainfall. Soil moisture content affects microsclerotia survival, root infection, and disease development. It has been found that microsclerotia can survive in dry soils for prolonged periods, but is unable to survive in saturated soils for more than a week. Such interaction between drought and charcoal root rot has also been shown in *Phaseolus vulgaris* (common bean) under laboratory conditions. It has been reported that drought conditions in England and Wales have resulted in higher incidences of common scab caused by *Streptomyces scabiei* in *Solanum tuberosum* (potato).

Infection occurs for 6 weeks after the start of tuber initiation and dry soils facilitate rapid infection of the fungus on developing tubers. The amount of scab on a tuber's surface is directly related to the length of time that the plants are deprived of irrigation. The timing of drought occurrence also affects the severity of scabs on surface of tubers and it was found that drought during early stages of tuber development resulted in more scabs.

Research showed that irrigating fields to as high as 90% field capacity effectively suppresses common scab. Given that a number of drought–pathogen stress combinations have a net negative influence on crop yields, it is important to devise strategies for improving crop performance under these stresses. A promising way of doing so is to identify measurable parameters or traits that are affected by combined stress conditions, which can be modified favorably to improve crop productivity under combined stress conditions. In the section below, we highlight some key traits that can be used for crop improvement under combined drought and pathogen infection.

POTENTIAL TRAITS FOR SCREENING GENOTYPES FOR TOLERANCE TO COMBINED DROUGHT AND PATHOGEN INFECTION

Root System Architecture

Root system architecture (RSA) acts as a major interface between the plants and several biotic as well as abiotic factors and enables the plants to circumvent the environmental challenges by sensing and responding to them. The length and density of primary as well as lateral roots play a crucial role in drought stress tolerance. Development of high root length density (RLD) along with increased root diameter in response to drought stress confers drought tolerance in rice. For example, rice lines with low RLD show reduced drought tolerance. High RLD favors improved plant growth under drought conditions as it provides access to moisture present at deeper soil depths.

Likewise, under drought stress, *Zea mays* (maize) with high RLD and few lateral roots had high plant water status, increased leaf photosynthesis, stomatal conductance, and increased overall growth, compared to plants with low RLD and more lateral roots. The presence of fewer but longer lateral roots results in enhanced rooting depth thereby increasing water acquisition from deeper layers of soil which helps in improved plant performance under drought.

Interestingly, RSA also plays a significant role under pathogen infection in plants. In an evaluation of virulence of *Pythium debaryanum* and *Pythium ultimum* (causal agents of root rot) on *T. aestivum*, plants with high root length had less fungal infection. In contrast, infection of *Rhizoctonia solani* (causal agent of root rot) on *S. lycopersicum* caused reduction in total root length, number of root tips, and magnitude of root branching, which compromised water exploration from deep soil layers and consequently the shoot growth. Thus there seems to be a correlation between RLD and the extent of pathogen infection by root infecting fungi. Thus we speculate that increasing the RLD of plants might help in reducing pathogen infection.

Combined drought and root infecting pathogens cause greater damage to plants as both stresses appear to additively disrupt the RSA. For example, under drought conditions, *Fusarium solani* f. sp. *phaseoli* (causal agent of root rot in beans) infects the roots of *P. vulgaris* in deep layers of soil and affects water absorption. As a result of infection, accessibility to water present at deeper soil profiles is compromised under drought conditions, leading to severe reductions in plant growth.

Under combined stress, the time of occurrence of pathogen infection or drought stress has a significant effect on the net impact. *Phytophthora cryptogea* (causal agent of root and crown rot) infection on drought stressed *Carthamus tinctorius* (safflower) resulted in severe root rot disease development and a marked reduction in fresh weight of roots, compared to the conditions where drought stress followed the infection. Similarly, *Phytophthora parasitica* (causal agent of root rot) infection on drought stressed *S. lycopersicum* resulted in greater disease severity, represented by an increase in the number of brown roots, reduced root length, and low fresh weight, compared to pathogen infection followed by drought stress.

Drought stress induced increase in root growth and exudation of amino acids (such as alanine and proline) and carbohydrates (such as pentose and glucose) are known to be responsible for enhanced root rot disease development on drought stressed plants. Root exudates serve as nutrients for the growth of soil borne pathogens. These drought induced changes in the host physiology enhance pathogen infection by directly attracting more pathogens as well as intensifying the existing infection on plant roots.

Additionally, pathogen infection modulates the composition of root exudates. Tomato roots infected with *F. oxysporum* f. sp. *radicis-lycopersici* exhibited decreased exudation of citric acid, but increased secretion of succinic acid as compared to the non-infected roots. Moreover, co-infection with bio-control bacterium *Pseudomonas fluorescens* WCS365 resulted in less disease and more secretion of succinic acid. Identification and analysis of exudates commonly secreted under drought and pathogen stress, and attempts toward manipulating the secretion of these exudates by inhibiting or over-expressing the secretory pathways may help in conferring tolerance to drought and pathogen infection.

However, studies in this direction needs to be done to prove the suitability of this approach. A number of studies have demonstrated no influence of drought stress on pathogen infection induced root damage in plants. For example, in an assessment of the effect of drought stress on infection by *Gaeumannomyces graminis* (Sacc.) var. *tritici* (causal agent of root rot) in wheat under low and severe drought stress conditions, found that infection under both the drought levels caused similar reduction in root dry mass. In addition, carbon assimilation rate and root decay were also found to be reduced similarly under both drought intensities, indicating that increasing drought intensities had little effect on disease development. Furthermore, *Pythium irregulare* and *R. solani* infection on *T. aestivum* cultivars under drought stress did not result in any significant change in root lesions inflicted by pathogen infection compared to infection on well-watered plants.

Taken together, in most cases plant survival under concurrent drought and pathogen infection is compromised if RLD is affected as it influences the acquisition of water. Plants with the ability to maintain high RLD may perform better under combined drought and pathogen infection. Considering the role of RLD in both drought tolerance and pathogen infection, this trait can be utilized as a potential morpho-physiological trait for selecting cultivars with resistance to combined drought and pathogen stress.

Additionally, root phenotyping tools can be exploited for screening plants with combined stress tolerance. Several studies have reported quantitative trait loci (QTLs) associated with RSA under drought stress. For example, a constitutive QTL, designated as root-abscisic acid 1 (ABA1), associated with root traits like branching, diameter, angle, and total dry mass has been identified in maize. Similarly, a QTL for ABA-induced reduction in lateral root growth and size has been identified in *A. thaliana*. Moreover QTL (ARR2.1) for root rot resistance and tap root diameter (TD2-1) are correlated and increase in tap root diameter was related to enhanced resistance. Hence, we expect that some QTLs associated with efficient RSA may also be used in breeding programs to develop combined drought and pathogen resistant crops.

Leaf Pubescence

Trichomes (leaf hairs) are modified epidermal cells found in uni- or multi-cellular, branched or unbranched, and glandular or non-glandular forms all over the surface of a plant. Though number and types of trichomes are genetically controlled, the environmental conditions also determine their pattern of occurrence. Plants grown in semi-arid environments maintain water levels by foliar absorption of water with the help of trichomes. Trichomes entrap water droplets and it has been shown that *Phlomis fruticosa* (Jerusalem sage) leaves with trichomes in mesophyll cells absorb dew deposits, which results in decreased water potential of drought stressed leaves, compared to the leaves of *Hedera helix* without trichomes.

In addition, photosynthetic performance of hairy leaves is greater than that of non-hairy leaves under water stress conditions. In some cases, drought conditions also increase trichome production in plants as a means of adaptation. For instance, drought stressed *Sinapis arvensis* (wild mustard) plants had more trichomes compared to control plants of the same line. Studies have found that trichomes can serve as a barrier to infection by foliar pathogens.

For example, *Phytophthora infestans* (causal agent of late blight) infection in *S. tuberosum* is negatively correlated with the presence of glandular trichomes. Presence of trichomes can reduce the relative humidity at the leaf surface, which is unfavorable for the germination of fungal spores. Trichomes may also secrete exudates that possess anti-fungal activities. For example, exudates secreted by glandular trichomes present all over the plant surface of chickpea are shown to decrease infection by *Ascochyta rabiei* (causal agent of ascochyta blight) due to the anti-fungal properties of the exudates.

It was found that increased concentrations of exudates inhibited the conidial germination of *A. rabiei* while low concentrations promoted it. Identification of the pathways and genetic elements behind the glandular secretions from the trichomes under pathogen stress and their careful manipulation can enhance the resistance of plants.

In contrast to the above reports, trichomes in some plants may favor pathogen growth. For example, trichomes present on the leaf surface of common beans were reported to favor the growth of *P. syringae*. As trichomes retain water, exudates released from the broken cuticle at the base of trichomes might favor microbial growth. Similarly, *A. thaliana* mutant *gl1* (GLABROUS1) plants with less trichome density were found to be tolerant to infection by necrotrophic fungus *B. cinerea* but *try* (TRYPTYCHON) mutants with more trichome density were found susceptible to infection.

Not many studies have been done to probe into the role of trichomes in pathogen infection. A closer understanding of plant–pathogen interaction at this interface may help in further unraveling the role of trichomes in enhancing pathogen infection. It is evident that plants produce more trichomes under drought to minimize transpiration. In many cases, the presence of glandular trichomes has been shown to provide tolerance against pathogen invasion. However, as mentioned above, there are exceptions to this rule.

Moreover, role of glandular trichomes and their secretory products under drought stress needs to be studied. Taken together, an extensive understanding of the nature of plant–pathogen interaction under drought stress would be needed in cases where trichomes enhance pathogen growth. Mapping for leaf pubescence related QTLs have been done for many plants like *Gossypium hirsutum* and *A. thaliana*. It can be hypothesized that enhancement of trichome production impart protection against combined drought and pathogen infection in many cases and trichomes can be considered as a potential morpho-physiological trait conferring tolerance to the stress combination. Moreover, identification of QTLs related to leaf trichome density and secretion under drought and pathogen infection can also help in breeding genotypes better adapted to the combined stress. It can also be utilized for exploring the genes and pathways regulating trichome production and secretion which can be suitably modified to confer enhanced resistance under combined stress scenarios.

Leaf Water Potential Regulation

A change in plant water potential is directly correlated to soil moisture level and is also affected by fungal and bacterial pathogens that disrupt the function of the plant vascular system. However, some traits related to maintenance of plant water potential are negatively affected by drought and pathogen stress. For example, plants close stomata under drought stress in order to reduce the transpirational loss of water. In contrast, infection by *Uromyces phaseoli* (causal agent of leaf rust) inhibits stomatal closure on *P. vulgaris* due to the toxins produced by the pathogen which indicates that pathogen infection in cases like this can compromise drought tolerance. Some pathogens may reduce plant water content even under sufficient soil moisture conditions. For example, *U. phaseoli* infection in *P. vulgaris* results in wilting at high soil water potential due to xylem damage, whereas uninfected plants experience wilting only under drought. Inhibition of stomatal closure by toxins secreted by *U. phaseoli*, disruption of cuticle layer and impaired stomatal resistance account for the increased water loss, which further reduce leaf water potential of plants under drought stress.

Similarly, demonstrated a marked reduction in shoot water potential, leaf turgidity and transpiration in *Vigna unguiculata* (cowpea) plants subjected to concurrent drought and *M. phaseolina* (causal agent of charcoal rot and stem blight) infection. showed that *V. vinifera* subjected to combined drought and *Xylella fastidiosa* (causal agent of leaf scorch) infection experience a significant reduction in leaf water potential and stomatal conductance, which aggravates the scorch symptoms more in drought stressed plants, compared to well-watered plants.

When *P. vulgaris* was exposed to simultaneous drought and *M. phaseolina* infection, high transpiration rate, decreased water potential and low stomatal resistance was observed in the stressed plants. Drought stress caused plants to produce carbohydrates which facilitated the growth and infection of *M. phaseolina*. In addition, it was found that varieties resistant to infection maintained high leaf water potential compared to susceptible varieties. In case of charcoal rot due to infection by *M. phaseolina* in *G. max*, it has been found that maturation of the sclerotia was induced only by the reduced leaf water potential due to drought stress. It was also found that symptoms appeared only after imposition of drought stress.

Likewise, found that post flowering drought stress caused a reduction in leaf water potential in *S. bicolor*. Plants with reduced water potential had longer *M. phaseolina* lesions, which directly resulted in a reduction of grain yield. had demonstrated that drought-resistant bean varieties showed resistance to *M. phaseolina* infection as well.

Taken together, leaf water potential can be influenced by both drought and vascular pathogens and improved water status of plants under drought conditions might correspond to improved pathogen as well as drought resistance. One of the factors defining plants response to vascular pathogen infection is the xylem vessel dimension; *V. vinifera* genotypes with smaller xylem diameter are known to be less susceptible to infection by fungal vascular wilt pathogens. Identification of QTLs related to xylem diameter and xylem pit anatomy can be helpful to identify mechanisms for tolerance against combined drought and pathogen infection. Thus, plant water potential can be used as an important morpho-physiological trait to screen plants resistant to combined drought and pathogen infection.

Cuticular Wax

Cuticle plays a vital role in protecting plants from drought stress and pathogen invasion. When stomata are closed under drought stress, a small amount of water is lost through cuticular layer.

Cuticular layer also acts as a barrier to pathogen infection as it is hydrophobic and devoid of moisture. The significance of cuticular layer has been studied under drought stress conditions. For example, drought stress led to an increase in the concentration of cuticular wax components such as alkanes, aldehydes, and ketones in *A. thaliana*, resulting in increased wax coverage in the stressed plants.

Under drought stress, the drought tolerant *T. aestivum* plants exhibit enhanced thickness of the cuticular layer while the susceptible varieties do not show any change in cuticle thickness. Likewise, the importance of cuticular wax has also been studied under pathogen infection. exposed wild type and glossy mutants of *Z. mays* (*gl4*) to *Clavibacter michiganensis* (causal agent of leaf blight and Goss's wilt of maize).

Compared to the wild-type, more bacterial colonies were observed on the *gl4* mutants, which had a thin cuticular layer due to an alteration in the wax biosynthesis pathway. Nutrient and water exudation through the weak cuticular layer might have encouraged the colonization of bacteria, leading to more pathogen growth in the *gl4* mutants. showed that bloomless (*bm*) mutants of *S. bicolor*, deficient in the synthesis of epicuticular wax and having a thin cuticular layer, were highly susceptible to infection by *Exserohilum turcicum* (causal agent of leaf blight) compared to the wild type plants. The rate of water loss was found to be high in the *bm* mutant compared to wild type plants.

This apparently suggests that the thickness of cuticular wax can be used as a trait to identify plants tolerant to *E. turcicum*. Plants without stomata and deficient in cuticular wax have been used to study the significance of cuticular wax under pathogen infection. showed that penetration of pathogen through stomata was more with astomatous fruits (cutin deficient, *cd* mutant) of *S. lycopersicum*. Only the *cd* fruits were found to be infected with *B. cinerea* (causal agent of gray mold) depicting a role of cutin in pathogen resistance.

Along with the cutin content, the composition and architecture of the wax layer also determines their role in defense. In the above study, it was also found that among the three cd mutants, cd1 which showed lack of microfissures, elevated level of amyrins and decreased levels of alkanes of chain length >30 showed maximum water loss and minimum susceptibility to *B. cinerea*.

Although, there are no studies showing the direct role of cuticular wax under combined drought and pathogen infection, the above evidence suggests its probable role in combined stress. Thus, plants produce more complex and thick cuticular wax layer in response to drought stress, which in turn might impart tolerance to pathogen infection.

Additionally, the composition and architecture of the wax layer is equally important in defining the role of cuticle in defense mechanism. Detailed investigation of the pathways that determine the composition and structure of cuticle layer may help in identifying targets which can be manipulated to impart improved resistance to plants against combined drought and pathogen infection. have found that QTLs for epicuticular wax, rate of water loss from excised leaves and harvest index co-located with QTLs associated with shoot and root-related drought resistance traits in rice. One example of such QTL is a region on chromosome 8 of rice.

Considering the importance of cuticular wax in providing resistance against many pathogen infection, identification of QTLs linked to wax content and disease resistance should also be done. Cuticular wax may be considered as a trait that can be used to screen plants tolerant to combined drought and pathogen infection. The measurement of wax content can be made by simple weight analysis by immersing leaves in chloroform and determining the wax content after chloroform evaporation. Thus, the trait can be efficiently utilized for large scale screenings of plants better adapted to combined drought and pathogen infection.

Canopy Temperature

Canopy temperature (Tc) has been used to measure drought stress tolerance of plants. Tc varies with each leaf under drought and pathogen infection, as stress induced drooping and curling of leaf reflects radiation differently. Tc plays a significant role in plant growth under drought stress. *T. aestivum* plants under drought stress were found to have high Tc and yielded less than irrigated plants. Plants which maintain low Tc under drought stress conditions possess high plant water status and thus are better adapted for drought stress.

Tc of *T. aestivum* plants infected with *Mycosphaerella graminicola* (causal agent of *Septoria tritici* blotch), can be positively correlated with disease occurrence as infected plants had higher Tc. The increase in Tc can be ascribed to the damaged cuticular layer due to infection by pathogens. Furthermore, a negative correlation was observed between Tc and leaf greenness as pathogen infection progressed. Thus, measurement of plant Tc can be used to identify both infected and un-infected areas.

The significance of Tc under concurrent drought and pathogen infection has been shown in some studies. When *B. vulgaris* (sugar beet) was infected with *Pythium aphanidermatum* (causal agent of root rot) under drought stress, increased Tc was observed in drought stressed plants as compared to control plants. Increased Tc can be attributed to infection induced root damage, resulting in interruption of water uptake and a reduction in plant water potential.

Infected plants had higher Tc as compared to the drought stressed, -uninfected plants. Similar increase in Tc was observed in *Gossypium* spp. infected with *Phymatotrichum omnivorum* (causal agent of *Phymatotrichum* root rot) under drought stress. Under concurrent drought and *M. phaseolina* (causal agent of charcoal rot) infection, increased leaf temperature and decreased stomatal resistance were observed in stressed *P. vulgaris*. Likewise, infection of *Sclerotinia minor* Jagger (causal agent of watery mold and soft rots) on thinned and un-thinned *Arachis hypogaea* (peanut) plants under drought stress.

Drought stressed un-thinned plant canopy showed increased disease severity compared to the thinned treatments. Thinning lead to a modification in plant canopy size, which affected the microclimate as well as Tc. High relative humidity and microclimate associated with the un-thinned canopy favored disease infection, whereas thinned canopy exhibited lower relative humidity and lesser disease infection.

The effects were further supported by a reduction in soil moisture content in the thinned fields, compared to the non-thinned ones; thinned canopy might have reduced canopy level humidity and increased transpiration rate as well as water uptake from soil, resulting in lower Tc and consequently lesser infections.

Thus increasing space between the plants can be utilized as an agronomic practice in areas with less soil moisture availability for providing resistance against combined drought and pathogen infection. It has been demonstrated that plants which regulate transpiration and gas exchange could maintain cooler canopies under drought conditions. Tc has also been found to increase with increasing number of dead leaves, further emphasizing the importance of transpiration and gas exchange in reducing Tc.

Overall, a combination of drought stress and pathogen infection shows a negative correlation with Tc. Plants which maintained normal Tc under these conditions did not compromise growth and yield. Thus, Tc, which is shown to be affected by both individual and combined drought and pathogen infection, can be considered as an important trait for assessing combined drought and pathogen tolerance of plants.

A simple measurement of Tc using infra-red thermometers can be implemented as an efficient means to screen genotypes better adapted to grow under combined drought and pathogen infection. Some QTLs for Tc has been found to be related to root development.

For example, QTL for Tc at chromosome 2B of *T. aestivum* is also the main QTL responsible for root developmental in wheat. An attempt to find the relation between Tc and Cephalosporium stripe disease (CSD) of winter wheat caused by *Cephalosporium gramineum* has been recently made. Although the authors could not find any significant relation between QTLs for Tc and disease severity, they suggest that a better evaluation of Tc might be helpful in proving the correlation between Tc and CSD resistance. As Tc is a good indicator of water status of plants investigation to find correlation between Tc and vascular diseases can also reveal mechanisms and QTLs for resistance to combined drought and pathogen infection.

GENETIC ENGINEERING STRATEGIES FOR BIOTIC AND ABIOTIC STRESS TOLERANCE AND QUALITY ENHANCEMENT IN CROPS

Biotechnology has been a leading example in many areas for more than two decades, right from the commercialization of the first ever transgenic crop in the form of 'Flavr-Savr' transgenic tomato with enhanced shelf-life trait. The first field trials of transgenic horticultural plants had been carried out in France and USA in 1986. Transgenic Flavr Savr tomato is the first successful example of genetically modified food crop and was approved for commercialization in USA in 1994.

The main resistant traits introduced into horticultural plants and already commercialized are insect-pest resistance (*Bt.* toxin gene) and herbicide tolerance while other important studies concern virus resistance, male sterility, etc. Among various genetically modified (GM) horticultural crops, GM papaya showing resistance to Papaya ring spot virus contributes to approx. 53% of the total share of GM horticultural crops cultivated globally. Herbicide tolerance trait is dominating the GM horticultural crop acreage followed by insect resistance and virus resistance traits (ISAAA).

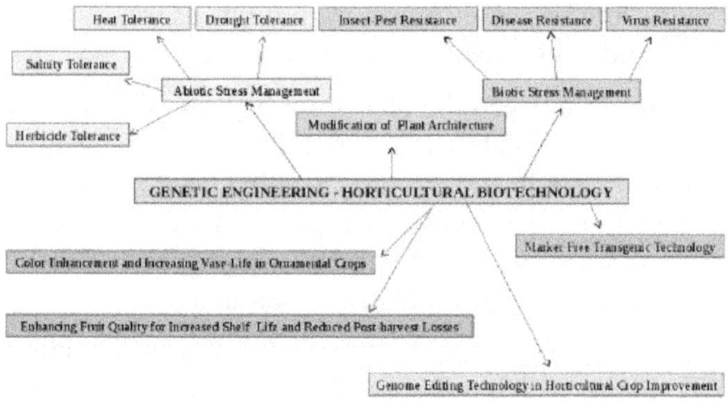

Fig. 1. Overview of various transgenic strategies targeting horticultural crop improvement

Also, the RNA interference (*RNAi*) technology has gained popularity in plant genetics and system biology during these days due to its stable transgene expression. The applications of this technology cover wide range from insect resistance, viral and disease resisance, drought, heat, salinity tolerance; developing designer flower colors by knocking down the expression of certain endogenous genes, increasing shelf-life, plant architecture modification etc. It is being used as a potential tool in tweaking the regulation of various metabolic pathways in plants and assigning functions to the genes involved, thereof. The most studied crop so far is tomato, but research activities had already been carried out in various horticultural crops such as fruits, vegetables and flowers. With the advancement of regeneration and genetic transformation protocols, extensive research efforts have been made to incorporate genes for various biotic and abiotic stress tolerance/resistance, enhancing shelf-life, modification of plant architecture and color and texture modification traits in a number of horticultural crops, which have been summarized and discussed in this review paper. An overview of various transgenic strategies targeting horticultural crop improvement is depicted in Fig. 1.

BIOTIC STRESS MANAGEMENT THROUGH TRANSGENIC APPROACH

Insect-pest Resistance

At present, insect-pest resistance is lacking generally in many crop plants. The use of chemical control measures is proving hazardous to the consumers and also not environmentally sustainable. From a grower's perspective, any genetic improvement that could reduce the cost of chemical application to combat pests would be of significant benefit. *Bt* (*Cry*) gene isolated from a soil bacteria *Bacillus thuringiensis* has proven highly effective in controlling various lepidopteran insects in a number of crops successfully. Insect resistance was firstly reported in tomato using *Bt.* gene in 1987. Transgenic *Bt.* tomato plants exhibited resistance against *Spodoptera litura* and *Heliothis virescens.* Fruit trees like transgenic persimmon carrying *cryI* gene were found resistant to *Plodia interpunctata* and *Monema flavescens.* Brinjal is among the highly consumed vegetables in Asia and specifically in Indian subcontinent. However, it is extremely damaged by a lepidopteran insect, i.e., *Leucinodes orbonalis.* Researchers transformed a synthetic *cry1Ab* gene coding for an insecticidal crystal protein (*ICP*) to brinjal (*Solanum melongena* cv. Pusa Purple Long) by co-cultivating cotyledons with *A. tumefaciens.* Gene expression was evaluated by double-antibody sandwich ELISA analysis. The transgenic lines displayed significant differences in the insect mortality in fruit bioassays.

It was suggested to express a very high level of insecticidal crystal protein to confer complete protection against *Leucinodes orbonalis* by employing some fruit-specific promoter for better inducible expression. Potato varieties engineered for resistance to Colorado potato beetle were in commercial production for several years and were technically and agronomically successful, allowing significant reductions in insecticide use.

Researchers transformed cauliflower var. Pusa Snowball K-1 with a synthetic *cryIA*(*b*) gene and the transgenic plants indicated the effectiveness of the transgene against infestation by diamondback moth (*Plutella xylostella*) larvae during insect bioassays. Researchers developed transgenic cabbage (*Brassica oleracea* var. *capitata*) line DTC 507 with a synthetic fusion gene of *B. thuringiensis* encoding a translational fusion product of *cry1B* and *cry1Ab* g-endotoxins to confer resistant against diamondback moth (*Plutella xylostella*), the most destructive pest of cruciferous plants across the globe. *Bt* cabbage plants expressing the fusion protein in mature leaves caused 100% mortality to all the four larval stages of diamondback moth.

Complete mortality of the neonate larvae had been observed within 24 h and within a period of 48 h in case of other three stages of larvae. *Bt* gene (*Cry1Ac*) has been successfully transformed and expressed in okra (*Abelmoschus esculentus*) for incorporating resistance against fruit and short borer (*Earias vittella*), which is the most serious insect-pest of this crop in Asia.

Okra is severely affected by *Earias vittella* and its larvae bore into pods and shoots of the plant and eat the internal tissues leading to withering of the plant and reduction in the market value of the pods. In insect bioassays, fruits from transgenic lines showed 100% larval mortality. Natural as well as synthetic insect resistance genes had been transferred into a number of horticultural crops for imparting resistance against various insect-pests.

Researchers transformed kiwifruit plant (*Actinidia chinensis*) with a synthetic chimeric gene *SbtCry1Ac* encoding for protein *btCry1Ac*, When the transgenic plants were screened for insect resistance in insect bioassays, an average of 75.2% *Oraesia excavate* inhibition rate was reported at 10 days of post-infection. This technology could be highly useful to protect yield losses of kiwifruit due to insect attack, which is an economically as well as nutritionally important fruit crop, offering a remarkably high vitamin C content.

Other genes such as protease inhibitors, trypsin inhibitors, lectins, etc. have also been employed for incorporation of insect-pest resistance in many crop species. Ding et al. developed insect-resistant transgenic Taiwan cauliflower *Brassica oleracea* var. *botrytis* cvs. expressing the *trypsin inhibitor* gene, isolated from local sweet potato. The transgenic plants expressed resistance to *Spodoptera litura* and *Plutella xylostella* in *in-planta* bioassays. Transgenic strawberry expressing cowpea (*Vigna unguiculata*) *protease trypsin inhibitor* (*CpTi*) gene under a constitutive promoter developed resistance against vine weevil (*Otiorhynchus salcatus*).

The *CpTi* transgenic lines reduced the frequency of survival of weevil larvae and pupae during insect bioassays. Gessler and Patocchi developed transgenic apple lines with *trypsin inhibitor* encoding *CpTI* gene from cowpea and *cry1A(c)* gene of *B. thuringiensis* for incorporation of resistance against codling moth pest. Almost all the genotypes of chrysanthemum are infested by two aphids, namely *Myzus persicae* and *Aphis gossypii*, lowering the flower quality and also transmitting viruses. Researchers transformed chrysanthemum genotype 1581 by *Agrobacterium tumefaciens*-mediated technique with *SAE* gene, under the control of chrysanthemum *RbcS* promoter to incorporate aphid resistance. The *protease inhibitor* sea anemone equistatin (*SAE*) has three domains for inhibition of both cysteine and aspartic proteases.

In another study, *Chrysanthemum morifolium WRKY48* (*CmWRKY48*) transcription factor over-expressing transgenic chrysanthemum plants were found to inhibit the population growth of aphids. Another category of plant pests, root-knot nematode (*Meloidogyne incognita*) causes severe yield losses in many horticultural crops. Genetic transformation of various *proteinase inhibitor* genes from plants is considered as the most potential strategy to prevent such yield losses.

Cysteine proteinases are involved in the digestion process of root-knot nematodes and binding of various *cystatins* to the active sites of *proteinases* inhibits their activity by proteolytic digestion. Researchers developed transgenic plantain (*Musa* sp.) cv. Gonja manjaya plants expressing a maize *cystatin* gene that inhibits the digestive *cysteine proteinases* and a synthetic peptide that disrupts nematode chemoreception.

The best level of resistance exhibited by the transgenic plants against the major pest species *Radopholus similis* was 84% for the *cystatin*, 66% for the peptide and 70% for the dual defense. In another study, Papolu et al. developed transgenic brinjal plants expressing a modified rice *cystatin* (*OC-I-D86*) gene under a root-specific promoter, *TUB-1* for inducing resistance against root-knot nematode (RKN).

Transgenic plants were confirmed for gene integration and expression using PCR, Southern and Western blotting, ELISA and qPCR assays. When one transgenic line (single copy event) was challenged with root-knot nematode, 78.3% inhibition rate in reproduction of root-knot nematode had been reported. In an earlier study, transgenic banana plants expressing the same *cystatin* gene (*OC-I-D86*) exhibited 70% resistance to the migratory endoparasite, *R. similis*. Researchers had also reported a partial resistance (67%) against *M. incognita* in transgenic potato roots expressing same gene.

Root lesion nematode (RLN), *Pratylenchus penetrans*, is one of the main pests of lily producers, particularly in USA, where lily (*Lilium longiflorum*) cv. 'Nellie white' assumes a great economic importance as cut flowers and constitutes one of the most valuable species. Vieira et al. developed transgenic lilies over-expressing the *OC-I-D86* gene which displayed an enhanced resistance to root lesion nematode infection by means of nematode reduction up to 75%. The transgenic lily plants also exhibited an increased biomass and better growth performance additionally as compared to non-transformed plants.

DISEASE RESISTANCE

The next major constraint limiting the production of horticultural crops is different diseases caused by pathogenic fungi, bacteria and viruses. Conventional breeding seems to have limited application due to non-availability of resistant gene(s) in gene pool of a particular crop. One of the main targets of genetic transformation is to improve tolerance or to incorporate resistance in plants against different pathogens. Genetic engineering of disease resistance in crops has become popular and valuable in terms of cost and efficacy.

For imparting resistance against bacterial and fungal diseases, various genes like *chitinase*, *glucanase*, *osmotin*, *defensin*, etc. are being transferred into various horticultural crops world over. Various glycolytic enzymes encoded by genes like *chitinase*, *glucanase*, *PR proteins*, etc. inside the plant cells have cell wall degrading capabilities, which attract their use for developing transgenic plants for incorporation of resistance against fungal pathogens.

Among different strategies used for genetic engineering for disease resistance, the employment of systemic acquired resistance (SAR)-related genes is of paramount importance. SAR is long lasting and often associated with local and systemic accumulation of salicylic acid (SA) and induced expression of many genes including pathogenesis-related (PR) genes. A gene for a PR protein from tomato (*PR-5*) had been expressed in transgenic sweet orange and regenerants showed increased tolerance to *Phytophthora citrophthora*.

Researchers introduced *Arabidopsis thaliana*-derived *NPR-I* gene into tomato. Transgenic tomato plants developed enhanced heat tolerance and resistance against tomato mosaic virus (*ToMV*). The transgenic lines also conferred significant level of resistance to bacterial wilt (BW) and Fusarium wilt (FW) along with moderate degree of enhanced resistance to gray leaf spot (GLS) and bacterial spot (BS). Malnoy and Aldwinckle developed transgenic apple lines over-expressing *MpNPR1-1* (ortholog of *AtNPR1*), which exhibited broad spectrum resistance against *V. inaequalis*, *Gymnosporangium juniper-virginianae*, a causative agent of cedar apple rust and *Erwinia amylovora*, which causes fire blight. The Fuji apple, the most popular and most widely cultivated variety of apple in China, is highly susceptible to powdery mildew disease. Apple powdery mildew, which is caused by *Podosphaera leucotricha*, damages leaves and young fruits, thus leading to huge yield losses.

Researchers introduced *Malus hupehensis* derived *NPR* (*MhNPR1*) gene into 'Fuji' cultivar of apple for development of resistance against powdery mildew disease. *NPR1* gene plays a key role in regulating salicylic acid (SA)-mediated SAR in plants. *MhNPR1* gene induced the expression of *MdPRs* and *MdMLO* genes which interact with powdery mildew as revealed by RT-PCR and the transgenic apple plants expressed enhanced resistance to powdery mildew disease. Over-expression of *AtNPR1* gene in tomato and carrot plants also exhibited resistance to bacterial and fungal pathogens.

Commercial sweet orange cultivars are suffering from this deadly disease. Similar findings were reported by over-expression of a *Vitis vinifera NPR1,1* (*Vv NPR1,1*) gene which conferred to enhance resistance against powdery mildew in grapevine. In the United States, Huanglongbing (HLB) is a very serious disease of citrus, which is associated with a phloem-limited bacterium *Candidatus liberibacter asiaticus* (CLas).

Researchers over-expressed an *Arabidopsis thaliana NPR1* gene under a constitutive promoter *CaMV35S* and also under a phloem-specific *Arabidopsis SUC2* (*AtSUC2*) promoter in sweet orange cultivar 'Hamlin' and 'Valencia'. *NPR1* gene is involved in the induction of expression of several native genes involved in the plant defense signaling pathways. The transgenic plants exhibited reduced disease severity and a few lines remained disease-free even after three years of planting in a high-disease pressure field site.

Another category of genes imparting disease resistance is *Chitinases,* which are glycosyl hydrolases that catalyze the degradation of chitin, an insoluble, linear β-1,4-linked polymer of *N*-acetyl glucosamine, a cell wall component of various bacteria and fungi, and thus code for pathogen resistance. *Chitinase* gene has been transferred to a number of crops for harboring fungal resistance. In carrot, the tobacco class I *ChiC* gene has shown resistance against *Botrytis cinerea, RCC2* gene, a rice *chitinase* displayed enhanced resistance to *Sphaerotheca humuli* in transgenic strawberry plants.

In another study, Yamamoto et al. transformed a rice *chitinase* gene (*RCC2*) into the somatic embryos of grapevine cv. Neo Muscat and reported an increased resistance level against powdery mildew fungus, *V. necator*. Schestibratov and Dolgov also developed transgenic strawberry plants expressing *thaumatin II* gene from *Thaumatococcus danielli* and reported some level of resistance to *B. cinerea* during in vitro assays. Researchers expressed a *chitinase* gene from *Phaseolus vulgaris*, a *glucanase* or a *thaumatin-like protein*, both from *Nicotiana tabacum* and a combination of both in strawberry cv. 'Pajaro'.

Two transgenic lines expressing *chitinase* gene showed enhanced tolerance to *Botrytis cinerea*. Khan et al. attempted to confer resistance to early blight of potato, caused by *Alternaria solani*, transformed a *chitinase* gene, *ChiC*, isolated from *Streptomyces griseus* strain HUT 6037, along with a bialaphos resistance (*bar*) gene into potato. The herbicide-resistant transgenic potato plants demonstrated enhanced resistance against *Alternaria solani* in in vitro bioassays. However, in another study, Moravcikova et al. reported that the high level of expression of cucumber class III *ChiC* gene in potato could not enhance resistance against the phytopathogenic fungus, *Rhizoctonia solani* (causing black scurf disease in potato) to any considerable level.

Researchers had also expressed bacterial *chitinase* (*chi B*) gene in litchi cv. Bedana, which, however, showed low level of *chitinase* activity and only partial resistance against *Phomopsis* sp. pathogen had been reported in transgenic plants. Various endo-chitinases genes such as *CHIT42* and *CHIT33* from *Trichoderma harzianum* had also been successfully transformed and expressed to impart increased fungal tolerance in potato, apple, broccoli, carrot and lemon.

Researchers developed transgenic tomato plants over-expressing a wheat *chitinase* gene, *chi194*, under the control of maize *ubiquitin 1* promoter. The transgenic tomato lines showing higher expression of *chitinase* activity were found to be highly resistant to Fusarium wilt disease of tomato caused by *Fusarium oxysporum f.* sp. *Lycopersici*.

In another study, transgenic litchi (*Litchi chinensis*) plants containing a rice *chitinase* gene were developed to increase the antifungal response. The transgenic lines exhibited higher *chitinase* activity and disease resistance than the non-transformed plants. Guava wilt disease caused by a soil borne fungus *Fusarium oxysporum f.* sp. *psidii* is emanating as a serious threat to guava growers throughout the entire globe. To control this disease, Mishra et al. transferred a *Trichoderma-endochitinase* gene into guava (*Psidium guajava*). *In vitro* pathogen inhibition assay and spore germination assay revealed that the crude extract of the transformed plants inhibited the germination of fungal conidia and were resistance to wilt disease.

A number of defense mechanisms were evolved in plants over thousands of years to overcome pathogen attack/infection and the role of many genes or various pathways has been investigated and identified. The disease resistance conferred by *glucanase* gene may be attributed to solubilizing elicitors from the fungal cell walls which induce production of antifungal phytoalexins. Researchers also proposed the role of *glucanase* in the induction of the transcription of a plant defense gene, *phenylalanine ammonia lyase* in response to fungal attack.

Further, *glucanase* is a hydrolytic enzyme, which breaks down the cell wall component, glucan of many necrotrophic fungal pathogens. An increased level of resistance in transgenic potato plants expressing soybean *glucanase* gene against *Phytopthora infestans* has been reported due to the increased *glucanase* activity. Transgenic kiwifruit over-expressing soybean β-*1,3 glucanase* gene exhibited a six fold increased enzyme activity leading to a decrease in the disease lesion area caused by the gray mold fungus, *Botrytis cinerea*. Almost all the cultivated varieties of brinjal are susceptible to wilt disease caused by *Verticillum dahliae* and *Fusarium oxysporum*, which cause considerable yield losses annually.

To generate wilt disease resistance in brinjal, Singh et al. transformed alfalfa *glucanase* gene coding for an acidic *glucanase* into brinjal cv. Pusa Purple Long. The selected transgenic lines, confirmed with DNA and protein blotting techniques, showed enhanced level of resistance against these wilt causing fungi with a delay of 5–7 days in disease development as compared to the non-transgenic plants. Sometimes, it has been found that the transgenes were capable of inducing disease resistance trait, but have altered the plant growth processes due to the use of a constitutive promoter. In a study, Mercado et al. expressed β-*1,3-glucanase* gene *bgn13,1*, isolated from *Trichoderma harzianum* in strawberry under the control of *CaMV 35S* promoter. The transgenic lines showed reduced anthracnose symptoms (from 61.0 to 16.5%) in leaf and crown than control plants after inoculation with *Colletotrichum acutatum*. However, most of the transgenic lines displayed stunted phenotype and reduced yield due to the reduction in number of fruits per plant and a reduced fruit size.

The use of various antimicrobial proteins coding genes like *defensins* had been advocated for combating a large class of fungal and bacterial pathogens. Defensins represent a class of antimicrobial peptides which play an important defensive role against fungi, bacteria and protozoa, but are non-toxic to mammalian cells and plants. Defensin genes encoded proteins react by creating certain pores in the fungal hyphal membrane and, thus, disturb the ion-influx–outflux and kill the fungal pathogens.

Researchers reported an enhanced level of resistance in transgenic tomato expressing *Capsicum annum defensin* gene against various fungal pathogens. A bell pepper *J1 defensin* gene was also reported to confer resistance against anthracnose disease of mango, which is caused by *Colletotrichum gloeosporioides*. Protein extract from processed embryos of transgenic mango cv. 'Ataulfo' inhibited the growth of *C. gloeosporioides, Aspergillus niger* and *Fusarium* spp. Transgenic banana plants over-expressing two defensin genes—*PhDef1* and *PhDef2* had been found resistant against *Fusarium oxysporum* f. sp. *cubense*.

In addition to that, genetic transformation of many non-plant antimicrobial compounds like cercopin, attacin, phytoalexins had been reported to enhance resistance level in plants expressing them. Transgenic apple expressing maize Leaf color (*Lc*) gene exhibited resistance to fire blight and scab diseases. In another study, transformation of a biotin binding protein and a *proteinase inhibitor* gene from *Nicotiana alata* exhibited resistance against light brown apple moth disease. Constitutive expression of a fungus-inducible *carboxy esterase* gene (*PepEST*) under *CaMV35S* promoter was reported to increase the anthracnose disease resistance in transgenic pepper (*Capsicum annum*). *PepEST* gene expression in fruits leads to disease resistance development by generation of H_2O_2 and expression of pathogenesis-related (*PR*) genes, which encode for a number of small proteins having antimicrobial activity. On infection of the anthracnose fungus, *Colletotrichum gloeosporioides* on the transgenic fruits of pepper cv. Nokkang, a higher level of expression of *PR* genes, namely *PR3, PR5, PR10* and *PepThi* was reported than the non-transgenic plants. Further, a lower rate of disease occurrence (30%) was reported in the transgenic fruits than in the wild-type plants.

Various types of polyamines including putrescine, spermidine and spermine play a key role in imparting tolerance/resistance to both biotic and abiotic stresses. Hazarika and Rajam transformed tomato cv. Pusa Ruby with a human *S-adenosyl methionine decarboxylase* (*samdc*) gene, which is involved in the biosynthesis of polyamines viz. spermidine and spermine.

The transgenic tomato plants synthesized higher level of polyamines and also expressed enhanced level of resistance against *Fusarium oxysporum* causing wilt disease and *Alternaria solani*, the early blight causing fungus. In addition to that, the transgenic lines also expressed better tolerance to a variety of abiotic stresses including high temperature, drought, salinity and chilling stress. Shin et al. developed transgenic chilli pepper plants (*Capsicum annum* cv. Nockwang) with *Tsi1* (tobacco stress-induced 1) gene via *Agrobacterium tumefaciens*-mediated gene transfer technique using cotyledon and hypocotyl explants. The protein product of *Tsi1* gene has got some role in regulating stress-responsive genes and pathogenesis-related (*PR*) genes.

The transgenic chilli plants expressed enhanced resistance to pepper mild mottle virus, cucumber mosaic virus, a bacterial pathogen—*Xanthomonas campestris* pv. *Vesicatoria* and a fungal pathogen—*Phytophthora capsici*. Genetic transformation of *Vf* gene, imparting scab disease resistance caused by *Venturia inaequalis* under *CaMV 35S* promoter in apple, had been found to impart scab resistance in susceptible cultivars of apple in a number of studies.

Banana (*Musa* sps.) is one of the most important fruit crops being cultivated in about 120 countries across the globe. India is the largest global producer of banana. *Banana Xanthomonas* wilt (BXW), which is caused by *Xanthomonas campestris* pv. *musacearum*, is considered as one of the most destructive diseases of this fruit crop, particularly in East and Central Africa. In a study, Namukwaya et al. expressed *plant ferredoxin like protein* (*Pflp*) gene under the control of *CaMV35S* promoter in transgenic banana cv. 'Sukali Ndiizi' and 'Nakinyika' to develop resistance against BXW disease. In bioassay studies, 67% of the transgenic lines were found resistant to BXW and did not show any disease symptoms, while the wild-type plants expressed severe symptoms of wilting. In another study, grapevine rootstock of *V. berlandieri* × *V. rupestris* cv. Richter 110 had been transformed with an *Agrobacterium* oncogene-silencing gene to develop crown gall-resistant lines.

Researchers reported a reduced severity of Pierce's disease and pathogen mobility in transgenic grape cv. Freedom by the over-expression of an *rpfF* gene (from *Xylella fastidiosa*), which codes for the *synthase* for diffusible signal factors (*DSFs*). Researchers transformed *Vitis vinifera* Thompson Seedless grape with a *stilbene synthase* gene, *VqSTS6*, isolated from Chinese wild-type *V. quinquangularis* accession Danfeng-2 under a fruit-specific promoter to develop resistance against powdery mildew disease, caused by *Uncinula necator*.

The transgenic plants synthesized enhanced quantity of trans-resveratrol and other stilbene compounds as compared to the control plants and expressed enhanced resistance to powdery mildew fungus. It has been found that *VqSTS6* gene is involved in resveratrol biosynthetic pathway in grapes and, thus, plays a key role in imparting protection against invading pathogens. Researchers reported antisense expression of the peach *MLO* gene in strawberry (*Fragaria* × *ananasa*) conferred cross-species resistance to *Fragaria*-specific powdery mildew. RNA-interference (*RNAi*) technology being used recently is quite successful in controlling various bacterial and fungal diseases in plants by switching off the expression of certain endogenous genes.

In one such study, transgenic tomato plants expressing *hpRNA* constructs against *Agrobacterium iaaM* and *ipt* oncogenes were found to be resistant to crown gall disease. Recently, researchers reported an increased level of resistance in grapevine to powdery mildew by *RNAi*-mediated silencing of the susceptible (*S*-gene) *MLO-7*.

VIRUS RESISTANCE

In fruit crops, the *coat protein*-mediated approach to engineer virus resistance has been in application to introduce resistance against various viral diseases. Papaya is grown in many tropical countries, but its cultivation is being threatened by Papaya Ring Spot Virus (PRSV), a disease that is considerably lowering its yield. Using biotechnological interventions, the *coat protein* gene of the virus has been transferred to papaya to confer PRSV resistance.

Since 1998, GM papayas have been cultivated in Hawaii, USA, which had shown considerable resistance to PRSV. PRSV-resistant transgenic papaya varieties 'SunUp' and 'Rainbow' have now occupied >80% shelf-space in the US market. Strawberry is susceptible to various devastating fungi, bacteria and viruses. Finstad and Martin developed transgenic strawberry plants expressing a *coat protein* (*cp*) gene from strawberry mild yellow edge potexvirus (*SMYELV-CP*) and these lines conferred resistance to the virus.

In another study, researchers developed transgenic chilli pepper plants with a *coat protein* gene (*CMVP0-CP*). Three independent transgenic events, which were earlier highly tolerant to CMVP1 pathogen, were also found to be tolerant to CMVP0 pathogen. The production and productivity of watermelon (*Citrullus lanatus*) have been affected considerably by two viruses, namely Zucchini yellow mosaic virus (ZYMV) and papaya ring spot virus type W (PRSV W) worldwide. In an attempt to get rid of these two viruses altogether, Yu et al. transformed three watermelon cultivars, namely 'Feeling', 'China rose', and 'Quality' with chimeric construct containing truncated *ZYMV coat protein* (*CP*) and *PRSV W CP* genes via *Agrobacterium tumefaciens*-mediated gene transfer technique. Two completely immune transgenic lines of 'Feeling' cultivar had been obtained during greenhouse bioassays where these two lines showed complete resistance to *ZYMV* and *PRSV W* and no virus accumulation was detected by Western blotting from these transgenic lines.

Also, transgenic papaya plants with the mutated *replicase* (*RP*) gene from PRSV showed high resistance or immunity against PRSV in the field. Borth researchers developed transgenic banana (cv. Dwarf Brazilian) plants resistant to banana bunchy top virus (BBTV) by transforming four gene construct derived from the *replicase associated protein* (*Rep*) gene of the Hawaiian isolate of BBTV. The transgenic plants showed no bunchy top symptoms, while the non-transgenic plants expressed bunchy top symptoms. Researchers transformed lily cv. 'Acapulco' plants with a defective cucumber mosaic virus *replicase* gene and four transgenic lines were found to show enhanced level of virus resistance.

RNAi technology has been found successful to impart resistance to various viral diseases in plants. The expression of a self-complementary hairpin RNA under the control of *rolC* promoter controlled the systemic disease spread caused by plum pox virus without preventing local infection. Using a hairpin RNA gene silencing strategy, transgenic poinsettia plants resistant to Poinsettia Mosaic Virus have been developed.

Researchers developed transgenic plants of tomato with *AC4* gene-*RNAi* construct and the transgenic plants were found to show the suppression of tomato leaf curl virus activity. Transgenic banana plants expressing *siRNA* targeted against viral replication initiation (*Rep*) gene were developed by Shekhawat et al., which showed high level of resistance to BBTV infection.

ABIOTIC STRESS MANAGEMENT THROUGH TRANSGENIC APPROACH

Abiotic stresses such as heat, drought and salinity are the major environmental constraints affecting production and productivity of almost all horticultural crops. Conventional plant breeding has not been proved that much successful in addressing abiotic stress mitigation so far. The reason might be that the traits are controlled by a number of genes present at a quantitative trait locus (QTL). To combat the negative effects of various abiotic stresses, it is pre-requisite to identify potential candidate genes or QTLs (gene networks) associated with broad-spectrum multiple abiotic stress tolerance. Various abiotic stresses including drought, high temperature, salinity, frost and flood, etc. adversely affect overall crop growth and productivity by affecting the vegetative and reproductive stages of growth and development. These stresses generally trigger a series of physiological, biochemical and molecular changes in the plants which often result in damage to the cellular machinery. These changes include the disruption of cellular osmotic balance leading to dysfunctional homeostasis, ion distribution and oxidative stresses which cause denaturation of integral proteins of plants.

Plants respond to such stresses in a variety of mechanisms which trigger the cell signaling process, transcriptional controls and production of a number of stress conditions related tolerant proteins, antioxidants and osmotic solutes to maintain homeostasis and to protect and repair the damaged integral proteins. Generally, plants which are stress sensitive are unable to synthesize such compounds under stress conditions and, thus, are rendered liable to various stresses which hamper their overall growth.

A number of genes have been identified in a number of plants/organisms, closely or distantly related, which code for the synthesis of these stress protecting compounds and thus can be targeted for genetic transformation into sensitive genotypes. Such genes have been classified into three categories as: (a) genes which code for the synthesis of various osmolytes such as mannitol, glycine betain, proline, and heat shock proteins, (b) genes responsible for ion and water uptake and transport like aquaporins and ion transporter, etc. and (c) genes regulating transcriptional controls and signal transduction mechanism, examples *MAPK, DREBI,* etc. Research on genetic modification of various horticultural crops for improved abiotic stress tolerance has been explored.

DROUGHT TOLERANCE

Various genes controlling signaling and gene regulatory pathways offer certain key targets for genetic engineering for abiotic stress tolerance. Transcription factors (*TFs*) that regulate or switch on the expression of a number of genes involved in imparting abiotic stress tolerance in plants have been proposed as the most efficient targets for genetic transformation. These transcription factors include *DREB1* gene family, *Myb* gene family, etc. Tsai-Hung et al. transformed tomato plants with a DNA cassette containing an *Arabidopsis* C repeat/dehydration-responsive element binding factor 1 (*CBF1*) cDNA and a *nos* terminator, driven by a cauliflower mosaic virus *35S* promoter. These transgenic tomato plants were more resistant to water deficit stress than the wild-type plants.

Researchers reported improved tolerance to cold and drought stress in transgenic apple by the over-expression of a cold-inducible *Osmyb 4* gene from rice, which codes for a *TF* belonging to *Myb* family. The over-expression of *DREB1b TF* gene had also been reported to induce cold tolerance and drought tolerance in transgenic grapevine. Chrysanthemum is one of the leading ornamental cut flowers across the globe and its production is severely hampered by various environmental conditions.

Drought stress harms this crop to the maximum extent by retarding its growth. *WRKY* transcription factors (*TFs*) work as positive or sometimes negative regulators in various abiotic stress responses in plants. Fan et al. transformed a *CmWRKY1 TF* derived from *Chrysanthemum morifolium* and over-expressed it in chrysanthemum cultivar 'Jinba'. It was found that *CmWRKY1* regulates an ABA-mediated pathway by suppressing the expression levels of various genes including *PP2C*, *ABI1* and *ABI2*, and activating the expression levels of genes like *PYL2*, *SnRK2-2*, *ABF4*, *MYB2*, *RAB18* and *DREB1A* in a positive regulation. The transgenic plants displayed increased drought tolerance in polyethylene glycol (PEG) stress as compared to control plants. Also, multiple abiotic stress tolerance in banana had been reported by the over-expression of *MusaWRKY71* gene, which is a very potential abiotic stress-responsive *WRKY TF* gene, cloned from *Musa species* cv. Karibale Monthan.

A bacterial *mannitol-1-phosphate dehydrogenase* (*mtlD*) gene driven by the constitutive cauliflower mosaic virus (*CaMV*) *35S* promoter was transferred into tomato plants which provides improved abiotic stress tolerance in the transformed plants (Khare et al.). Drought (polyethylene glycol in medium) and salinity (sodium chloride in medium) tolerance tests revealed that transgenic lines exhibited a higher tolerance for abiotic stresses than non-transformed plants. To impart tolerance to various abiotic stresses in potato, researchers transformed a potato-derived gene *StnsLTP1* into potato (*Solanum tuberosum* cv. Desiree) using *Agrobacterium tumefaciens*-mediated genetic transformation method. Under stress conditions, transgenic potato lines displayed enhanced cell membrane integrity by reduced membrane lipid peroxidation activity and H_2O_2 content, comparatively.

Also an increased level of antioxidant enzyme activity with enhanced accumulation of ascorbates and upregulation of various stress-related genes including *StAPX*, *StCAT*,

StSOD, etc. was reported in transgenic potato plants. In an attempt to improve abiotic stress tolerance in mulberry, a very important plant of silk industry, *Hva1* gene encoding for late embryogenesis abundant protein from barley, was transformed by *Agrobacterium tumefaciens*-mediated method. LEA proteins comprise a group of hydrophillins, which are induced as a response to dessication in seeds and are also stimulated under various abiotic stress conditions like dehydration, salinity, chilling or high temperature stress in vegetative tissues of plants.

The transgenic lines displayed an enhanced level of tolerance to drought, salinity and cold conditions than normal plants as quantified by free proline, membrane stability index (MSI) and PS II activity. Glycine betaine plays an important role in drought stress tolerance by scavenging oxidative stress-inducing molecule (free radicals) and it also protects the photosynthetic system in plants. Cheng et al. transformed *choline oxidase* gene (*CodA*) isolated from *Arthrobacter globiformis*, which is involved in the biosynthesis of glycine betaine, into potato cv. 'Superior' under an oxidative stress-inducible *SWAP2* promoter for inducing drought stress tolerance trait.

Under water-stress conditions, transgenic potato plants showed expression of *codA* gene and an accumulation of glycine betaine with a higher leaf water potential as compared to the non-transformed plants. In the stress-recovery treatment, transgenic potato plants displayed a stronger antioxidant activity, higher chlorophyll content, more efficient photosynthesis and better recovery, comparatively. Plant micro RNA (*miRNA*) regulates several developmental and physiological phenomena inside the plants including drought responses. Researchers transformed tomato with an *miR 169* family member, *Sly-miR169c*, which can effectively down-regulate the transcripts of the target genes—three nuclear factor Y subunit genes (*SINF-YA1/2/3*) and one multidrug resistance-associated protein gene (*SIMRP1*), which are down-regulated under drought stress. The transgenic tomato plants over-expressing *Sly-miR169c* displayed reduced stomatal opening, reduced leaf water loss and transcription rate with enhanced drought tolerance traits.

HEAT TOLERANCE

Under heat stress, many reactive oxygen species (ROS) such as hydrogen peroxide (H_2O_2) and superoxide are produced inside the plant cells, leading to various kinds of physiological disorders in plants which affect crop growth and productivity. These ROS denature enzymes and damage various cellular components inside the plant cells. Tolerance to heat stress is straightway correlated with the increased capacity of plants to scavenge ROS.

Thus, it is very important to scavenge ROS to maintain normal growth and metabolism of plants. Plants have developed a variety of mechanisms to combat ROS by the production of various enzymatic systems like superoxide dismutase (SOD) to remove superoxide ions, glutathione reductase (GR) and peroxidase to scavenge peroxide ions (H_2O_2), etc. Thus, over-expression of ROS scavenging enzymes in plants via genetic transformation offers a much potential strategy to overcome heat stress.

Researchers reported over-expression of cytosolic ascorbate peroxidase (*cAPX*) gene improved tolerance to heat stress in transgenic apple. Wang et al. developed transgenic tomato plants which over-expressed *cAPX* gene with enhanced tolerance to heat (40 °C), In field tests, detached fruits from field grown transgenic tomato plants showed enhanced resistance with the exposure to direct sunlight as compared to the fruits from wild-type (non-transgenic) plants.

Over-expression of *Cu/Zn superoxide dismutase* (*Cu/Zn SOD*) gene (derived from *Manihot esculenta*) under an oxidative stress inducible promoter *SWPA2* in potato led to enhanced heat stress tolerance.

Cu/Zn SOD is an ROS scavenging enzyme and, thus, helps in quenching of free radicals released under heat stress in plants. Transgenic plants expressed enhanced tolerance to 250 μM methyl viologen and the visible damage due to heat stress was around 25% in the transgenic plants as compared to the non-transgenic/wild-types, which were destroyed completely under heat stress.

The non-enzymatic methods involve the production of a variety of chemical compounds including polyamines, carotenoids, ascorbic acid, tocopherol, etc., which directly react with ROS, scavenge them and thus provide protection to the plants against heat stress. Polyamines play an important role in imparting thermal stress tolerance in plants. *S-adenosyl-I-methionine decarboxylase* (*SAMDC*) is one of the key regulatory target enzymes in polyamines biosynthesis. Cheng et al. over-expressed *SAMDC* cDNA, isolated from *Saccharomyces cerevisae*, in tomato plants for enhanced polyamines production.

Transgenic lines produced 1.7–2.4-fold higher levels of spermidine and spermine with enhanced antioxidant enzyme activity and better protection of membrane lipid peroxidation as compared to wild-type plants, leading to enhanced tolerance to high temperature stress (38 °C). Over-expression of heat shock proteins in plants has been proposed as one of the potential strategies to combat heat stress. HSPs function as molecular chaperons, who are involved in correct protein folding, assembly, translocation, degradation and they also provide stability to integral proteins and cell membranes under heat stress.

Researchers over-expressed *CgHSP70* gene conferring for heat tolerance in chrysanthemum. The transgenic lines exhibited an increased peroxidase (POD) activity, higher proline content and reduced malondialdehyde (MDA) content. Proline is an important osmoprotectant that protects cells from damage under heat stress and transgenic plants were better able to tolerate heat stress than wild-type plants.

SALINITY TOLERANCE

Salinity or salt stress is one of the most prevalent abiotic stresses that severely affect the quality and quantity of different horticultural produce. Around 20% of the world irrigated agricultural land is affected with salinity problem. Salinity tolerance is a complex mechanism governed by many genes. Plants which are exposed to abiotic stress conditions produce several pathogenesis-related proteins to compensate the adverse effect of stress conditions. *Osmotin* is one of the important pathogenesis-related proteins, which is produced by the plants to combat various biotic and abiotic stresses.

Researchers over-expressed tobacco *osmotin* gene in strawberry (*Fragaria* × *ananasa* Duch.) and found that the transgenic strawberry plants exhibited tolerance to salt stress. Chilli plants are not easily amenable to tissue culture and genetic transformation, thus limiting the scope of genetic improvement for various biotic and abiotic stresses.

Researchers could successfully improve the tolerance of chilli pepper (*Capsicum annum* L.) plants by the ectopic expression of tobacco *osmotin* gene via *Agrobacterium tumefaciens*-mediated gene transfer technique. T_2 generation of transgenic pepper plants revealed enhanced levels of chlorophyll, proline, glycine betaine, ascorbate peroxidase (APX), superoxide dismutase (SOD), glutathione reductase (GR) and relative water content (RWC) in biochemical analysis and survived in salinity level up to 300 mM NaCl concentration. In comparison to other horticultural crops, citrus species are the most sensitive to soil salinity, which greatly limit growth and productivity of citrus crops across the globe. Cervera et al. transformed Carrizo citrange, an excellent rootstock of citrus with a yeast-derived halotolerance gene, *HAL 2*, which is involved in salt tolerance mechanism. *HAL2* gene is involved in the methionine biosynthetic pathway and confers tolerance to lithium and sodium ions. It encodes for a salt-sensitive *biphosphate nucleotidase*, which is required for sulfate accumulation. The transgenic lines expressing *HAL2* protein showed improved tolerance to salinity than the wild-type plants. Tomato is considered as one of the most important vegetable crops worldwide for the commercial value it offers.

Researchers developed transgenic tomato plants expressing tolerance to chilling and salt stress by incorporation of *cytosolic ascorbate peroxidase* (cAPX) gene, derived from pea (*Pisum sativum* L.). *Ascorbate peroxidase* plays a key role in quenching hydrogen peroxide (H_2O_2) in plant cells, thus providing protection against oxidative injury induced by chilling and salt stress. The transgenic plants showed better seed germination rate (26–37%) than the wild type (3%) when the seeds were placed at 9 °C for 5 weeks. *APX* activity was found 10–25 folds higher in transgenic plants under salinity stress (200–250 mM) conditions, thus ensuring minimum damage to the leaves comparatively.

Various abiotic stresses including salinity, chilling and oxidative stresses are the critical factors limiting the cultivation and productivity of sweet potato (*Ipomoea batatas*), a root vegetable crop. It has been observed that the increased production of glycine betaine in plant cells improves their tolerance level towards these stresses. Fan et al. transformed sweet potato cv. Sushu-2 with a chloroplastic *betaine aldehyde dehydrogenase* (*SoBADH*) gene from *Spinacia oleracea*, which is involved in the biosynthesis of glycine betaine.

The over-expression of *SoBADH* gene in transgenic sweet potato improved tolerance towards salinity, oxidative stress and low temperature by providing protection against cell damage by maintaining cell membrane integrity, stronger photosynthetic activity, reduced ROS production and activation of ROS scavenging mechanism. To enhance the tolerance of tomato plants to salinity stress, researchers transformed a strawberry *D-galacturonic acid reductase* (*GalUR*) gene into cherry tomato (*Solanum lycopersicum*) lines to increase the ascorbic acid content. Transgenic tomato plants enriched with high fruit ascorbic acid contents had been found more tolerant to abiotic stress induced by viologen, NaCl and mannitol as compared to the wild-type plants. The transgenic events could survive at a salt stress up to 200 mM and also showed higher expression levels of antioxidant genes including *APX* and *CAT*, responsible for imparting additional capabilities to the transgenic plants for salt tolerance. Under high salinity stress conditions, ion-homeostasis within the plant cells get disturbed altering the overall metabolism. Bulle et al. developed transgenic chilli pepper (*Capsicum annuum*) plants expressing wheat Na^+/H^+ antiporter gene (*TaNHX2*) to develop tolerance towards salinity stress. Transgene integration and expression were confirmed by PCR, Southern hybridization and RT-PCR in T_1 generation.

In biochemical assays, transgenic lines gave enhanced levels of proline, chlorophyll, superoxide dismutase, ascorbate peroxidase, relative water content and reduced level of H_2O_2 and malondialdehyde as compared to the non-transformed plants under salt stress conditions. Over-expression of *TaNHX2* gene has already been evaluated in tomato to combat salinity stress. To improve salt tolerance in bottle gourd, Han et al. transformed a bottle gourd line 'G5' with *Arabidopsis thaliana*-derived *H^+-pyrophosphatase AVP* gene. The *AVP1*-expressing transgenic lines exhibited an improved salt tolerance and maintained higher relative water content under salt stress regime in glasshouse. When watermelon plants were grafted onto the transgenic bottle gourd root stock, they also exhibited greater salt tolerance generating higher biomass and photosystem II quantum yields.

HERBICIDE TOLERANCE

Herbicide tolerance in bedding plants can be expected to significantly reduce the cost of weeding in a landscape environment. The herbicide glyphosate is a potent inhibitor of the enzyme 5-enolpyruvylshikimate-3-phosphate synthase (*EPSP*) in higher plants. A complementary DNA (cDNA) clone encoding *EPSP synthase* was isolated from a complementary DNA library of a glyphosate-tolerant *Petunia hybrida* cell line (MP4-G) that overproduces the enzyme. This cell line was shown to overproduce *EPSP synthase* messenger RNA as a result of a 20-fold amplification of the gene.

A chimeric *EPSP synthase* gene was constructed with the use of the cauliflower mosaic virus *35S* promoter to attain high-level expression of *EPSP synthase* and introduced into petunia cells. Transformed petunia cells as well as regenerated transgenic plants were found tolerant to glyphosate. Transgenic pineapple plants transformed with the *bar* gene for bialaphos resistance were developed and evaluated for tolerance to herbicide Basta. Seven months after transfer to the field, plants were found tolerant to 1600 ml/rai of the herbicide Basta® X (stock concentration 15% w/v glufosinate ammonium), this being twice the dose recommended for field application of the herbicide. Transgenic plants tolerant to Glufosinate ammonium should facilitate more effective weed control in pineapple plantations without damage to the crop.

ENHANCING FRUIT QUALITY FOR INCREASED SHELF-LIFE AND REDUCED POST-HARVEST LOSSES

Excessive softening is the main factor limiting fruit shelf-life and storage. Transgenic plants modified for the expression of cell wall modifying enzymes have been used to investigate the role of particular activities in fruit softening during ripening. Fruit ripening has been modified by altering the activity of cell wall enzymes such as polygalacturonases that are involved in tissue softening and deterioration. The biosynthesis of ethylene, the fruit ripening hormone, has also been blocked in several ways to delay fruit ripening. Calgene Inc., USA (1994) developed the first commercialized transgenic plant, a long shelf-life tomato by the suppression of *polygalacturonase* (PG) gene by antisense strategy. PG gene encodes for *polygalacturonase* enzyme which degrades pectin, the major component of fruit cell wall. Researcher has given the brand name 'Mac Gregor' to its transgenic tomato and the fruits of this plant had enhanced shelf-life for approximately 2 weeks longer without softening.

The Flavr Savr tomatoes have improved flavor and total soluble solids (TSS), in addition to the enhanced shelf-life. However, this Flavr Savr variety was withdrawn from the market three years later because of its disease susceptibility and lack of productivity. The plant hormone ethylene is involved in senescence in many flowers and fruits and their vase life can be extended by either blocking ethylene biosynthesis or ethylene reception.

Later on, other tomato varieties with increased shelf-life were developed through antisense RNA inhibition of *ACC synthase* or *ACC oxidase*, two ethylene precursors. Delayed leaf senescence has been achieved in tobacco and petunia by manipulation of cytokinin biosynthesis. Researchers at the Horticultural Research International, the United Kingdom, have identified the genes which control the taste, smell and color of strawberries. As a result, it would now be possible to create super strawberries that taste sweeter using transgenic approaches. Blackheart is a fruit defect caused by exposure of pineapples to higher temperatures which stimulates *polyphenol oxidase* (PPO) activity.

Researchers cloned a *PPO* gene from pineapple fruits under conditions that produce blackheart. The *PPO* gene has been silenced in transformed plants and transgenic plants are under field evaluation. Also, Park et al. demonstrated that fruit from tomato plants expressing *Arabidopsis thaliana* H^+/cation exchanger (*CAX*) gene has more calcium (Ca^{2+}) and prolonged shelf-life when compared to controls. Nambeesan et al. expressed a yeast *spermidine synthase* (*ySpdSyn*) gene under constitutive (*CaMV35S*) promoter and fruit-ripening specific (*E8*) promoter in *Solanum lycopersicum* (tomato).

The *ySpdSyn* transgenic fruits had a longer shelf-life, reduced shriveling and delayed decay symptom development in comparison with the wild-type (WT) fruits. Crop maturity indicated by the percentage of ripening fruits on the vine was delayed in a *CaMV35S-ySpdSyn* genotype, with fruits accumulating higher levels of the antioxidant lycopene. Notably, whole plant senescence in the transgenic plants was also delayed compared with wild-type plants. Sajeevan et al. over-expressed *AtSHN1*, a transcription factor associated with epicuticular wax biosynthesis to increase leaf surface wax load in mulberry. It has been proposed that leaf surface waxes contribute for cuticular resistance, protect mesophyll cells from dessication and, thus, reduce post-harvest water losses. Transgenic mulberry plants expressing *AtSHN1* gene displayed dark green shiny appearance with increased leaf surface wax content and showed significant improvement in leaf moisture retention capacity even 5 h after the harvest as compared to wild-type plants.

Recombinant-DNA technology also finds its multifaceted applications in improvement of the nutritional quality of fruits. To improve the vitamin C accumulation in fruits and make them more health supportive to the consumers, researchers used *RNAi* technology to silence the expression of a mitochondrial APX gene in tomato fruit. *Ascorbate peroxidase* (APX) oxidizes the reduced L-ascorbic acid in ascorbate metabolic pathway and, thus, reduces vitamin C content. The mitochondrial *APX* enzyme activity in transgenic lines was reduced by 0.4–0.8 times than the non-transformed plants and this led to 1.4–2.2 times increase in ascorbic acid content in tomato fruits.

Altering the fruit texture leading to softening of fruits is a desirable trait from the processing point of view. Researchers transformed tomato cv. Pusa Uphar with a fruit-specific *expansion* gene, *LeEXP1* under the control of a fruit-specific promoter *LeACS4*. The over-expression of *LeEXP1* gene resulted in enhanced fruit softening, as determined by force required to rupture the fruit pericarp than the non-transgenic plants with increased red coloration of fruits at different stages of ripening. This trait may find usefulness in tomato processing industry. Behboodian et al. silenced the expression of *ACC oxidase* (*ACO1*) gene in tomato using *RNAi* approach to lower down ethylene production, which is involved in regulating fruit ripening and flower senescence. The transgenic tomato lines harboring the *hpRNA-ACO1* gene exhibited lower ethylene production and a longer fruit shelf-life of 32 days as compared to 10 days for wild-type fruits. Further, fruits of transgenic tomato demonstrated reduced level of firmness loss as a result of decrease in *pectin methylesterase* (PME) and *polygalacturonase* (PG) activities.

COLOR ENHANCEMENT AND INCREASING VASE LIFE IN VARIOUS ORNAMENTAL CROPS

Genetic engineering techniques had so far limited impact in ornamental horticulture field. However, ornamental horticulture and particularly floriculture are very well suited to the approach of genetic engineering technology. The primary focus on color modification is important in cut flowers because flower color is an important driver of new variety development. Ornamental plants generally have eye-catching attraction and aesthetic importance. Altering flower color or plant architecture has been an important area of work for all the floriculturists. Several ornamental plants, including carnation, rose and gerbera, have been engineered for modified flower color.

The primary color imparting pigments present in flowers are a wide class of flavonoids, carotenoids and betains, which are the targets of genetic engineering for color alterations/modifications. Flower color modification venture includes modifying the flavonoid biosynthetic pathway through the introduction of new genes, over-expression of certain key regulatory genes or by silencing the expression of those target genes by co-suppression strategy, antisense gene technology or by *RNAi* technique. Research has been focused on the manipulation of either anthocyanins (red and blue colors) or carotenoids (yellow and orange colors), with the intent of creating a wider range of flower colors than occurs naturally, as well as to produce natural dyes for industrial purposes.

The first application of genetic engineering to modify flower color led to the production of an orange pelargonidin-producing Petunia variety, which produced flowers with pale brick color. This was achieved by the expression of *dihydroflavonol-4 reductase* (*dfr*) gene from maize in a petunia line. Chalcone synthase (*Chs*) is another gene which had been used for production of pink, white and variegated flowers (sense and antisense genes were used) in petunias, chrysanthemum, gerbera and roses.

Transgenic violet carnations have been successfully produced with the introduction of *F32 52 H* gene from *Petunia hybrida* which encodes a flavonoid required for the biosynthesis of delphinidin. The vase life of flowers can be altered by manipulating the biosynthesis of ethylene. The enzymes *ACC synthase* and *ACC oxidase* are encoded by genes *acs* and *aco,* respectively, and both have been cloned from many species including carnation. Transgenic carnation having antisense *aco* gene had been produced and exhibited longer vase life. Flowers of the transgenic carnation plants exhibited low climacteric ethylene production and a markedly delayed petal senescence.

Till date, the only genetically modified product commercialized on a significant scale are the color modified carnations developed in a joint venture by Suntory Ltd. and Florigene Ltd. Florigene is selling transgenic Moon series carnations engineered for dark violet–purple color around the world. The varieties are developed in Australia and flowers are produced primarily in South America for marketing in the United States and Japan.

In another study, fruit-specific *RNAi*-mediated suppression of a photomorphogenesis regulatory gene (*DETI*) was reported to enhance the carotenoid and flavonoid content in tomatoes. In 2009, transgenic blue roses had been developed by two companies, namely Florigen Ltd. and Suntory Ltd. in Australia. A package of three genes was transferred to red rose plants as; a synthetic *RNAi* gene that switches off the red rose dihydroflavonol reductase (DFR) gene, a *delphinidin* gene from blue pansy and a DFR gene from iris that had an affinity for producing delphinidin.

The resultant roses exclusively accumulated delphinidin in the petals, and the flowers had blue hues, not so far achieved by hybridization breeding. Worldwide, chrysanthemum is considered as one of the most economically valuable flowers. Research efforts have been focused on improving various traits in chrysanthemum including flower color and architectural variants. Chrysanthemums are typically used as cut flowers or potted plants. Obtaining bright red and blue colored flowers has always been the charm of chrysanthemum growers. In an attempt to develop red colored flowers, researchers down-regulated *CmF32 H* gene using *RNAi* approach and over-expressed the *Senecio cruentus F32 52 H* (*PCFH*) gene in chrysanthemum. The transgenic chrysanthemum plants developed bright red flowers and exhibited a significantly increased cyanidin content.

MODIFICATION OF PLANT ARCHITECTURE

Genetic engineering has succeeded in relevance to modifying plant architecture to satisfy the instinct of beauty of mankind. Ornamental plants are being grown for the purpose of beautifying, embellishing or improving human environments. Differential expression of a number of diverse genes in different backgrounds has led to diversification in plant forms.

Reduction in height is a major consideration during commercial production of chrysanthemum. In an attempt to reduce plant height, reesearchers ectopically expressed a tobacco *phytochrome B1* (*PHY-B1*) gene in 'Iridon' chrysanthemum under the control of *CaMV35S* promoter. The transgenic plants developed a short stature as compared to the wild-type plants looking similar to that caused by the application of a commercial growth retardant.

The leaves of transgenic plants developed more intense color inferring the biosynthesis of higher levels of chlorophyll pigment. Later in 2003, Petty et al. reported decrease in chrysanthemum plant height by introducing *Arabidopsis GA insensitive* (*gai*) gene. In another study, researchers transferred *CAG* gene in antisense orientation into *Chrysanthemum morifolium* and observed alteration in gynoecium and androecium to corolla-like tissues. Khodakovskaya et al. developed transgenic chrysanthemum by transforming *ipt* (*isopentyl transferase*) gene, a cytokinin biosynthetic gene, in an attempt to change the plant architecture.

The transgenic plants were found shorter in height with more number of flowers, but with smaller size and late flowering period than the wild-type plants. In another study, researchers reported lessened branches formation in transgenic chrysanthemum transformed with *DgLsL* gene in antisense orientation. Dwarfed transgenic plants of petunia had also been developed by transforming *gai* (gibberellic acid insensitive) gene from *Arabidopsis thaliana.*

Further, researchers reported that *Ls*-like sense gene expression in transgenic carnation resulted in lack of axillary branches formation. Cultivars of many crops having shorter stems generally exhibit higher harvest index, offering additional commercial advantage to the growers. In another study, the ectopic expression of a *PHYA* gene in potato plants significantly inhibited stem elongation and increased the harvest index through hypersensitivity to far red (FR) light.

MARKER-FREE TRANSGENIC TECHNOLOGY

Generally, the methods of genetic transformation employ selection markers such as antibiotic resistance genes or herbicide tolerance genes for the selection of desirable transgene expression in transformed cells. However, except for the role as a selectable marker, these genes do not have any relevant function inside the plant cell and, thus, they exert an extra burden on the plant genome. Also, the constitutive expression of these genes encoded proteins affects the plant metabolism in a negative way. Further, use of marker genes, particularly those coding for antibiotic resistance, has been facing a strong criticism and opposition, particularly in edible crops including fruits and vegetables.

Developing marker-free plants or finding out suitable alternatives of antibiotic or herbicide tolerance genes has been proposed with the hope of increasing consumers' acceptance for genetically modified crops. A set of new technologies has been developed which involve the elimination of marker genes during transgenic plants development, which come under the recent molecular advent as 'Marker-free transgenic technology' or 'Clean-gene technology'. Such technologies would be helpful to minimize biosafety concerns during biosafety research trials and the transgenic products would fetch wider consumer acceptance. Site-specific recombination has been suggested by many workers as a potential strategy.

Researchers proposed chemically inducible site-specific recombination systems as valuable tools for excision of transgenes when their expressions are not required. De Vetten et al. suggested the use of marker-free gene construct for genetic transformation of potato followed by polymerase chain reaction (PCR)-based selection of transformed cells for identification of transformants. Co-transformation with two gene constructs followed by segregation of marker gene and gene of interest in segregating generation has been explored.

Researchers devised a strategy to eliminate public concerns regarding proliferation of antibiotic and herbicide resistance genes into the environment by constructing a super binary vector having two T-DNA systems to generate marker-free transgenic chrysanthemum plants. The vector system was designed having two T-DNA regions—one having *hygromycin phosphotransferase* (*hpt*) selectable marker gene and the other T-DNA containing *â-glucuronidase* (*uidA*) gene, placed adjacent to each other with no intervening region. A total of 17 *hpt*-resistant/*gus* positive T_0 chrysanthemum plants were evaluated for segregation in T_1 generation and among those, approximately 15.7% carried the transgene. Suitability of a twin T-DNA system for generating marker-free transgenic plants has been established.

For the development of selectable marker-free transgenic plants, Multi-Auto-Transformation (MAT) vector system, which combines positive selection using *isopentenyl transferase* (*ipt*) gene, the first enzyme of cyokinin biosynthesis, with a site-specific recombination and removal system offers a very potential tool. Marker gene is removed from the transformed cells by the mechanism of homologous recombination. The MAT vector backbone is composed of yeast site-specific recombination R/RS system to excise the DNA fragment and the *ipt* gene cloned between two directly oriented recombination sites. The MAT vector system has been employed in a number of crops to develop marker-free transgenic plants including *Antirrhinum majus*, citrus, *Kalanchoe blossfeldiana* and *Petunia hybrida*. Researchers used MAT vector in which *ipt* gene was used as a selection marker and *Wasabi defensin* (WD) gene, isolated from *Wasabia japonica* as a target gene, to transform tomato plants. The marker-free transgenic tomato plants exhibited enhanced resistance against a number of fungi including *Alternaria solani*, *Botrytis cinerea*, *Fusarium oxysporum* and *Erysiphe lycopersici*. Also, *phosphomannose isomerase* (PMI) gene derived from *E. coli* had been developed as an efficient positive selection marker for apple transformation, which induced the capability to grow on mannose supplemented medium in the transformed cells. Further, plastid engineering has also been advocated as one of the most viable techniques to avoid transgene spread to other related crop (non-target species).

Cisgenic crops represent a step towards a new generation of genetically modified crops. Development of genetically modified crops, which do not possess any selectable marker (e.g., antibiotic resistance or herbicide tolerance) gene in the end product and also, if the inserted gene is derived from the same organism/plant, would be a welcome step to increase consumers' acceptance for that product and to minimize the environmental risks associated with genetically modified crops. In this direction, Vanblaere et al. developed cisgenic apple plants by inserting the endogenous scab resistance gene *HcrVf2* under the control of its own regulatory sequences into the scab susceptible apple cultivar 'Gala' using R/RS vector system to develop marker-free transgenic plants. Dhekney et al. also used cisgenic approach to develop disease-resistant apple.

GENOME EDITING TECHNOLOGY IN HORTICULTURAL CROP IMPROVEMENT

Recent technology relies on certain engineered endonucleases (EEN) that cleave DNA in sequence-specific manner due to the presence of a sequence-specific DNA-binding domain. These *endonucleases* recognize specific DNA sequence and, thus, efficiently and precisely cleave the target genes. The double-strands breaks (DSBs) of DNA result in cellular DNA repair mechanisms, including homology-directed repair (HDR) and non-homologous end joining breaks (NHEJ), leading to gene modification at the target sites in the genome of plants. Generally, this technology employs three types of engineered endonucleases viz. Zinc finger nucleases (ZFNs), TALENs and CRISPR/CaS for site-specific cleavage. The CRISPR/Cas9 system has originated from bacteria and archae. CRIPSR/Cas9 is comparatively easy to prepare, affordable and can be better upscaled than ZFNs and TALENs. Cas9-induced double strand breaks in the plant genome are repaired by non-homologous end joining (NHEJ) method. During this repair, small insertions or deletions may occur disturbing the open reading frame of a protein or introduces a stop codon.

Zinc finger nucleases (ZFNs) and transcription activator-like effector (TALENs) technologies of genome editing affected from the disadvantages of high technical complexity and low efficiency. On the other hand, the clustered regularly interspersed short palindromic repeats (CRISPR)-associated protein 9 (CRISPR/Cas9) technology has revolutionized genome editing by overcoming the disadvantages of ZFNs and TALENs due to a high efficiency, low cost involvement, simplicity and versatility.

CRIPSR/Cas9 genome editing technology will probably avoid the current GM regulations mechanisms as the Cas9 protein-guide RNA complexes get rapidly decomposed in the regenerating cell cultures and thus will broaden the utility of this technology with greater global acceptance levels in comparison to the transgenic technology. A gene-edited crop does not necessarily contain any transgene and, thus, does not require very stringent regulation and thus such crops may find quick acceptance among consumers.

Precise genome editing offers a wonderful technology to decipher plant gene functions and in improvement of crop plants. CRISPR/Cas9 editing tools have been efficiently applied in a number of horticultural crops including tomato, petunia, citrus, grape, potato and apple for gene mutation, repression, activation and epigenome editing.

Genome editing in plants has been revolutionized with the development of CRISPR/Cas9 technology. Recently, this technology has found application in developing resistance against many viruses. Researchers could successfully develop virus resistance in cucumber using Cas9/subgenomic RNA (sgRNA technology) to disrupt the function of recessive *eIF4E* (eukaryotic translation initiation factor 4E) gene. Homozygous T_3 transgenic cucumber plant which was targeted to both eIF4E sites developed immunity to cucumber vein yellowing virus (Ipomovirus) infection and resistance to the potyviruses—Zucchini yellow mosaic virus and Papaya ring spot mosaic virus-w. In a recent study, Malnoy et al. reported the direct delivery of purified CRISPR/Cas9 ribonucleoproteins (RNPs) targeting to bring mutagenesis in *MLO-7*, a susceptible gene in order to increase resistance to powdery mildew in grape cultivar Chardonnay and *DIPM-1*, *DIPM-2* and *DIPM-4* genes in apple cultivar Golden delicious to increase resistance to fire blight disease.

Researchers demonstrated the usefulness of genome editing technology, CRIPSR/Cas9—as a powerful tool to effectively create knockout mutations in watermelon. They targeted *CIPDS* (*phytoene desaturase*) gene for mutagenesis for developing albino phenotype. All the transgenic watermelon plants harbored *CIPDS* mutations and developed clear or mosaic albino phenotype, indicating that CRIPSR/Cas9 system has 100% genome editing efficiency in transgenic watermelon lines to introduce new functions. Induction of parthenocarpy has always been desired in horticultural crop plants for various industrial purposes and for eating quality. Researchers demonstrated a CRISPR/Cas9 system-based breeding strategy to generate parthenocarpic tomato plants. Using CRISPR/Cas9 system, they could effectively introduce 100% somatic mutations into *SlIAA9*—a key gene controlling parthenocarpy in T_0 tomato plants.

Regenerated T_0 mutants showed morphological changes in leaf shape and seedless fruit, which is a characteristic of parthenocarpic tomato. In a very recent study, Kishi-Kaboshi et al. reported for the first time gene editing in chrysanthemum using CRISPR/Cas9 system and developed transgenic chrysanthemum plants expressing the yellowish-green fluorescent protein (*CpYGFP*) gene from *Chiridius poppei*. Two sgRNAs were selected to target different positions in the *CpYGFP* gene and they obtained transgenic calli containing mutated *CpYGFP* genes (CRISPR-*CpYGFP*-chrysanthemum). Finally, the CRISPR-*CpYGFP*-chrysanthemum shoot containing a mutation in the *CpYGFP* gene was obtained.

CONCLUSION

The applications of recombinant-DNA technology or genetic engineering in crop improvement are immense to solve the problem of global hunger as population is increasing day by day with depriving sustainable intensification. However, horticultural crops have got less attention in this area so far. In contrast to the increasing global adoption of biotech field crops, biotechnology has limited commercial success to date in horticultural crops including fruits, vegetables, flowers and landscape plants. At this juncture of time, we cannot ignore the potential of this technology for the genetic enhancement of our horticultural crops to combat various production constraints like biotic or abiotic stresses and fruit quality improvement. Transgenic technology provides a potential technique for genetic enhancement using desirable trait of interest in plants. There is a need to address various regulatory obstacles for commercial release of various transgenic crops so that the real benefit of this wonderful technology may reach the consumers, the end users.

After the advent of Next Generation Sequencing (NGS) technologies, many horticultural crops including strawberry, papaya, grapevine, sweet orange, mango, etc. have been sequenced, which has now solved the problem of lack of genomic information and thus facilitated the target gene/site to be modified using genome editing technology. This has also improved the breeding efficiency as various genes/QTLs coding for various horticulturally important traits have been identified. In addition to that, transcriptome sequences of a number of horticultural crops are now available in public databases. This vast information will assist in identifying various genes governing various important traits and will help in identifying the target sites for genome editing and genetic transformation.

6

NUTRITIONAL COMPOSITION AND ANTIOXIDANT PROPERTIES OF FRESH PRODUCE

Harvesting: Time of harvesting is determined by the degree of maturity.

- For cereal and pulses, a distinction is made between maturity of stalks (straw), seed pods and seeds with respect to storage and preservation.
- In fruit and vegetables, the maturity indices such as size, development of colour, days from full bloom to harvest (DFFB), degree of ripeness, tenderness in vegetables, starch iodine ratings etc are taken into consideration.
- Harvesting before maturity, increases the risk of loss through moulds and decay.

Pre-Harvest Drying: Pre-harvest drying is generally practiced only in case of cereals and pulses. Extended pre harvest drying ensures good preservation.

Sorting, Grading and Packaging: These operations are used for fruit and vegetables.

- Sorting is done to remove diseased, bruised, rotten, bird eaten fruit or vegetable.
- To select only uniform and healthy produce.
- Grading of produce is then done on the basis of size and colour before packaging.
- Packaging in appropriate containers is done after wrapping individual fruits.

Transport: Collection and initial transportation of the produce depends on the place and condition of grading, packaging and storage space.

- Great care is needed in transporting the produce from orchard to grading and packing house and then to the distant market.
- Storage in godown and cold stores should be done carefully to prevent bruising.
- Over loading and rough handling should be avoided.
- During stacking in the trucks, proper aeration should be provided to the produce.
- Gentle handling is required during loading and unloading of the fresh produce.

Post-harvest drying: Post harvest drying is practiced in case of agricultural crops. The duration of drying of grains depend upon weather and atmosphere conditions during drying in open conditions.

- Drying for a longer time also causes loss thus drying for longer period should be avoided.
- If the grain is not dry enough, it is spoiled by moulds and may rot during storage.
- Moreover, if the grain is too dry, it becomes brittle and can crack after threshing, during hulling and milling.

Winnowing: The broken grains can be removed with the husks during winnowing and are also more susceptible to certain insects (beetles and weevils). Besides, excessive drying leads to loss in weight.

Threshing: If crop is threshed before it is dry enough, this operation will be incomplete. Further, if grain is threshed when it is too damp and immediately heaped up or stored in storage room or bags, it will be much more susceptible to attack from micro-organism, thus limiting its preservation.

Storage: Long term storage space should be adequate, effective and hygienic. In closed structures like granaries, warehouses, cold stores, CA stores, hermatic bins etc, the control of cleanliness, temperature and humidity is particularly important. Damage caused by pests (insects, rodents and moulds) can lead to loss of food value in quality as well as quantity.

Processing: In food grains excessive hulling or threshing can also result in grain losses, for examples hulling of rice can suffer cracks and lesions. The grains not only become unfit for consumption but also become vulnerable to insects such as the rice moth (Corcyra cephalonica). In fruit and vegetables both quantitative and qualitative losses occur during different operations of processing include peeling, trimming, washing and blanching, heat processing, filling and packaging.

Marketing: Marketing is the last operation of post harvest system. Although, produce can be marketed at different points in the agro-food chain particularly from the growing place (orchard), stores, wholesale markets and retail markets. In fresh fruit and vegetables, excessive delays in marketing of the produce at wholesale markets during gluts leads to losses in quantity as well as quality. Thus, the duration between harvesting, storage and marketing should be as short as possible.

Post harvest losses: With respect to agricultural commodities, the loss is defined as a measurable reduction in foodstuffs which may affect either quantity, quality or both. Post harvest losses arise from the fact that freshly harvested agricultural produce is a living entity that breathes and undergoes changes during post harvest handing.

- Loss in form of damage, is the visible sign of deterioration and can only be partial e.g., chewed grain, birds eaten fruit/vegetable etc.
- Damage restricts the use of the product, while loss makes it use impossible.
- Food loss refers to total modification or decrease of food in quantity or quality and makes it unfit for human consumption.

TYPES OF LOSSES

- Quantitative loss: It refers to a loss in terms of physical substance i.e. reduction in weight and volume which can be assessed and measured.
- Qualitative loss: It refers to the food and reproductive value of the products. Broadly it includes loss in quality, nutrients, vitamins, appearance, seed viability and commercial loss.
- Direct and indirect loss: Direct losses occur when the disappearance of food stuff is caused by leakage, pilferage or consumption by pests (insects, rodents, birds), loss in weight whereas indirect losses occur when a reduction in quality leads to the consumer's refusal to purchase.

POST HARVEST LOSSES IN HORTICULTURAL CROPS

The value chain in post-harvest management of horticultural crops mainly comprise of pre-harvest factors, harvesting, market preparation (pre-cooling, sorting, grading, packaging and on-farm storage), transportation, storage, value addition/processing and by-product waste management (Table 1). Factors responsible for post harvest losses in the value chain are discussed as under:

Pre-harvest Practices

a. Varietal improvement: Good returns can only be obtained from a good quality raw material and selection of varieties is most essential. In horticultural crops, there is range of genetic variation in composition, quality and post-harvest life potential.

- Selection of carrot and tomato cultivars with comparatively high carotenoid and vitamin A contents.
- Onion and tomato cultivars with high TSS and longer shelf-life.
- Potato cultivars with low sugars.
- Watermelon cultivars with higher sugar content and firmer flesh.
- Spur type apple cultivars with better colour and firmness.
- Grape cultivar with high sugar content.
- Ginger with less fibre.
- Pineapple cultivars with higher content of ascorbic acid, carotenoids and sugars.

Thus, there is a need for development of cultivars for specific traits having longer shelf life and suitability for processing.

- Cultural practices: Proper nutrition, irrigation prior to harvest and other cultural operations in the field affect the quality of harvested produce.
- Pre-harvest treatments: Pre-harvest treatments like cultural practices, nutrition, irrigation and chemical sprays of fruit and vegetable in the field affect the post-harvest shelf-life of the commodity. Chemical sprays of fungicides, calcium

chloride, bio-pesticides etc used as pre-harvest treatments have been utilized to reduce the post-harvest losses in different fruits.

Harvesting

1. **Maturity indices:** Maturity at harvesting is the most important determinant of storage life and final fruit quality. Fruit picked either pre-mature or too late, are more susceptible to post-harvest physiological disorders and diseases than those picked at proper stage of maturity. Maturity indices developed for mangoes (Dashehari, Chausa, Langra and Mallika), guava, date palm, apple, aonla, Nagpur mandarin, acid lime, sweet oranges, pomegranate, grapes, ber, litchi, capsicum and stone fruits can be used to harvest fruits at optimum maturity. Many leafy vegetables and immature fruit vegetables like cucumber, green beans, peas and okra attains optimum eating quality prior to reaching maturity, but are often harvested late and consequently result in low quality produce.
2. **Method of harvesting:** The methods of harvesting (manual or mechanical) significantly affect the post-harvest quality of fruits and vegetables. Mechanical injuries (bruising, surface abrasions and cuts) during harvesting can accelerate loss of water and vitamin C resulting in increased susceptibility to decay causing pathogens. Most fresh fruits and vegetables in India are harvested by hand. However, commodities meant for processing like tomatoes and potatoes can be mechanically harvested. Different fruits require different methods of harvesting.
   - Mangoes are manually harvested with the use of bamboo poles to which a net is attached and are lowered to the ground in a basket.

Different mechanical harvesters developed for harvesting of fruits are:

- Mango harvester.
- Bael fruit picking platform.
- Strawberry harvester cum holding tool.
- Sapota harvester.
- Semi-mechanized hydraulically operated oil palm harvesting device.
- Ber and jamun fruit harvester along with fruit saver device.

Harvesters of different sizes with respect to economic feasibility and affordability of farmers are required to be developed in view of difficult hilly terrains in most of the fruit growing areas.

Market preparation

a. Pre-cooling: Immediate cooling of the produce to its optimum storage temperature for removal of field heat is the most effective and well established tool for extending storage life of fresh horticultural perishables. It is now considered as the most accepted treatment but rarely used in actual practice in India.

   - Protocols optimized for pre-cooling of mango, apple, banana and stone fruits etc need to be refined further for commercial adoption.

- Packing the produce with crushed or flaked ice to provide rapid cooling along with high RH.
- Keeping produce under shade can also be used in farmers' fields.

b. Ripening: Considerable information exists on the physical and chemical changes that occur during fruit ripening, particularly with respect to polysaccharide hydrolysis and role of ethylene etc.

   - Artificial ripening of banana with 500 ppm etheral dip or spray and keeping at 18-20°C has been found to impart uniform ripening in Cavendish group of banana. The treatment appears to be good substitute for smoking or carbide treatment presently practiced by many wholesalers in artificial ripening of mango and banana.
   - Ripening in sapota is hastened by the application of etheral, 2, 3, 5-TP and maleic hydrazide.
   - Ripening chambers are needed to develop for checking the use of banned chemicals like calcium carbide.
   - Very few large scale enterprises use ripening chambers for uniform ripening of banana yet such facilities are beyond the reach of small growers.

c. Sorting/grading: Mostly, manual grading of fruits on the basis of size is done in India. Growers generally employ trained packers who also grade the fruit on the basis of their experience.

   - Few mechanical graders developed in the country like expanding belt type grader, mechanized sorting, washing, wax coating and size grading unit for Nagpur mandarin, mosambi and kagzi lime.
   - Mechanical grader for black pepper and raw cashewnut.
   - Onion halving and spherical fruit grader.
   - Mango fruit grader and sapota fruit grader.
   - All these graders need to be popularized among the farmers after taking into account their affordability.
   - Development of complete packaging line with size grader, de-saping unit, hot

water treatment unit, washing, sponging, waxing, dehydration and packaging for mango can be used to reduce losses and to ensure good returns.

TABLE 1: IDEAL VALUE CHAIN FOR POST HARVEST HANDLING OF HORTICULTURAL CROPS

Issue	Unit operation	Remarks
Harvesting	Improved cultivars	ultivars with better attributes, shelf life and suitable for processing
	Cultural practices	Pre-harvest cultural practices like nutrition, irrigation, water sprays helps in uniform colour and size development
	Pre harvest treatments	Pre harvest treatments to increase colour, firmness and improve storage quality.
	Maturity indices for intended use	Maturity indices for fresh market, short and long distance marketing, storage and processing
	Time of harvesting	Harvesting during cool hours (evening and early morning) improve storage life
	Harvesting methods and tools (manual/mechanical)	Mechanical harvesting is suitable for bulk handling
Market preparation	Pre-cooling	Pre cooling to remove field heat and improve storage quality
	Ripening	Ripening chamber for mango, banana, papaya etc for uniform ripening and alternative for banned calcium carbide
	Sorting/ trimming	Sorting/trimming to separate undesirable part and to select healthy and uniform produce
	Grading	Grading with respect to size, colour and quality
	Post harvest treatment	
	Washing/cleaning	Washing of fruit/vegetable to remove any dust, dirt or debris on the surface, soil in case of tuber crops.
	Polishing, waxing, pest or fungus treatments	Helps to impart shine to the produce, slows down respiration and checks spoilage

Issue	Unit operation	Remarks
	Coating, SO2 fumigation ethylene scrubber Curing	Treatments helps to extend shelf-life by checking respiration, ripening and spoilage
		Extends storage life in onion and potato, etc
	Packaging/wrapping	
	Polyethylene cushioning, wrapping in poly-	Cushioning/wrapping reduces bruising impact during trans-
	propylene, waxed paper etc	portation and also extends storage quality
	Packaging boxes	Corrugated fibre board box (CFB), wooden box, plastic lined CFB
		Reusable plastic crates for processable produce.
	On farm storage	Storage at the farm and packing house.
Transport	Time of transport	Immediately after harvesting, loading during night (cool hours)
	Loading/stacking	Load gently to avoid bruising and impact
	Transport with or without refrigeration	Refrigerated transport is preferred for perishables
Storage	Cold storage, CA storage,	Preferred for better storage quality of perishables
	Zero Energy Cool Chamber	Suitable for short term storage near farm/orchard
	Handling at destination	Unload produce at sanitized place to avoid losses
Value addition/Processing	Utilize market surplus	Processing for manufacture of value added products
	Semi finished products manufacturing	Raw slices, pulps and juices at growing area/production catchments to reduce losses during transportation and their utilization for finished product manufacturing.
	Drying	•Solar tunnel drier or solar cum gas fired drier for continuous drying and safe storage of dried products to prevent insect damage and spoilage

Issue	Unit operation	Remarks
	Osmotic drying	•Popularization for commercial adoption for selected fruits• Development of method for left over syrup utilization protocol•
	Carbonated juice & beverages	•Fruit juice based carbonated beverages from citrus, apple, grapes etc to replace aerated waters
	High flavonoids rich products	• Development of method to utilize peel rich in polyphenols in product manufacture
	Pickling	• Mechanization for adoption at production catchments
	Minimally processed fruits/ vegetables	• Development, refinement and popularization of the technology
Issue	**Unit operation**	**Remarks**
	Finished products like jam, jelly, candies/preserve, puree, sauce, ketchup, canned and bottled products etc	•Mechanization for commercial adoption at production catchments • Juice, squash, ready to serve (RTS) drinks, nectar, appetizers etc
	Juices and Beverages	
	Concentrated juices/pulps	• Development of protocol and equipment for juice concentration for cottage scale entrepreneurs
	Extruded products	• Development of processes for fruit and vegetable based extruded products
	Fermented products	• Utilization of marketable surplus for preparation of fruit based low alcoholic beverages
By-product and waste utilization	Industrial alcohol, pectin, dietary fibre, oil, starch, candies, vinegar, biocolour, bio-flavonoides	• Utilization of food industries waste and crop residue like pomace, peel, seed/pit and waste for value addition and checking environmental pollution

d. Post-harvest treatment: Post harvest treatments are applied to fruits before packaging and storage to improve colour, appearance, firmness and to extend the shelf-life. Various post harvest treatments includes:
   - SO_2 fumigation of litchi fruits followed by citric acid dip, surface drying and packaging in CFB box for retention of peel colour.
   - $CaCl_2$ dip in apple and stone fruits.
   - Degreening of citrus fruits.
   - Gamma radiation (60-90 Gy) of cured onions to check sprouting.
   - Vapour heat treatment, fumigation, bio-pesticide for control of fruit flies in fruits like mango, papaya etc.

e. On-farm storage: On-farm storage is required during gluts and in remote and inaccessible areas in the country. Low cost evaporative cool chambers like zero energy cool chamber, evaporative cooled structures and household evaporative coolers can be used to keep the produce in a good condition for short periods.

f. Packaging: Packaging is an integral element in the marketing of fresh horticultural produce as it provides an essential link between producer and consumer.
   - Fruits and vegetables are generally packed in the field without any pre-treatment.
   - Some are even transported without any packaging like tomatoes.
   - Due to large volumes of non-edible portions of vegetables such as cauliflower, peas etc are transported to wholesale markets from the field.
   - Removal of these non-edible portions prior to marketing would reduce both transportation costs and environmental pollution.
   - Traditional forms of packaging such as bamboo baskets, wooden baskets and gunny bags for packing of many fruits like citrus, mango, cull apples and vegetables like beans, peas, onion, garlic, okra, potato, etc are widely used.
   - Corrugated fibre board boxes are also used for packaging apples, oranges, grapes, stone fruits, sapota etc to reduce load on the forests.

Transportation

Almost all horticultural produce are transported in non-refrigerated trucks to distant markets leading to loss in quantity as well as quality. Temperature management during transportation of fresh fruits and vegetables over long distances is critical. Some of the important points needed to be kept in mind during transportation are:

- Loads need to be stacked to enable proper air circulation, in order to facilitate the removal of heat from the produce as well as to dissipate the incoming heat from the atmosphere.
- Simple intervention like immediate transportation of produce after packaging and loading.

- Movement during cool hours preferably during night.
- Optimum stacking provisions for proper aeration in the truck.
- Cushioning to check vibrations impact etc for safe transportation.
- Transportation in refrigerated vehicle can also be used to reduce post harvest losses of highly perishable produce.

Storage

Storage losses of fruits and vegetables are high due to large variation in temperature and humidity between growing fields to the place of marketing in case of distant marketing. The lowest temperature that does not cause chilling injury is the idle storage temperature for fresh fruits and vegetables. Optimum refrigerated storage requirement for different fruits have been worked out as 1.7-3.3°C for apples, 12.8°C for banana, 0-1.7°C for grapes, 8.3-10°C for guava, mango & pineapple and 5.5-7.2°C for oranges. When combined with low temperature, controlled or modified atmospheres appreciably retards respiratory activity and delays softening, senescence and changes in quality of stored product.

Processing

a. Primary processing: Simple primary processing operations like sorting, trimming, grading, washing, surface drying and packaging can be used to prepare fruit and vegetables for immediate marketing. The available equipment and technologies for various unit operations of primary processing include farm level fruit and vegetable washing machine, basket centrifuge, shrink packaging of fruit and vegetable and hydro cooler-cum-washer for fruits and vegetables, vegetable dryer, tamarind dehuller and deseeder, pomegranate aril remover cumin cleaner-cum-grader, turmeric washing and polishing machine etc.

b. Secondary Processing

1. Drying: Drying is the oldest and cheapest method of preservation of horticultural produce. Significant information is available on the use of solar drier for drying of fruits, vegetables, plantation crops and spices. However, medicinal and aromatic plants can be dried in solar drier, poly tunnel solar drier or mechanical drier. Pre-treatments of fruits like peeling, slicing, blanching, sulphuring, etc, are used for preparing fruit and vegetables for drying.
2. Osmotic dehydration: Osmotic dehydration consisting of partial removal of water by dipping in sugar syrup prior to washing in a mechanical dehydrator is now a standard accepted practice for preparation of intermediate moisture products with acceptable sensory qualities. Some fruits not otherwise fit for drying owing to inherent high acid and astringent taste can also be dried by using this technique. Fruits suitable for osmotic dehydration include pineapple slices, mango slices, banana slices, sapota, apricot, apple and grapes etc.
3. Processing of lesser utilized fruits: Large quantities of lesser utilized horticultural crops like bael, aonla, jack fruit, aloe vera etc cannot be consumed in fresh form without processing. They are known for many therapeutic/medicinal and

nutritive properties. Processing of such crops can play an important role in satisfying the demand for nutritious, delicately flavoured and attractive natural foods of high therapeutic value.

- Bael fruit having hard shell and mucilaginous texture is popularly consumed as a dessert fruit in India.
- Kokum and hill lemon are not acceptable in the fresh form owing to their high acidity, while fresh aonla has a strong astringent taste.
- The products prepared from such fruits include unripe mango drink,(mango pana) high fiber and anti-oxidant rich.
- Cashew apple powder fortified with 2% cereal flour.
- Bael products like ripe bael drink, squash, RTS, (ready to serve) drink jam and jelly, bael dry powder.
- Aonla preserve, candy, shreds, chayawanprash, squash, dehydrated powder, aonla beverages, toffees etc.
- Aloe vera gel and beverages.

iv. Value addition: The fruit and vegetables which can not be sold in the fresh market can be utilized for preparation of different value added products. The value added products include juice, concentrate, fruit based carbonated juices, canning, pulp extraction, pickling, chutney and sauce making, preserves and candies, beverages like squashes, ready to serve (RTS) drinks and appetizer etc from different fruits and vegetables.

v. Fermented products: Production of alcoholic drinks like cider, wine, vermouth, vinegar etc is now an accepted practice for utilization of different fruits.

- Manufacture of champagne (sparkling wine), still wine and brandy from grapes is commercially practiced in the country.
- Other fermented beverages include cider, wine and vermouth from apple, plum, apricot, wild apricot, peach, strawberry, banana etc.

vi. By product waste utilization: Fruit processing plants generate large volume of by-products in the form of pomace, seeds, stones/pits skin, peel which is thrown as a waste. Though such left over produce still contain good proportion of nutrients which can be utilized to prepare large number of value added products for industrial uses.

- Pomace can be used for extraction of pectin, dietary fibre and industrial alcohol.
- Oil extracted from fruit stones/seed left after processing of stone fruits pulp can be used for cooking of foods, pharmaceutical and cosmetic purposes.
- Vinegar extracted from mango peel.
- High fiber containing biscuits from aonla and apple pomace.
- Peel oil, pectin powder, peel candy and animal feed are some of the citrus peel products.
- Oil and fiber from oil palm.

AREAS FOR RESEARCH AND DEVELOPMENT

- Selection and development of improved cultivars for better post-harvest quality.
- Development and adoption of integrated pre and post-harvest treatment protocol (irrigation, water and chemical spray, ethylene management, 1-MCP, fumigation, vapour heat treatment) for different fruit and vegetables.
- Development of protocol for harvesting of produce (maturity indices, time of harvest, method and machineries) for use in different orchard situations.
- Optimization of method for artificial ripening of fruits.
- Development of cost effective mechanical graders (based on size and colour) cum polisher of different size and capacities for different horticultural commodities.
- Optimization of methods for pre-cooling and cool chain handling of horticultural perishables.
- Modification in existing transport vehicle using evaporative cooled system for safe transport of perishables.
- Development and adoption of shrink wrapping/packaging, cushioning and modified atmosphere packaging technology for selected produce.
- Refinement and commercial adoption of evaporative cooled chambers for use in growing areas, retail and whole sale outlets.
- Standardization of post-harvest treatment like irradiation, vapour heat, biopesticides, fumigation etc control of pest and diseases in fresh and stored products.
- Development and integration of protocol for primary processing and market preparation (sorting, grading, washing, surface drying and packaging) of selected fruits, vegetables, medicinal and aromatic plants.
- Standardization of process and equipment for drying of horticultural produce in poly tunnel solar dryer using different modifications like solar, electrical and fuel for uninterrupted drying.
- Establishment of processing units for manufacturing semi-finished products (pulps, slices, raw juices) for their later use in processing industry.
- Development of process and machinery for minimal processing of organically processed products, juice concentrates extruded products, fruit wines, brewed vinegar, flavonoid rich products and health foods from horticultural produce.
- Standardization and adoption of technology for manufacture of pectin, fibre, industrial alcohol, starch, natural colourants and oil extraction from food industries waste (peel, pomace and seed kernels).
- Development of HACCP protocol for fresh and processed horticultural produce.
- Estimation, monitoring and control strategies for pesticide residue, heavy metal and toxins in fresh and processed products for food safety.

MATURITY INDICES, HARVESTING AND POST HARVEST HANDLING OF FRUITS AND VEGETABLES

Maturity

It is the stage of fully development of tissue of fruit and vegetables only after which it will ripen normally. During the process of maturation the fruit receives a regular supply of food material from the plant. When mature, the abscission or corky layer which forms at the stern end stops this inflow. Afterwards, the fruit depend on its own reserves, carbohydrates are dehydrated and sugars accumulate until the sugar acid ratio form. In addition to this, typical flavour and characteristic colour also develop. it has been determined that the stage of maturity at the time of picking influence the storage life and quality of fruit, when picked immature like mango develop white patches or air pockets during ripening and lacking in normal brix acid ratio or sugar acid ratio, taste and flavour on the other hand if the fruits are harvested over mature or full ripe they are easy susceptible to microbial and physiological spoilage and their storage life is considerably reduce. Such fruits persist numerous problems during handling, storage and transportation. Therefore, it is necessary or essential to pick up the fruits or vegetables at correct stage of maturity to facilitate proper ripening, distant transportation and maximum storage life.

Horticultural Maturity

It is a developmental stage of the fruit on the tree, which will result in a satisfactory product after harvest.

Physiological Maturity

It refers to the stage in the development of the fruits and vegetables when maximum growth and maturation has occurred. It is usually associated with full ripening in the fruits. The Physiological mature stage is followed by senescence.

Commercial Maturity

It is the state of plant organ required by a market. It commonly bears little relation to Physiological maturity and may occur at any stage during development stage.

Harvest Maturity

It may be defined in terms of Physiological maturity and horticultural maturity, it is a stage, which will allow fruits / vegetables at its peak condition when it reaches to the consumers and develop acceptable flavour or appearance and having adequate shelf life.

Table 2: Criteria of maturity for harvesting fruits and vegetables

Fruit	Physical	Chemical
Mango	Olive green colour with clear lenticels, shoulder development size sp. gravity, days from fruit set.	Starch content, flesh colour
Banana	Skin colour, drying of leaves of the plant, brittleness of floral ends, angularity of the fruit, and days from emergence of inflorescence.	Pulp/peel ratio, starch content
Citrus	Colour break of the skin from green to orange, size	Sugar/acid ratio, TSS
Grapes	Peel colour, easy separation of berries, character- istic aroma	TSS 18-12 Thompson seedless, 12-14 for Bangalore Blue, 14-16 for Anab-e-shahi
Apple	Colour size	Firmness as measured by pres-sure tester
Papaya	Yellow patch or streaks.	Jelliness of the seed, seed colour

Vegetables are harvested at harvest maturity stage, which will allow it to be at its peak condition when it reaches the consumer, it should be at a maturity that allows the produce to develop an acceptable flavour or appearance, it should be at a size required by the market, and should have an adequate shelf life. Time taken from pollination to horticultural maturity under warm condition, skin colour, shape, size and flavour and abscission and firmness are used to assess the maturity of the produce.

TABLE 3: TIME TAKEN FROM POLLINATION TO HORTICULTURAL MATURITY

Sl. No.	Vegetables	Time to harvest Maturity (days)
1.	Ridge gourd	5 -6
2.	Squash	7 - 8
3.	Brinjal	25 - 40
4.	Okra	4 - 6
5.	Pepper (green stage)	45 - 55
6.	Pepper (red stage)	60 -70
7.	Pumpkin (mature)	65 - 70
8.	Tomato (mature green)	35 - 45
9.	Tomato (red ripe stage)	45 - 60
10.	Peas	30 - 35

Skin Colour

Loss of green colour in citrus and red colour in tomato.

Shape, Size and Flavour

Sweet corn is harvested at immature stage, smaller cobs marketed as baby corn. Okra and cow pea are harvested at mature stage (pre fiber stage). In chilli, bottle gourd, bitter gourd, cluster beans maturity is related to their size. Cabbage head and cauliflower curd are harvested before un pleasant flavour.

Abscission and Firmness

Musk melon should be harvested at the formation of abscission layer. In cabbage and lettuce should be harvested at firmness stage.

FACTORS AFFECTING MATURITY

1. Temperature: Higher temperature gives early maturity, e.g. Gulabi (Pink) grapes mature in 100 days in Western India but only 82 days are enough in the warmer Northern India.
2. Lemon and guava takes less time to mature in summer than in winter. Sun-scorched portions of fruits are characterized by chlorophyll loss, yellowing, disappearance of starch and other alcohol insoluble material, increase in TSS content, decrease in acidity and softening.
3. Soil: Soil on which the fruit tree is grown affects the time of maturity, e.g. Grapes are harvested earlier on light sandy soils than on heavy clays.
4. Size of planting material: This factor in propagated fruits affects fruit maturity, e.g. In pineapple, the number of days taken from flowering to fruit maturity was more by planting large suckers and slips than by smaller ones.
5. Closer spacing: Close spacing of hill bananas hastened maturity.
6. Pruning intensity: It enhanced the maturity of Flordasun and sharbati Peaches.
7. Girdling: Process of constricting the periphery of a stem which blocks the downward translocation of CHO, hormones, etc. Beyond the constriction which rather accumulates above it. In Grape vines it hastens maturity, reduces the green berries in unevenly maturity cultivar and lowers the number of short berries. It is ineffective when done close to harvest. CPA has an additive effect with girdling. Number of days from flowering to harvest can be calculated for some commodities by use of the degree-concept.

Heat Units

Harvest date of newly introduced fruits in a widely varying climate can be predicted with the help of heat unit. For each cultivar the heat requirement for fruit growth and development can be calculated in terms of degree days: Maturity at higher temperature is faster as the heat requirement is met earlier. This heat unit helps in planning, planting, harvesting and factory programmes for crops such as corn, peas and tomato for processing.

MATURITY OF FRUITS AND VEGETABLES

Banana

The fruit is harvested when the ridges on the surface of skin change from angularity to round i.e. after the attainment of 3% full stages. Dwarf banana are ready for harvest within

11- 14 months after planting while tall cultivars takes about 14-16 months to harvest. Peel colour change from dark green to light green the remaining style ends were dry, and brittle and fruits were less angular in shape.

Guava

TSS acid ratio, specific gravity and colour are determined the maturity in guava, e.g.

Allahabad safeda	35.81
Apple colour guava	26.39
Chittidar guava	28.13
Lucknow	49 -34.25
Specific gravity	Less than I
Colour	Light green to yellow.

Ber

In ber maturity is judged by colour (yellow), specific gravity (less than 1) and TSS

Pomegranate

Sugar percentage should be 12-16% and acid percentage 1.5—2.5%, variety Ganesh harvest when seed colour becomes pink. In this stage TSS 12.5% and sugar acid ratio 19.5%.

Bael

It takes one year for fruiting after flowering. It is the fruit which ripen after one year of flowering. April start harvesting and may end it start in flowering.

Mango

This can be judged when one or two mangoes ripen on the tree are fall on the ground of their own accord. This process of fallen is known as tapaca specific gravity 1,01—1.02 and TSS 10-14%.

TABLE 4 MATURITY INDICES OF VEGETABLE CROPS

Root, bulb and tuber crops	Maturity indices
Radish and carrot	Large enough and crispy
Potato, onion and garlic	Tops beginning to dry and topple clown
Yams, bean and ginger	Large enough
Fruit Vegetables	
Cowpea, snap bean, sweet pea, winged bean	Well filled pods that snap readily
Lima bean and pigeon pea	Well filled pods that are beginning to lose their greenness.
Okra	Desirable size reached and the tips of which can be snapped readily
Snake gourd	Desirable size reached and thumbnail can still penetrate flesh readily

Root, bulb and tuber crops	Maturity indices
Egg plant, bitter gourd, slicing cucumber	Desirable size reached but still tender
Tomato	Seeds slipping when fruit is cut, or green colour turning pink
Muskmelon	Easily separated from vine with a slight twist leaving clean cavity (full slip stage).
Watermelon	Dull hollow sound when thumped
Flower vegetables	
Cauliflower	Curd compact
Broccoli	Bud cluster compact

Harvesting

The goals of harvesting are to gather a commodity from the field at the proper level of maturity with a minimum of damage and loss, as rapidly as possible and at a minimum cost. This is achieved through hand-harvesting in most fruit, vegetable and flower crops.

Hand Harvesting

Hand harvesting has a number of advantages over machine harvest. People can accurately determine product quality, allowing accurate selection of mature product. This is particularly important for crops that have a wide range of maturity and need to be harvested several times during the season. Properly trained workers can pick and handle the product with a minimum of damage. Many fresh-market products have a short shelf life if they are bruised or damaged during harvest and handling. The rate of harvest can easily be increased by hiring more workers. Hand- harvesting also requires a minimum of capital investment. The main problem with hand harvesting is labor management. Labor supply is a problem for growers who cannot offer a long employment season. Labor strikes during the harvest period can be costly. In spite of these problems, quality is so important to marketing fresh- market commodities successfully that hand harvesting remains the dominant method of harvest of most fruits and vegetables and for all cut flowers.

Effective use of hand labor requires careful management. New employees must be trained to harvest the product at the required quality and at an acceptable rate of productivity. Employees must know what level of performance and must be encouraged and trained to reach that level.

Mechanical Harvesting

Mechanical harvest is currently used for fresh-market crops that are roots, tubers, or rhizomes and for nut crops. Vegetables that are grown below ground (radishes, potatoes, garlic, carrots, beets and others) are always harvested only once and the soil can be used to cushion the product from machine caused mechanical injury. Tree nuts and peanuts are protected by a shell and easily withstand mechanical handling. A number of products destined for processing such as tomatoes, wine grapes, beans, peas, prunes, peaches and some leafy green vegetables are machine harvested because harvest damage does not significantly affect the quality of processed product. This is often because the product is processed quickly after harvest. These crops have also been amenable to new production techniques and breeding that allow the crop to be better suited to mechanical harvest.

The main advantage of mechanical harvest equipment is that machines can often harvest at high rates. Tree nut harvesters, e.g. attaching a shaking mechanism to the tree and remove most of the nuts in few seconds. The nuts are either caught on a fabric-covered frame or picked up from the ground by other machines. This allows an orchard to be harvested very quickly compared to handshaking with poles. Machine harvest also reduces management problems associated with workers. The commodity must be grown to accept mechanical harvest.

Demerits of Mechanical Harvesting

Machines are rarely capable of selective harvest. Mechanical harvesting will not be feasible until the crop or production techniques can be modified to allow one time harvest. Harvesting machines often causes excessive product perennial crops eg. Bark damage from a tree shaker. The harvesting machines are quite expensive.

Post Harvest Handling

Being living organs, fruits and vegetables continue to respire even after harvesting when they have a limited source of food reserves. In addition to degradation of respiratory substrates, a number of changes in taste, colour, flavour, texture and appearance take place in the harvested commodities which make them unacceptable for consumption by the consumers if these are not handled properly. Post harvest technology starts immediately after the harvest of fruits and vegetables. The whole process of processing the commodities is categorized as Handling of fresh produce. Post harvest Technology of fresh fruits and vegetables combines the biological and environmental factors in the process of value addition of a commodity.

Precooling

Precooling (prompt cooling after harvest) is important for most of the fruits and vegetables because they may deteriorate as much in 1 hr at 32°C. In addition to removal of field heat from commodities, precooling also reduces bruise damage from vibration during transit. Cooling requirement for a crop vary with the air temperature during harvesting, stage of maturity and nature of crop.

There are many methods of precooling viz, cold air (room cooling, forced air cooling), cold water (hydro-cooling), direct contact with ice (contact icing), evaporation of water from the produce (evaporative cooling, vacuum cooling) and combination of vacuum and hydrocooling (hydrovac cooling). Some chemicals (nutrients/growth regulators/fungicides) can also be mixed with the water used in hydrocooling to prolong the shelf life by improving nutrient status of crop and preventing the spread of post harvest diseases.

Washing, Cleaning and Trimming

Before fresh fruits and vegetables are marketed various amounts of cleaning are necessary which typically involves the removal of soil dust, adhering debris, insects and spray residues. Chlorine in fresh water is often used as disinfectant to wash the commodity. Some fungicides like Diphenylamine (0.1 - 0.25%) or ethoxyquin (0.2 - 0.5%) may be used as post harvest dip to control the disorders, eg. Apple superficial scald. For cleaning of some fruit type vegetables (melons, brinjals, tomatoes, cucumber) they should be wiped with damp cloth. Many vegetable need trimming, cutting and removal of unsightly leaves or other vegetative parts.

Sorting, Grading and Sizing

Sorting is done by hand to remove the fruits which are unsuitable to market or storage due to damage by insects, diseases or mechanical injuries. The remainder crop product is separated into two or more grades on the basis of the surface colour, shape or visible defects. For example, in an apple packing house in India 3 grades viz. Extra Fancy, Fancy and standard may be packed for marketing. The fourth "cull" grade is meant for processing. After sorting and grading, sizing is done either by hand or machine. Machine sizers work on two basic principles: weight and diameter. Sizing on the basis of fruit shape and size are most effective for spherical (Oranges, tomato, certain apple cultivars) and elongated (Delicious apples and European pears or of non-uniform shape) commodities, respectively.

Curing

Curing is an effective operation to reduce the water loss during storage from hardy vegetables viz, onion, garlic, sweet potato and other tropical root vegetables. The curing methods employed for root crops are entirely different than that from the bulbous crops (onions and garlic). The curing of root and tuber crops develops periderms over cut, broken or skinned surfaces wound restoration. It helps in the healing of harvest injuries, reduces loss of water and prevents the infection by decay pathogens.

Onions and garlic are cured to dry the necks and outer scales. For the curing of onion and garlic, the bulbs are left in the field after harvesting under shade for a few days until the green tops, outer skins and roots are fully dried.

Waxing

Quality retention is a major consideration in modem fresh fruit marketing system. Waxes are esters of higher fatty acid with monohydric alcohols and hydrocarbons and some free fatty acids. But coating applied to the surface of fruit is commonly called waxes whether or not any component is actually a wax. Waxing generally reduces the respiration and transpiration rates, but other chemicals such as fungicides, growth regulators, preservative can also be incorporated specially for reducing microbial spoilage, sprout inhibition etc. However, it should be remembered that waxing does not improve the quality of any inferior horticulture product but it can be a beneficial adjunct to good handling.

The advantages of wax application are:

- Improved appearances of fruit.
- Reduced moisture losses and retards wilting and shrivelling during storage of fruits.
- Less spoilage specially due to chilling injury and browning.
- Creates diffusion barrier as a result of which it reduces the availability of 02 to the tissues thereby reducing respiration rate.
- Protects fruits from micro-biological infection.
- Considered a cost effective substitute in the reduction of spoilage when refrigerated storage is unaffordable.
- Wax coating are used as carriers for sprout inhibitors, growth regulators and preservatives.

The principal disadvantage of wax coating is the development of off- flavour if not applied properly. Adverse flavour changes have been attributed to inhibition of O_2 and CO_2 exchange thus, resulting in anaerobic respiration and elevated ethanol and acetaldehyde contents. Paraffm wax, Carnauba wax, Bee wax, Shellac, Wood resins and Polyethylene waxes used commercially.

Packaging

Proper or scientific packaging of fresh fruits and vegetables reduces the wastage of commodities by protecting them from mechanical damage, pilferage, dirt, moisture loss and other undesirable physiological changes and pathological deterioration during the course of storage, transportation and subsequent marketing. For providing, uniform quality to packed produce, the commodity should be carefully supervised and sorted prior to packaging. Packaging cannot improve the quality but it certainly helps in maintaining it as it protects produce against the hazards of journey. Striking developments have been in the field of packaging of horticultural produce and the gunny bags, grasses and stem leaves used so far for packaging are now being replaced by a variety of containers such as wooden boxes, baskets woven from bamboo or twigs, sack/jute bags and corrugated fibre board (CFB) boxes.

Storage

A number of storage techniques (ground storage, ambient storage, refrigerated storage, air cooled storage, zero energy storage, modified atmospheric storage, hypobaric storage and controlled atmosphere storage) are being used for fruits and vegetables depending upon the nature of the commodity and the storage period intended.

7

VASE LIFE IMPROVEMENT OF FLOWERS AND FOLIAGE

INTRODUCTION

Vase life refers to the length of time that a bouquet or any bunch of cut flowers retains their appeal and aesthetic value, especially when sitting in stagnant water.

Delicate flowers like roses tend to wither within a few days, so they have a short vase life, whereas s have a longer vase life, lasting up to several weeks if cared for properly.

There are many ways people extend the vase life of their freshly cut flowers.

- Cut flower quality is affected by a wide range of features, like flower number, size, shape, colour (patterns), fragrance, uniformity of blooming, leaf and stem colour, plant shape and developmental stage, and absence of pests and diseases. Postharvest performance involves preserving these specific features for as long as possible. Although these features are caused by a range of physiological processes, they depend mostly on water balance, carbohydrate availability, senescence and pest and disease control. Although there is a clear role for genotype, preharvest growing conditions, and postharvest storage, are just as important: these conditions can be optimised to improve vase life. Of special interest is the long-term storage of flowers, for reefer transport. This way of transport allows a more sustainable product chain.
- Postharvest techniques to achieve greater cut flower longevity must accomplish two seemingly conflicting purposes i.e. the promotion of bud growth and development of the plant part to full opening and retardation of metabolic processes leading to senescence. Many researchers have shown that an appropriate chemistry of vase solution can reduce rate of respiration and senescence and extend the longevity of cut flowers. These preservatives are composed of carbohydrates such as sugar, inhibitors of ethylene biosynthesis such as silver thiosulphate (STS) and aminoethoxyvinyl glycine and germicides e.g. 1-methyl-cycloproprene, chlorine dioxide and 8-hydroxyquinoline sulphate.

- The ethylene sensitivity of cut flowers drives the search for the appropriate vase solution. For instance, is an ethylene-sensitive flower and if preservatives are not used its vase life is normally shortened from about 12-14 days to 7 days. As such, commercial producers widely use STS to extend the vase life of cut flowers due to its remarkable efficacy on vase life extension. However, concern about potential heavy-metal contamination of the environment by STS waste solution has increased the search for alternative methods of controlling flower senescence in s. Most cut flower dealers have used many unorthodox substances, but the results are inconclusive. Many natural products such as essential oils, lemon juice, apple and other fruit extracts have been evaluated for vase life extension of cut flowers because they are environmentally friendly and possess antimicrobial properties.
- Compost obtained from decaying organic matter is known to have antimicrobial properties and humic substances that protect and provide nourishment to plants in many ways. It is therefore proposed that compost can be used a vase solution. Compost stability also defined as resistance of compost organic matter to degrade is important as immature compost may lead to depletion of oxygen or nitrogen and release of phytotoxic compounds, which can adversely affect cut flower if used as vase solution. This suggests that properly formulated mature and stable compost tea mixed with other natural additives can potentially be a substitute for synthetic chemical preservative solutions. This review highlights some of the influential factors affecting cut flower postharvest longevity, aesthetics and metabolism. Recent development and use of organic vase solution are also highlighted.

FACTORS AFFECTING VASE LIFE

Vase life refers to the time period for which a cut flower retains its appearance in a vase. Vase life is determined based on attributes such as diameter and length of florets, the opening of flowers, changes in fresh weight, diameter or length of stem or pedicel, senescence pattern, the color of petals, total longevity and foliage folding. Several efforts have been directed at the evaluation of pre- and post-harvest factors affecting vase life of cut flowers some of which are described below.

Preharvest Conditions Affecting Vase Life

It has been speculated by the authors of several studies that preharvest environmental factors alter the carbohydrate status in cut flowers, which in turn affect their vase life. Despite support for this hypothesis, literature information on effects of preharvest factors on vase life of cut flowers are not comprehensive. However, previous studies on light, nutrient solution, temperature and water relations effects on cut flowers are documented. For instance, it is generally accepted that higher light intensity conditions help to produce optimal carbohydrate content in horticultural crops and extend flower life although effects may vary with genotype. Supplementary lighting and elevated carbondioxide levels have led to longer vase life in (*Dianthus caryophyllus*), chrysanthemum (*Dendranthema grandiflora*) and rose (*Rosa* spp.) while elevated light intensity is known to affect stomatal density which in turn affects gas exchange or photosynthetic rate.

In contrast to high light intensity, elevated temperatures can deplete carbohydrate reserves and decrease keeping the quality of flowers by increasing respiration rate. Some reports indicated that temperatures as low as 12o-15°C or as high as 27°C during three weeks before harvest decreased the longevity of cut flowers. The possible interaction of temperature with other factors complicates the understanding of temperature effects on vase life. Low temperature can also increase pigmentation such as flavonoids and anthocyanins. For instance, 'Golden Wave' roses kept at low temperatures (i.e. less than 15°C) developed a greenish tint.

This was probably due to incomplete conversion of chloroplasts to chromoplasts, which was thought to be caused by low carbohydrate levels. It is usually assumed that if the fertility and the structure of the growing medium are within optimal range for plant growth, there will be little or no effect on the longevity of cut flowers. Another preharvest factor is salinity stress. Halevy & Mayak reported that high levels of soil salinity relative to that in the plant sap will cause exosmosis and thus, reduce plant water uptake.

Postharvest Conditions Affecting Vase Life

Postharvest physiology of cut flowers deals with the functional processes in plant material after it has been harvested or removed from its natural growing environment until it is utilized or deteriorated. Darkness or low light intensity during shipping of cut flowers and foliage can reduce flower quality due to leaf and flower abscission.

Moreover, dark-held plants stored at low temperature when moved to ambient temperatures undergo a number of stress changes such as rapid loss of chlorophyll, proteolysis, loss of catalase activity and increased membrane permeability. At harvest, water and materials transport in the detached plant part is interrupted leading to reduced xylem water potential below the water potential of adjacent cells. Thus, water is withdrawn from cells of the cut flower to cause shrinkage and wilting. Also, the quality of vase water is important for maintaining the vase life of cut flowers and foliage.

Pure clean water with limited amounts of salt and alkalinity is a requisite. The presence of salts, basic ions, and fluoride in water reduce vase life as salts and basic ions are responsible for changes in pH that can affect osmotic potential. Fluorine in water causes softening of tissues, leaf browning and wilting. It was found that addition of small amounts of fluorine (i.e. ~1 ppm) in vase solution had injurious effects on rose cut flowers. Some stems of cut flowers may exude sap, various cations, anions and organic acids that can be harmful to flowers themselves, which suggested that the vase water will need regular replacement. Addition of cobalt ions to vase solution partially closed stomata, inhibited vascular blockage in rose stem and increased the water uptake by cut flowers.

On the other hand, addition of magnesium and calcium ions in vase solution did not affect flower vase life. Storage of cut flowers is economically important as it enables the producers to distribute and transport the product to its destination and to meet the demand of buyers. But an exponential increase is seen in respiration of cut flowers with an increase in storage temperature. The temperature outside the physiological norm can cause damage to cut flowers by freezing injury that results in the immediate collapse of tissue or by heat injury that can heat up tissues above the thermal cell death point.

Cooling prior to packaging and transportation reduces metabolic changes such as enzymatic activity, slows down the maturation of flowers, and reduces ethylene production, metabolic processes and improves longevity. The close association of flower respiration during storage and vase life suggest the potential use of controlled atmosphere technology to extend vase life. There are two controlled systems i.e. modified atmosphere packaging (MAP) and controlled atmosphere packaging (CAP) systems.

Cut flowers and foliage can be enclosed in a gas impermeable package to create an inside environment that can modify carbon dioxide, oxygen, nitrogen, water vapor and trace gases. To provide an optimum atmosphere for increasing the storage length and quality of produce, a MAP is created by altering the normal composition of air (i.e. 78% nitrogen, 21% oxygen, 0.03% carbon dioxide and traces of noble gases). Under CAP conditions, the atmosphere is modified from that of the ambient atmosphere, and these conditions are maintained throughout the storage duration. MAP uses the same principles as CAP.

However, it is used on smaller quantities of cut flowers and the atmosphere is only initially modified. MAP and CAP for packaging fresh cut products reduce polyphenol oxidase and ethylene production and reduced amount of oxygen and elevated carbon dioxide reduces respiration. However, oxygen concentration below a critical level for supporting aerobic respiration may result in the glycolytic conversion of pyruvate into acetaldehyde and ethanol, which leads to tissue fermentation and browning.

Many postharvest diseases such as botrytis rot, mildews and other molds affect cut flowers. The wounded stem-end quickly depletes oxygen in the vase solution due to the increased rate of cellular respiration, which also provide perfect growing conditions for microbes. Microbial cells are responsible for plugging the vascular system, enzymatic damage to vascular tissue, disruption of cell membrane integrity by releasing toxic metabolites and enhance the production of ethylene.

ETHYLENE AND FLOWER LONGEVITY

Role of ethylene in Senescence

The Role of endogenous ethylene in triggering petal senescence is well known. Flowers can be classified as climacteric or non-climacteric based on ethylene sensitivity that can vary from one species to another. In , it is observed that ethylene is first produced in the pistil and then acts on the petals. This induces the expression of genes for ACC synthase, ACC oxidase and cysteine proteinase resulting in auto-catalytic ethylene production. The physical signs of ethylene response include in-rolling of the petals and wilting of flowers.

Ethylene Biosynthesis

Ethylene biosynthesis is under strict metabolic pathway. The sequence of the pathway of ethylene biosynthesis is defined as methionine → S-adenosyl methionine (SAM) → 1- aminocyclopropane -1-carboxylic acid (ACC) → ethylene. It was observed that ACC content was very low in freshly harvested cut flowers, but it rose rapidly with the onset of senescence.

The lifespan of cut flowers is often terminated by the abscission of petals or by petal wilting. In many species, these processes are regulated by ethylene through changes in endogenous levels. Ethylene production changes during plant development and in relation to physiological status.[24] Ethylene production is induced during certain critical stages of growth such as abscission of leaves and senescence of flowers along with other external factors such as mechanical wounding, environmental stress and certain chemicals like auxins.

METABOLIC CHANGES IN CUT FLOWERS

Wounding and Emboli Formation

Suberization is the general response to mechanical damage of plant tissue in which a hydrophobic polymeric material called suberin attaches to cell walls. Suberin can be found in intercellular spaces between xylem parenchyma cells adjacent to wounded vessels and is responsible for compartmentalization after wounding. As such, suberization can be a barrier to water uptake. Upon cutting off a stem, water column in plants breaks resulting in the drawing of air into xylem vessels which constricts water flow in the xylem. Also, plant tissue blockage termed as emboli is formed immediately upon cutting the stem.

Emboli result in a temporary reduction in water uptake but may become permanent if the rate of transpiration exceeds the water conductance of the wounded stem. The reversion of air emboli present in xylem vessels partly results in a positive water balance and restoration of water uptake after harvest of cut flowers and foliage. The use of techniques such as detergent dips or vase solutions, low pH, re-cutting of the stem under water and immersing the stem in deep water containing biocide may help in improving rehydration and the vase life of recalcitrant cut flowers like Heliconia (*Heliconia rostrate*).

Water Balance and Weight Change

The most common cause of short vase life in cut flowers was found to be water stress following the termination of steady flow of water at harvest. Undesirable appearance and short vase life of cut flowers may result from reduced water uptake by xylem blockage due to microbial growth, deposition of gum and mucilage in the lumen of xylem vessels and the presence of air emboli in vascular tissues. This can cause competition for water among the various tissues in the cut flower tissues. Water imbalance due to continuous loss of water through transpiration may result in drooping and pre-mature wilting of flowers making them unacceptable to consumers.

Thus, the appropriate water balance and turgor level, for example, are necessary for the development of buds to bloom. Water loss also enhances premature ripening, senescence, browning, chilling injury and thus, limits the vase life of cut flowers and foliage. More importantly, the vase life of cut flowers is determined by the rate of transpiration through stomata and the type and nature of the solutes present in the vase solution. It is generally known that the sugar in vase solution helps to reduce transpiration rate.

Senescence and Abscission

The cells of senescing organs undergo gradual and orderly disassembly. Plant growth hormones, especially cytokinin appear to control the monocarpic senescence process. Ethylene is known to stimulate changes in membrane lipids of leaves and petals, which results in the rolling of petals, drying of sepals and corollas, wilting of flower and change in color. Earlier, van Doorn & Stead reported increased protease activity and respiration, and reduced lipid fluidity of membranes in petals of cut flowers undergoing senescence.

On the stem of cut flowers, there are several leaves that also suffer the degenerative process of senescence. These reports suggested that leaf senescence leads to several metabolic changes including increased activity of protease, nucleases and chlorophyllases; and can be triggered by stressful conditions such as high salinity, low light conditions and removal of roots.

Cellular Leakage

Two major metabolic and biochemical changes that occur in senescing petals are increases in respiration and hydrolysis of cell components. Respiration mediates the release of for example energy, oxidative breakdown of carbon compounds, synthesis and degradation of carbohydrates, proteins, lipids, organic acids, vitamins and pigments. Most of these occurrences are undesirable to the cut flower. Cell hydrolysis includes membrane rupturing, increase in cytoplasmic debris, invagination of tonoplast, endocytosis of cytoplasmic contents, the disappearance of microtubules, reduction in cytoplasmic volume, and change in proton flux across the plasma membrane and increase in peroxisomes.

These lead to a dramatic increase in the leakage of several cellular molecules including amino acids, sugars, inorganic ions and anthocyanins in addition to an increase in the activity of petal 1-ACC synthetase, which is a precursor for ethylene production, and the disintegration of tonoplast and mitochondria.

Change in Flower Color

Carotenoids constitute yellow, orange and red pigments whereas anthocyanins and other related compounds are responsible for the red, purple and blue coloration of plant tissues. Anthocyanins often mask carotenoids and chlorophyll and are red at acidic pH and blue at basic pH. This gives rise to a phenomenon called 'blueing' that results in a shift of coloration to blue with aging due to depletion of sugars. Cellular pH is important in the regulation of metabolism, and due to the presence of large vacuoles, usually, the cellular pH is highly acidic (i.e. < 5). To maintain cytoplasmic pH, metabolic processes consume or produce protons in the plasmalemma and in the tonoplast. In senescing petals, disassembly of chloroplasts is a programmed process and thylakoid proteins are immobilized. Thus, loss of photosynthetic activity of cells begins with the orderly disassembly of thylakoids.

ROLE OF CARBOHYDRATES IN VASE SOLUTION

Carbohydrates have numerous roles in plants i.e. serving as photosynthetic precursors required for growth, respirable substrates, osmoregulators and sometimes as osmoprotectants. Additionally, carbohydrates can act as cellular signals by controlling gene expressions. Sugar added to vase solution is also known to extend the vase life in several cut flowers as it helps to maintain the respiration rate of floral tissues. Sugar content in plant organs change throughout plant development and some plant organs such as developing flowers act as a sink. With the cutting of a stem, there is a complete reduction of sucrose import to sink tissues of cut flowers.

This leads to the alteration of carbohydrate metabolism. Hence, the use of sucrose in the vase solution has a significant effect on the physiological status of floral tissues and the endogenous levels of sucrose. As such, the addition of sucrose helps in accelerating flower opening and delay flower senescence. Sucrose addition to vase solution was followed by an increase in the length of inflorescence, opening of flower heads and prolonging of the vase life of spikes of *Liatris spicata* (L.) Willd. It was observed that sucrose did not extend the vase life by improving solution uptake but by delaying senescence.

Despite the importance of sucrose as a substrate for respiration and as an osmolyte, other sugars and sugar alcohols (or polyols), including raffinose, trehalose, mannitol and sorbitol have the added function of protecting plants from stress. The polyols mannitol and sorbitol have two crucial functions. They act as both osmotic adjusters and protective compounds.

Therefore, it is believed that these sugars and polyols could be beneficial when added to vase solutions or used as pulse treatments following the harvest of cut flowers. Studies with vegetative cuttings indicated that higher levels of endogenous carbohydrates reduce sensitivity to ethylene, but do not reduce ethylene production. Therefore, in cut flowers, increased endogenous carbohydrates may not only increase vase life by increasing respirable substrate but might also decrease ethylene sensitivity.

Compost

Composting is the biological decomposition and stabilization of organic substrates. During the aerobic composting process, the elevated temperature at the thermophilic stage sterilizes the final compost product by making it free of pathogens and plant seeds. The products of aerobic composting are mainly carbon dioxide, water and heat whereas the products of anaerobic decomposition are mostly methane, carbon dioxide and low molecular weight organic acids and alcohols. Application of compost is known to reduce weeds and fungal diseases in crop production systems.

Brown and Tworkoski reported that growth of the brown rot fungus (*Monilinia fructicola*) was significantly slower on compost substrate. Stable compost can reduce plant pathogens and improve plant resistance to diseases. It also contains valuable nutrients like nitrogen, phosphorus and other essential micronutrients. The properties of compost are known to vary according to the feedstock used and the process of preparation of the compost. But compost formulation as a vase solution for cut flowers has rarely been reported in the literature.

Compost as Replacement of Synthetic Chemical Vase Solution

Compost is known to supply nutrients as well as beneficial organic compounds to plants. Compost is rich in biomass of microorganisms and high contents of total extractable nutrients (16 mg nitrogen/g, 9 mg phosphorus/g and similar levels for most other nutrients). Several organic and inorganic substances such as humic acid, sugar, and amino acids are present in compost extracts and these promote plant growth.

Compost tea is recently being used to control various plants diseases such as rose (*Rosa spp.*) powdery mildew, grape (*Vitis vinifera*) mildew, leaf anthracnose, peach (*Prunus persica*) leaf curl, cherry (*Prunus cerasus*) brown rot and apple (*Malus pumila*) scab. The exact mechanism is unknown, but it is thought that compost tea contains antibiotic compounds that suppress these plant pathogens.

Humic acids are a mixture of naturally occurring organic compounds present in composts, soils, natural waters and sediments and are generally derived from the decomposition of animal and plant residues. Humic acids promote plant chlorophyll content and photosynthesis. Humic substances serve as energy source for beneficial soil organisms, which release complex polysaccharides (i.e. sugar based compounds).

These compounds help in the improvement of carbohydrate contents of plants and can thus, influence the quality and life of cut flowers. This is because the postharvest life of cut flowers has been shown to be dependent on the carbohydrate status of the flowers.

Humic acids influence antioxidative defense mechanisms by stimulating catalases (CAT) to improve the postharvest quality of cut flowers. This agreed to observation made by Fan et al. who demonstrated that foliar application of humic acids increased chlorophyll content, rate of photosynthesis, flower size and flower fresh weight.

Carbohydrates in Compost

Organic matter has been reported to contain organic materials such as lignin, proteins, lipids, cellulose and non-cellulosic carbohydrates (hemicellulose, starch, and mono- and oligosaccharides) that are metabolized along different pathways. Presence of hydrolyzable carbohydrates (acidic and neutral) has been reported in compost. Hexose carbohydrates constituted the largest proportion of neutral sugars with glucose being the most abundant. It was also reported that the proportion of acidic sugars to total carbohydrates was higher in well-cured compost. The number of hexoses and pentoses in compost feedstock samples and humic and fulvic acids were reduced during the composting process. D-glucitol was the main sugar alcohol in compost but its relative amount decreased during composting while the amounts of all the other sugar alcohols increased. The compost of these carbohydrates in vase solution may influence cut flower vase life.

Pathogens in Compost

It is presupposed that well degraded, mature and cured compost does not contain any harmful microbial populations. However, an ecotoxicological tests will be required to ascertain compost safety, especially when used as a vase solution in households. Pathogen control and degradation of toxic organic compounds by composting are important public health measures.

It is assumed that during the composting process when the temperature rises between 45° to °70C, most of the bacterial pathogens and weed seeds are killed. However, the compost must be pathogen free to ensure profitable use of the product. A major concern is the potential for human pathogens to grow in the compost tea.

Compost Amendment of Growing Medium to Enhance Vase Life of Cut Flowers

Several experiments have been conducted using compost, which indirectly indicated that compost can improve the vase life of cut flowers. Compost has been used during the growth stages of flowers to increase the vase life of cut flowers. A study was carried out using different combinations of farmyard manure, poultry manure, sand, leaf compost and coconut coir in equal proportions to determine their effects on the morphological parameters and vase life of tuberose (*Polianthes tuberosa*).

The highest values of floral diameter, number of flowers per spike and shelf life were recorded for the combined sand and leaf compost medium. These components gave better quality cut flower production with maximum vase life compared to all the other medium treatments. Postharvest vase life was also studied by applying foliar humic acid fertilizer extracted from compost chrysanthemum flowers. Seedlings of chrysanthemum were sprayed with equal volumes of distilled water, inorganic compound fertilizer (i.e. nitrogen-phosphorus-potassium (NPK) at 16:6:20 (%)) and organic foliar humic acid (FHA) fertilizer every 15 days.

Application of FHA resulted in increased chlorophyll content, photosynthesis rate, soluble sugars and soluble proteins in leaves of chrysanthemum. Also, there was an increase in vase life, flower size and fresh weight of cut chrysanthemum flowers. The vase life was 61% and 33% greater in humic acid spray than in distilled water and NPK treatment, respectively. Another similar experiment also showed that 1.25 ml FHA (8%)+10g/m NPK (17:17:17) under greenhouse conditions increased vegetative growth, flower production and vase life of tulip (*Tulip* sp.) 'Triumph'. Tamrakar revealed that treatments of vermiwash, cow urine, and GA_3 have a significant effect on vegetative growth, flowering and vase life of gladiolus (*Gladiolus grandiflorus*) 'Candyman'.

These showed that water extract of compost tea may have the potential to extend the vase life of cut flowers, but detailed investigation is required prior to recommendation.

Potential of Compost Tea as Vase Solution

Compost tea is a water extract of matured and well cured compost that has been used for centuries since the Roman Empire. Compost tea is a sustainable, economic and feasible way to efficiently utilize nutrients in compost. It is rich in nutrients and beneficial microbial communities and prepared in an oxygenated environment. Anaerobic conditions are minimized to avoid the production of organic acids, which are detrimental to plants.

There is growing interest in the use of compost tea in integrated pest management as it inhibits several diseases due to the presence of high concentration of beneficial microorganisms. It is well acknowledged that compost tea has the potential to supplement the use of synthetic chemical pesticides with many environmental, health and economic benefits. Compost tea contains many microbial communities that enhance plant growth and suppress disease-causing pathogens including bacteria (*Bacillus*), yeasts (*Sporobolomyces* and *Cryptococcus*) and fungi (*Trichoderma*) as well as chemical antagonists such as phenols and amino acids. Compost tea can be tailored for its desired use. For example, a compost tea can be specifically brewed for use as soil organic matter builder, a disease suppressant, or a nutrient source.

MECHANISM OF ACTION OF COMPOST TEA

Compost acts as a food source for the antagonists that compete with plant pathogens. The mechanism by which compost tea works is not well understood but seems to depend on host-pathogen interaction and mode of application. These biological interactions are complex and can happen through the following non-exclusive mechanisms:

- Antibiosis: It is an antagonistic association between two organisms in which one is adversely affected. Some beneficial organisms can produce compounds that inhibit the growth of disease-causing organisms by producing antibiotics toxic to them. For example, bacteria *Pseudomonas* fluorescens strain CHAO produces hydrogen cyanide, 2,4-diacetyl phloroglucinol, and pyoluteorin which directly interfere with the growth of various pathogens;
- Competition: when beneficial microorganisms are present in a growing medium they tend to out-compete pathogenic bacteria or fungi for food sources and infection sites leaving no food for disease-causing organisms. Even if the disease-causing organisms start to grow, they cannot penetrate the tissues of the plant;
- Induced resistance: some beneficial microbes colonizing plant roots or foliage confer resistance to a plant by turning on plant genes that increase the plant tolerance to infection by pathogens;
- Parasitism: certain beneficial microbes can feed on specific pathogens, for example, *Trichoderma* species have been shown in numerous studies to secrete enzymes that digest the cell wall of some fungal root pathogens;
- Nutrients: compost tea also contains soluble nutrients that perform different key functions. These nutrients feed the beneficial organisms that already exist within

the compost tea, so they grow faster and are healthier and can perform disease suppressive functions. They also help in retaining the nutrients in the medium for plant uptake.

ORGANIC SUBSTANCES USED TO EXTEND VASE LIFE

Germicides

Synthetic germicides such as silver nitrate ($AgNO_3$) and 8-hydroxyquinoline citrate (8-HQC) improve water uptake by preventing the blockage of xylem vessels.[43] Organic antimicrobial compounds have been reported as an alternative biocide for the floriculture industry. Malic acid along with sucrose proved to be effective in controlling bacterial population in vase life preservative solution. This also limits ACC-oxidase activity in cut flowers.

Antimicrobial compounds were found in methanolic and ethanolic leaf extracts of *Jatropha curcas*, *Andrographis paniculata*, and *Psidium guajava*. These were found to be effective against microbes in vase solution of cut Mokara Red orchid flowers. The solution with leaf extract of *A. paniculata* had the lowest bacterial count compared to the other extracts. Thus, the leaf extracts can serve as a potential solution to reduce the microbial population and can be potentially utilized to extend the vase life of flowers.

Acidifiers

A major cause of reduction of vase life in flowers is water flow interruption due to microbial proliferation. Acidifying agent such as citric acid is used to lower the pH of vase solution. Citric acid was reported to increase the vase life and the fresh weight of cut roses. In chrysanthemum, the application of citric acid increased vase life, petal water content, initial fresh weight, and marketable value. Thus, citric acid can be used as a natural, cheap, safe and biodegradable compound that can be efficiently used as an alternative to synthetic treatments to prolong vase life of cut flowers.

Sucrose treatment alone without germicides promoted bacterial proliferation and led to a shortening of vase life. However, the addition of citric acid was effective as an antimicrobial compound and was readily and inexpensively obtained from lemons, lime and oranges.

Also, salicylic acid proved to increase chrysanthemum initial fresh weight, leaf relative water content and petal water content by 23, 49 and 73% respectively, as compared to the controls. At a concentration of 300 ppm, salicylic acid remarkably increased chrysanthemum vase life to 22 days.

Salicylic acid extension of vase life was attributed to its plant regulation and anti-stress properties. The effect of pre- and post-harvest application of salicylic acid was also reported by Soleimany-Fard et al. on cut alstroemeria (*Alstroemeria* sp.) flowers. About 3.0 mM salicylic acid improved the vase life, fresh weight, water uptake, water balance, total chlorophyll content, and suppressed water loss in alstroemeria 'Tampa' compared to control (distilled water) treatment.

Coagulants

Potassium aluminum sulfate (alum) has proved to be an effective coagulant that prolongs the vase life and increase the final weight of Sampaguita flower (*Jasminum sambac*). Alum is synthetic sulphate salt of aluminum and excess use can result in some negative impacts caused by aluminum. There are several other organic coagulants available like polyamines and melamine. However, literature information pertaining to their use for cut flower vase life extension is limited.

Hormones and Growth Regulators

Several growth regulators and hormones are known to extend the postharvest life of different flowers. Kinetin is known to increase water uptake and petal growth in leafless flower shoots of cut roses. It also maintained petal turgidity for an extended period. Gibberellins are also important growth regulators that increase the postharvest quality of cut flowering shoots of *Alstroemeria*, *Euphorbia fulgens* and *Narcissus tazetta*. Methyl jasmonate also improved the vase life and size of fresh cut Peony (*Paeonia officinalis*) flowers.

Herbal Extracts

Various herbal extracts like piper (*Piper magnificum)*, Annona (*Annona reticulata)*, curcuma *(Curcuma longa)*, tobacco (*Nicotiana tabacum*) and galangal (*Alpinia galangal*) are also known to control microbes isolated from the holding solutions of cut roses. Piper extracts at 0-10% (w/v) concentrations showed complete inhibition of microbial growth in vase solutions, delayed bent neck, reduced ethylene production and maintained freshness of leaves.

Essential oils are organic, safe, environmentally friendly and contain biodegradable compounds and have been proved to be the most suitable alternative for the prolongation of vase life in several cut flowers. Trials were conducted on to determine the effects of essential oils from thyme (*Thymus vulgaris* L.), summer savory (*Satureja hortensis* L.) and ajwain (*Carum copticum* L.). It was observed that all the extracted oils prolonged vase life.

Summer savory oil showed the highest effect by increasing the fresh weight to almost 100% in comparison to distilled water used as a control. Another nine essential oils extracted from anise (*Pimpinella anisum*), cumin (*Cuminum cyminum*), geranium (*Pelargonium graveolens)*, common lavender (*Lavandula angustifolia*), sage (*Salvia officinalis*), sweet basil (*Ocimum basilicum*), cinnamon (*Cinnamomum zeylanicum*), blue gum (*Eucalyptus globulus*) and lemon grass (*Cymbopogon flexuosus*) were tested for their antimicrobial effects on the vase life of cut roses by Shanan.

The study indicated that lavender, geranium and anise increased vase life, fresh weight and water uptake in addition reducing transpiration. They suppressed blockage of xylem vessels by reducing bacteria and fungi growth in the vase solution. Babarabie et al. studied rosemary essence along with cola, peppermint and apple extract and were found to be effective as natural ingredients for cut flower preservation.

IMPROVING ORNAMENTAL'S VASE LIFE THROUGH MOLECULAR APPROACHES

Floriculture has emerged as a viable diversification option in the agri-business. It is a rapidly expanding industry recording a growth rate of more than 15 per cent per annum in the last two decades. Flowers are highly perishable unlike other horticultural or agricultural crops. Owing to poor keeping quality, the post-harvest losses in floriculture are significantly higher than any other sector.

Although there has been significant increase in the area, production and productivity of flower crops in the past two decades, there is an urgent need to minimize the huge post-harvest losses in terms of the value of the produce which are estimated to be 30-40 per cent of farm value. The post-harvest losses become important especially when dealing with the export of fresh flowers to distant and foreign market.

The post-harvest life of many flowers is determined by the onset of petal senescence. The petals exhibit a characteristic 'in-rolling' behavior during senescence, and also in response to exogenously supplied ethylene. This 'in-rolling' can be delayed, or even prevented, by treatments (such as the application of silver ions or cytokinins) that block the biosynthesis or action of ethylene. The quality and post-harvest life of many plants and flowers are often reduced by the presence of ethylene in the environment and as plant hormone in plant. Generally, cut flowers are treated with different kinds of chemicals for increasing their shelf life.

We can minimize losses by applying post harvest handling techniques but these are not a permanent tactics for overcoming this problem. It might be possible by applying genetic-engineering approaches to reducing the ethyleneregulated senescence of flowers. The isolation of the genes of the ethylene signal-transduction pathway opens up many new possibilities for engineering longer-life flowers.

Enhanced vase life could be obtained by the introduction of resistance to ethylene or by the inhibition of expression of endogenous ethylene biosynthesis genes. Transgenic ornamental plants have potential to enhance leaf and flower longevity. To achieve the target of enhanced vase life, different biotechnological techniques have been used.

Ornamental horticulture, and particularly floriculture, is well suited to the application of genetic engineering technology. Genetic engineering allows the introduction of genes from outside the gene pool, and is precise, because a gene or genes targeted for a specific trait can be introduced. Biotechnology also shortens the time frame for new variety development.

Where phenotypic novelty is very important, as it is in ornamental horticulture, this is critical. Researchers have recently reviewed the potential for GM to improve leaf and flower longevity in transgenic ornamental crops.

Carnation

Carnation has been used for many years as a model system for studying the physiology of flower senescence. Petal senescence in carnation is an active process involving many biochemical and physiological changes. Carnations with longer vase life have been developed, but were not commercialized. Florigene Pty. Ltd. has developed transgenic carnation varieties which produce flowers with an enhanced vase-life, as a result of alteration of either ethylene biosynthesis or ethylene perception. The flowers from the transgenic plants do not require chemical treatment for maximum vase-life.

Ethylene production during senescence and ethylene sensitivity at a young flower age trigger flower senescence and have been considered in improving vase life in transgenic carnation. Transgenic carnations with delayed petal senescence have been created by silencing the ACO gene to down regulate ethylene production. Using antisense and sense suppression technology, several cultivars were transformed with ACC oxidase. The vase life of these transgenes was almost twice that of the original variety, increasing from 5 to 9 days. Co-suppression using ACC synthase (acs) has been successfully used to delay carnation flower senescence. To inhibit ethylene synthesis in carnation, researchers generated transgenic plants containing A. thaliana etr1–1 gene. Vase life of transgenic carnation flowers was increased 3 fold (senescence was delayed by 6–16 days) compared to the control plant. Vase life was even longer than thatmeasured in control flowers treated with chemicals such as with Aminooxyacetic Acid (AOA) or Silver thiosulfate (STS).

In carnation, autocatalytic ethylene production is caused by the expression of 1-aminocyclopropane-1-carboxylate (ACC) synthase and ACC oxidase genes and wilting is caused by the expression of the cysteine proteinase (CPase) gene during senescence after harvest. In order to improve vase life, sACO (sense ACC oxidase) or aACO genes (antisense ACC oxidase) were transferred by Agrobacterium-mediated transformation into potted carnation (*Dianthus caryophyllus* L. 'Lillipot').

Results showed that *in vivo* ACC oxidase activity in leaflet segments of cultured transgenic shoots was much lower than that in the non-transformed plant. Transgenic cut flowers had a prolonged vase life compared with control plants. Despite the fact that transgenic carnation exhibiting an improved vase life has not yet been commercialized, in the longer-term the incorporation of specific transgenic lines, most probably carrying a mutated ethylene receptor gene, may be a useful addition to a conventional breeding program.

Genetically engineered flowers produced by Florigene, an Australian-based biotechnology company. (a) So-called 'longlife' carnations have an ACC oxidase gene inserted in an antisense direction, which suppresses ethylene biosynthesis and leads to greatly extended postharvest petal life compared with non-transgenic control flowers.

Similar delayed senescence can also be achieved by supressing ASS synthase expression or by enhancing cytokinin biosynthesis through expression of an ipt transgene.

Rose

Studies were realized aiming to investigate the differential responses of flower opening to ethylene in rose cultivars. In these studies were cloned cDNA fragments of three Rh-ACSs (Rosa hybrida – ACC Synthase) and one Rh- ACO (Rosa hybrida – ACC Oxidase) genes that were designated as Rh-ACS1, Rh-ACS2, Rh-ACS3 and RhACO1 respectively.

Additionally, Northern-blotting analysis revealed that among three genes acs, ethylene-induced expression patterns of Rh-ACS3 gene corresponded to ACS activity and ethylene production in both cultivars. The results obtained suggest the existence of cultivars more sensitive to ethylene than others; and the changes of Rh-ACS3 expression caused by ethylene might be related to the acceleration of flower opening in the cultivars more sensitive cultivars and the inhibition in the more tolerants ones.

Petunia

Mutated ethylene receptor genes have frequently been used to convey resistance to ethylene in transgenic floral species. Bleecker et al. (1988) [3] showed that etr1-1 is a dominant mutation that confers ethylene insensitivity in other plants. Ethylene-insensitive petunia was created when they introduced etr1-1 under the control of the constitutive CaMV35S promoter. Transfection of petunia with etr1-1 placed under the control of floral-specific promoters FBP1 (floral binding protein) and AP3 (involved in floral organ development) resulted in a vase life of up to five times that of non-transformed flowers.

The transgenic petunia plants were insensitive to ethylene, and produced flowers that were larger and had a longer vase life than those from non-transformed plants. They found, however, that disease resistance was compromised.

Earlier, researchers observed reduced adventitious root formation in ethylene-insensitive transgenic petunia and concluded that ethylene plays an important role in the response of roots to environmental stimuli. More recently, researchers reported that transgenic ethyleneinsensitive petunias carrying the etr1-1 transgene had a decrease in pollen viability, root mass, seed weight, and seed germination. While seed germination and weight are maternally regulated, the transgene was found to be completely dominant in its effect on flower senescence.

Researchers have recently isolated PhEIN2, a petunia homolog of the Arabidopsis EIN2 gene, and constructed transgenic petunia plants with reduced PhEIN2 expression, and significantly delayed flower senescence. Some undesirable traits similar to those seen in transgenic etr1-1 plants were observed including inhibition of adventitious and seedling root hair formation, increased hypocotyl length, and premature death. A more dramatic reduction in ethylene sensitivity was achieved with the expression of both the etr1-1 and PhEIN2 transgenes.

Chrysanthemum

Normally, leaves of cut flowers show yellowness, even before the start of senescence. On the other hand, exposure to ethylene can also speeds up yellowing. This yellowing of leaves is a feature of senescence. This destroys flower attractiveness, diminishes quality and shortens vase life. Consequently, transgenic plants with reduced ethylene sensitivity i.e., chrysanthemum is anticipated to possess improved vase life. In transgenic *D. grandiflorum*, reduced leaf senescence has been proved very useful. Current studies have shown that yellowing of the leaves of some chrysanthemum cultivars is induced by ethylene. Researchers transferred mDG-ERS1 (etr1-4) into *C. morifolium* 'Sei-Marine' in order to reduce ethylene sensitivity in chrysanthemum.

Their results revealed the usefulness of the mDGERS1 (etr1-4) gene in conferring reduced ethylene sensitivity in chrysanthemum, and gave further support for the action of the DG-ERS1 gene in the perception of ethylene by chrysanthemum leaves.

Arabidopsis Thaliana

In a study to investigate the regulation of ACC synthase gene expression, the promoters of Arabidopsis thaliana ACS genes, AtACS4, AtACS5, and AtACS7, were fused to a GUS reporter gene, and the recombinant transgenes were introduced into Arabidopsis to produce three groups of AtACS: GUS transgenic plants. The data on histochemical and fluorometric studies revealed that promoters of AtACS4, AtACS5, and AtACS7 were all active in dark-germinated seedlings. AtACS5 had the highest promoter activity in leaves of 2-week-old lightgrown seedlings among the three AtACS genes studied. The promoter activities of all these AtACS genes were also found in the reproductive organs.

AtACS5 and AtACS7 were highly expressed in petals, sepals, carpels, stamens, cauline leaves, inflorescence stems, and siliques, while AtACS4 expression was undetectable in the petals of open flowers. Accordingly, each AtACS gene had a unique expression profile during growth and development. It appears that at any developmental stage or any growth period of Arabidopsis, there is always a member of AtACS multigene family that is actively expressed.

- For instance, transgenic carnations produced 90% less ethylene than non-transformed plants, transgenic begonia flowers had longevity similar to non-transformed flowers treated with silver thiosulfate, and antisense ACC oxidase transgenic torenia had a vase life up to 3.5 times that of non-transformed plants.
- Studies with clover showed that the suppression of ethylene activity occurred in transformed plants with the gene in anti-sense orientation. Thus flowers showed delayed petals senescence and a prolonged shelf life as compared to the control flowers.
- Transgenic torenia with an antisense ACC oxidase gene also produced more flowers per stem. Researchers reported that cut flowers of a transgenic carnation with an ACC oxidase cDNA in the sense orientation (sACO transgene) had a longer vase life than flowers of non-transformed carnation plants, and produced negligible amount of ethylene. The sACO transgene is thought to inhibit expression of ACO1, probably by co suppression in the gynoecium, which then suppresses ethylene production in all flowers of sACO-1 plants.
- In the case of pot plants, flower longevity on the plant is also important, as is the health of the leaves. Delayed leaf senescence has been observed in transgenic tobacco (*Nicotiana tabacum*) and petunia plants by introduction of genes that affect cytokinin synthesis.

- In one study, the delay to senesce was the result of expression of a gene for biosynthesis of cytokinins, the over production or external application of which are known to delay senescence. A strategy using a mutant ethylene receptor gene was also effective in petunia. Insertion of anti-sense ACC oxidase in transgenic torenia resulted in plants with flowers that had an extended vase-life and keeping quality has been improved in begonia.
- Success has been witnessed in Oncidium and Odontoglossum by mutating ethylene receptor gene.
- The main disadvantage of this strategy is that plants deprived of ethylene perception do not develop normally. Therefore, the use of organ and tissue-specific promoters that are inducible either developmentally or by a selective external stimulus is indicated.

8

POST-HARVEST MANAGEMENT OF SPICE, MEDICINAL, AND PLANTATION CROPS

Post-harvest management is a system of handling, storing, and transporting agricultural commodities after harvest. For some commodities such as coffee and cocoa, post-harvest activities may include drying and fermenting as well. The importance of post-harvest management has been established over the years—it strengthens the action chain that produces, transports, and processes food and all other related products that give sustenance to the world population.

Moreover, post-harvest management allows the effective planning of how to sustainably manage finite resources for the future. In other words, through effective post-harvest management, action can be done now so that the environment can sustain future generations.

During the post-harvest period, handlers and producers focus on preserving quality, quantity, and the safety of the commodities. For coffee and cocoa producers and traders, maintaining the integrity of the beans is important because it directly affects the price of the commodity. This means having to ensure that moisture, contaminants, and insects will not affect the quality of the commodities.

In the quality front, the major consuming countries like Europe and USA are demanding more and more quality compliance by the producing countries. To meet this challenge we have to equip ourselves to produce, process and market high quality spices, with internationally accepted food safety standards.

Hence, precautions has to be taken from the harvesting, primary processing which includes, washing, threshing, blanching, drying, cleaning, grading and packaging in order to meet the standards. The improvements in post harvest processing, mechanisation and value addition in spices in major spice crops are presented in this chapter.

IMPORTANCE OF POST HARVEST TECHNOLOGY

Fresh fruits and vegetables (F&V) have been part of human diet since the dawn of the history. The systematic nutritional value of the some F & V was recognized in the early

17th century in England. One example is the ability of the citrus fruit to cure scurvy, a diseases wide spread among naval personnel.

An example of the importance of the field to post-harvest handling is the discovery that ripening of fruit can be delayed, and thus their storage prolonged, by preventing fruit tissue respiration. The knowledge of the fundamental principles and mechanisms of respiration, leading to postharvest storage techniques such as cold storage, gaseous storage, and waxy skin coatings. Another well known example is the finding that ripening may be brought on by treatment with ethylene.

Fruits and vegetables are being rich in vitamins and minerals, known as protective foods. Due to their high nutritive valve, ready availability, and being inexpensive they make significantly contributes to human well-being. Realizing the worth of fruits and vegetables in human health ICMR recommend consumption of 120g of fruits and 280g of vegetables per capita per day.

- F&V are rich in ascorbic acid which have beneficial effects of wound healing and antioxidant. Dietary source of Vit. C is essential, since human beings lack the ability to synthesize it.
- Some F&V are excellent source of beta -carotene (provitamin A) which is essential for the maintenance of eyes health; and folic acid which prevents anemia.
- These also prevent degenerative diseases which are prevalent in people with sedentary lifestyle. Concern about obesity and coronary heart diseases have led to reduced levels of fat intake. Antioxidants, phenolic compounds and dietary fiber are considered to be beneficial in reducing risk of various cancers.
- Many F&V have neutraceuticals properties.

F & V provide variety in the diet through difference in colour, shape, taste, aroma and texture that distinguish from the other major food groups of grains, meats and dairy products. Sensory appeal of F&V is not confined to consumption but also has market value. Diversity in their colour and shape is used by traders in arranging product displays to attract potential purchasers; and chefs have traditionally used F & V to enhance the attractiveness of the prepared dishes or table presentations; to adorn meat displays and F & V carvings have becomes an art.

The ornamental provide sensory pleasure and serenity, derived from the colors, shape and aroma of individual species. Garden plants, cut flowers, foliage and flowering plants are increasingly used in exterior and interior decoration. Considerable commercial opportunities arise from their role in social, religious and economic ceremonies and special greeting occasion such as festivals, Valentine's day and others occasion.

NEED FOR POST HARVEST TECHNOLOGY

Species, F & V and ornamentals are ideally harvested based on optimum eating or visual quality. However, since they are living biological entities, they will deteriorate after harvest. The rate of deterioration varies greatly among products depending on their overall rate of metabolism, but for many it can be rapid. For example marketing chains where produce is transported from farm to end user within a short time period, the rate of PH deterioration is of little consequences. However, with the increasing remoteness of production areas from population centers, the time lag from farm to market is considerable. The deliberate storage of certain produce to capture better return adds to this time delay between farm and end user, by extending the marketing periods into times of shorter supply. Thus a modern marketing chain puts increasing demands on produce and creates the need for the PH techniques that allows retention of quality over an increasingly longer period.

POST HARVEST HANDLING OF SPICES

Black pepper

Harvest in Black pepper starts when one or two berries turn yellow. The spikes are nipped of by hand and collected in bags. Three models of pepper harvester were fabricated and evaluated on the basis of efficiency in the cutting action and easiness in operation. The most efficient and user friendly was the second model due to its light weight, easiness in operation and minimum loss. Recent advances in product diversification have necessitated harvesting of the berries at different stages of maturity. The primary processing in black pepper involves threshing, blanching, drying (sun drying or mechanical drying), grading and packing.

Threshing, Blanching, Drying

Traditional manual processing has given way to mechanization in majority of the processing stages. Mechanical threshers are used now to improve the quality of the product and to increase the efficiency of operations.

The quality of the black pepper can be improved by blanching, a simple treatment of dipping the mature berries taken in perforated vessel in boiling water for a minute before drying. Sun drying is the conventional method followed in order to bring the moisture content below 10%. The average dry recovery varies between 33-37% depending on the varieties. Recently, various types of dryers such as solar and mechanical dryers have been developed by different agencies which are highly efficient for drying pepper. Solar tunnel drying in a cylindrical tunnel shaped drying chamber of size 2 m × 3 m to a height of 2m for drying 100 kg pepper. A natural convection reverse air flow mechanical drier developed by Regional Research Laboratory, Thiruvananthapuram are used by the farmers and small scale industries for drying of black pepper. Models of varying capacities operated either electrically or by burning agricultural wastes can also be used for drying.

Cleaning and Grading

Cleaning and grading are basic operations that enhance the value of the produce and help to get higher returns. The dried pepper is cleaned to get rid of the extraneous matter such as dirt, stalks, leaves etc. Magnetic separator is used to remove metallic contamination such as iron fillings and stray nails. Vibratory conveyors with inclined decks in combination with air classification are used for efficient de stoning of spices. Broken pepper and light pepper grades are separated pneumatically. Pinheads which come along with garbled pepper are separated by sieving.

Cleaning on a small scale is done by winnowing and hand picking which removes most of the impurities. Grading of black pepper is done by using sieves and sifting black pepper into different grades based on size. TNAU has developed a hand operated cleaner cum grader suitable for cleaning and grading operations. At an operating speed of 25 rpm, the unit had a maximum effectiveness of 66.4% and capacity of 430 kg/h.

Packaging and Storage

Whole pepper is generally packed and transported in gunny bags and polyethylene lined double burlap bags. The bags are arranged one over the other on wooden pallets after laying polypropylene sheets.

Cardamom

Cardamom, known as the 'queen of spices', is one of the highly prized spices of the world. As the flowering continues over a long period several pickings are needed. Harvesting should be taken up only at a time when the capsules reached the maturity stage. Immediately after harvesting on each day, capsules are washed to get rid of dirt and treated with 2 % washing soda (Sodium Carbonate) for 10 minutes to retain green colour and reduce mould growth in RCC tanks. The treated capsules are spread in a single layer on portable drying trays for draining of wate and later for drying.

Curing

Cardamom curing or drying is the process in which the moisture content of green cardamom is reduced to 8 12%. There are mainly two types of drying viz. sun drying and artificial drying. In sun drying it is difficult obtain good green colour and requires 5 6 days or more depending up on the availability of sun light. Artificial drying can be done either by electrical or conventional flue pipe drier.

Garbling

The dried cardamom is then subjected to garbling. Traditionally this is achieved by rubbing the cardamom capsules against coir mat or wire mesh and winnowed to remove any foreign matter. An oscillatory type semi mechanical garbling unit developed, which has a capacity of 2 3 kg per batch whereas the rotary type garbler will take about 2 3 minutes to garble one batch of about 5 kg. Its capacity is 100kg/h and the efficiency was 98%.

Ginger

India and China are the world's largest producers and exporters of ginger. In India, domestic market prefers fresh green ginger for culinary use while two other types of dried ginger i.e. bleached and unbleached are also produced for export purpose. The fresh ginger immediately after harvest is subjected to washing, which is performed to remove dirt, residues of farm chemicals and other foreign materials.

Peeling and Drying

After washing the ginger rhizomes are subjected to peeling operation. Indigenously, peeling of ginger is done by scrapping with sharpened bamboo stick. The scrapped or peeled rhizomes are again washed well and dried in sun for a week or more and then rubbed again to give a polish. A mechanical brush type ginger peeling machine has been developed by Rajasthan Agricultural University. The peeling efficiency of the machine was 85% and the capacity was 200 kg/h. Another mechanical ginger peeler was developed with its peeling drum made of diamond cut mesh which has a peeling efficiency of 59%. The sun dried ginger is brown in colour, more or less irregular wrinkled surface and when broken, shows a dark brownish colour. However, it is reported that the quality of dry ginger cured in the brightsun is better than the quality of ginger cured in a closed oven with artificial heat.

Polishing

Polishing of dried ginger is done to remove the wrinkles developed during drying process. In the indigenous method the dried ginger is rubbed against a hard surface. However, hand or power operated polishers similar to turmeric polishers are also employed for the purpose of polishing dried ginger. In the case of hand operated polishers an output of 5-6 quintals per day of 8 hours is obtained with the help of two persons.

The dried ginger rhizomes are manually graded. The machines of various capacities to pulverize dried ginger from 25 kg per batch to continuous powdering of 2-3 t/day for large scale production are available. Turmeric Maturity of the crop is indicated by complete drying of the aerial plants including the base of the stem. The leaves and stem are cut close to the ground. Two days earlier to digging out the rhizomes, irrigation is given.

In case of ridge method of planting, plough is used to lift the rhizomes. Otherwise the rhizomes are dug out with manually. he harvested rhizomes have to be cured within 2-3 days after harvest for securing maximum out turn. In the turmeric growing regions of Tamil Nadu, mechanical harvesting of turmeric is being practiced. Tractor drawn turmeric harvester mounted on 35-45 hp tractor with a capacity to harvest 1.6 ha per day have been developed by TNAU. The tractor mounted harvester may be adjudged as the best mechanical harvester for saving of time and reduced the percentage of damaged rhizome apart from the removal of drudgery in harvesting of quality rhizome.

Post harvest Operations

The harvested turmeric rhizomes before entering into the market is converted into a stable commodity through a number of post harvest curing processes like boiling, drying, polishing and colouring. Curing of turmeric is taken up within 3 or 4 days after harvest. The fingers and mother rhizomes are separated and are cured separately, since bulbs take longer time to cook.

The recommended practice is to use clean water for boiling turmeric rhizomes in mild steel or galvanized iron pans and takes about 60-90 minutes. The cooked fingers are heaped on a cleaned drying floor and left undisturbed for 4-5 hours and later sun dried for 10-15 days by spreading in 5-7 cm thick layers on bamboo mat or on the drying floor.

Chillies

The harvested pods have to be kept in heaps either indoor or in shade away from direct sunlight for 2-3 days so as to develop uniform red colour. The ideal temperature is 22-25°C. Subsequent to this, pods should be dried in the sun by spreading them out on clean, dry mats, cemented or concrete surfaces/terraces etc. ensuring that the product does not get any contamination from the drying yard or premises. The drying surface/mats should not be smeared with mud/cow dung. Pods have to be spread out in thin layers for uniform drying with frequent stirring which is essential for preventing mould growth and discolouration. The material should be heaped and covered by clean gunny bags/tarpaulin at night. Unless the pods are dried properly they may lose their colour, glossiness and pungency. The safe moisture level of dried chilli pods is below 15 per cent. Instead of the conventional sun drying method, improved drying system could be used to ensure cleanliness and uniformity of colour of the product.

Storage Precautions

Well dried pods after removing plant parts and the foreign matter should be packed in clean, dry gunny bags and stored ensuring protection from dampness. Dunnage has to be provided to stack the packed bags to prevent moisture ingress from the floor. Care should be taken to stack the bags 50 to 60 cms away from the wall.

The product may be preferably sold soon after drying provided the farmer gets a remunerative price as long as storage may lead to deterioration. However, if cold storage facilities are used the product may be stored for 8- 1 0 months.

Insects, rodents and other animals should be effectively prevented from getting access to the premises where materials are stored. The stored product should preferably be exposed to sun periodically. If care is taken during all stages of cultivation, harvesting, post-harvest handling, processing, packing, storage and transportation by following good practices and sound methods we will be able to effectively prevent contamination and arrest deterioration of quality in any farm produce and ensure consumer satisfaction.

Conformance to the quality requirements of the buying countries, meeting the consumers' expectations and ability to maintain price competitiveness are the key factors which determine our survival in the international market. Improved productivity and quality will enhance the farmers income and increase the country's foreign exchange earnings.

Nutmeg and Mace

Nutmeg is one of the unique and costly spices in the world which yields two spices viz. nut and mace, each has its own aroma and taste. Its scientific name is Myristica fragans and it belongs to the family Myricaceae. Nutmeg and mace contain several chemical compounds that are known to have antioxidant, disease-preventing and health-promoting properties. Nutmeg also has many therapeutic applications in various traditional medicines for its anti-fungal, anti-depressant, digestive and carminative properties.

Nutmeg has huge demand in the West. The possible earnings from Europe will be multifold if quality standards are maintained. Western countries reject the Indian produce because of the presence of Aflatoxin, a poison from Aspergillus fungus. The permissible levels of Aflatoxin is 10 ppb (parts per billion) in Europe, 20 ppb in the U.S., and 30 ppb in the Indian market. Exploiting the export opportunities to European countries is the best available option for profitable farming. Since guaranteeing highest quality alone can ensure enhanced export, systematic post-harvest handling of the produce is imperative.

Nutritional Composition

Nut contains fixed oil trimyristin and many essential volatile oils like myristicin, elemicin, eugenol and safrole. These oils give a sweet aromatic flavour to Nutmeg. The other volatile oils are tineme, camphene, dipentene, cineoil, linalool, sabinene, safrole, terpeniol. Fixed oil content in the nutmeg varies from 25-40% and that of mace varies from 20-30%.

Post-Harvest Processing

Market value of nutmeg mace is fixed based on the colour and nutritional composition. Even though post-harvest processing includes decortication, husk removal, separation of nut and mace, the most important processing with respect to colour of mace is drying. The harvest season of nutmeg coincides with heavy rain in Kerala. Hence, nuts should be dried within 24 hours of harvest and moisture level should not exceed 10 per cent.

Conventional Drying

Nutmeg and mace are dried separately as the moisture content of wet nutmeg is higher than that of mace and time required for drying is different. Mace takes four to five days for drying and nutmeg takes 10 to 12 days depending upon the intensity of the sunlight. In sun drying, the disadvantages are the products are exposed to direct sunlight, no uniformity in drying, presence of mould and infections (Aflatoxin) and contamination. In the processing of mace, the red coloured aril is removed from the nutmeg that it envelops and is flattened out and dried for 10 to 14 days under shade; its colour changes to pale yellow, orange, or tan. The mace is detached, flattened and dried in sun on mats for 3-5 days.

Mechanical Drying

Mechanical drying is the solution to tide over the aflatoxin problem in nutmeg and mace. Mechanical and firewood-based driers can be used for drying both nutmeg and mace. In case of nutmeg drying, a temperature of 40°C may be adopted for about 6 to 8 hours for three days. Mace will take only 5 hours to dry at 40°C, but the colour will be yellowish which is not desirable. Maintenance of better quality, drying even in unfavourable climatic conditions at reduced drying time are the main advantages of mechanical drying.

Both nutmeg and mace need to be dried to a maximum moisture level of 10%. Artificial drying of nutmeg at 50°C, which takes 28 hours to reduce moisture content from 36% to 9% showed "case hardening" effect after drying. Hence slow drying is preferable in case of mechanical drying.

Pulsed Microwave Assisted Hot Air Drying of Mace

Since nutmeg is a spice of high commercial value due to its importance in many fields like food and medicine, the processing is very important. The drying of nutmeg in quick time with colour retention is still a challenge. Conventional methods of drying nutmeg is associated with many disadvantages. Normal sun drying or mechanical drying takes several hours to weeks for attaining the desired moisture content. Implementation of novel methods of drying and processing will lead to the production and processing of nutmeg with better quality attributes at a lesser cost.

Microwave-assisted convective hot air drying reduces drying time compared to convective drying and improves product quality compared to simple microwave drying. Pulsed microwave assisted hot air drying of mace takes two to three hours on an average. Preliminary post-harvest studies of nutmeg mace at Indian Institute of Crop Processing Technology in Thanjavur ensured that microwave drying helps to retain the red colour of the mace. Flavour component analysis of the raw, market, and microwave dried sample with best colour retention was done using GC-MS. Application of novel processing methods, such as pulsed microwave assisted hot air drying may be having the potential to solve this problem. More research should be done in this field to optimise a best drying method for nutmeg which will overcome all the disadvantages of the conventional drying systems.

Packaging and Storage

The dried mace can be well packed and stored in air tight containers or polybags which are approved for organic products with protection from sunlight so as to protect the red coloured pigment lycopene which is highly susceptible to oxidation in light. According to Agmark, nutmeg whole shall be packed in clean, sound and dry jute and cotton bags or in pouches made of food grade plastic. Nutmeg powder shall be packed in new, clean sound and dry containers made of tin, glass or in pouches made of laminated, extrusioned and metallised multilayer food grade plastic materials. The net weight of the nutmeg whole and nutmeg powder packed in a container shall be 25 gram, 50 gram, 100 gram, 500 gram, 1 kg and thereafter in multiples of 500 gram. Well packed nutmeg and mace can be stored normal temperatures in dry place. Moisture should be avoided so as to avoid mould attack and contamination with aflatoxin.

Potential Applications

- Nutmeg and mace are used to flavour drinks and foods such as cakes, sources, pickles and relishes.
- Used as an ingredient in ground spice mixtures like garam masala in Indian cuisine.

- Oil extraction, usually by hot manual pressing, produces nutmeg butter (also known as concrete or fixed oil or expressed oil). The butter is a highly aromatic, orange coloured fat which can then be processed into ointments and perfumes.
- Nutmeg jelly can be made from the pericarp.
- Nutmeg and mace oleoresins are extracted from the plant using organic solvents. They are produced for commercial flavourings and perfumes.
- Mace is typically used for seasoning seafood in West Indies.
- Ground nutmeg is mixed with petroleum jelly and warmed and used as balm (FAO, 1995).
- Medically, nutmeg is said to have stimulative, carminative and aphrodisiac properties.

Cinnamon

Cinnamon (*Cinnamom verum*) is an evergreen tree reaching to a height of 6-15m. The dried inner barks of these two species are the products for commerce. Its cultivation is more prevalent in hilly regions of the western ghats. Cinnamon tree may attain a height of about 10-15m, but in cultivation, it is generally cut back periodically. Two year old plants are cut during June-July to a height of about 15cm from the stump. This operation encourages the development of side shoots from the stump. Shoots are harvested from September-November. Usually coppicing is done in alternate years. The shoots having finger thickness and uniform brown colour are ideal for bark extraction. The stems are cut close to the ground when they are about 2 years old, as straight as possible, 1.0-1.25m length and 1.25 cm thickness. Such shoots are bundled after removing leaves and terminal shoots. Cuttings is followed by scraping and peeling operations.

Seed Spices

Out of 20 seed spice crops cumin, coriander, fennel, fenugreek, dill and ajwain contribute more than 95 per cent towards area and production. Cumin and fennel are dominant seed spices of Gujarat while coriander and fenugreek are important in Rajasthan. India exports raw as well as value added items to nearly 70 countries in the world and meet around 45% of the global demand and earns 361.5 crores of foreign exchange.

For exporting the seed spices, quality is the most important criterion. The quality of seed spices is assessed by mean of its intrinsic (Moisture, volatile oil, oleoresins content, major chemical constituents) as well as extrinsic (size, appearance, colour) quality. The produce must be safe, free from any health hazards substances and contaminants. The contaminants can be classified into three categories.

1. Physical contaminants Immature or shriveled seed, berries, insect infested product, presence of live or dead insect excreta of animals.
2. Chemical contaminants Added colour material, preservatives, antioxidants, fumigants, aflatoxin, pesticides/insecticide residue.
3. Microbial contaminants Presence of Salmonella, *E. coli*, yeast and mould.

Stage of Harvest

The crop of coriander matures in 90 to 135 days. The stage of maturity of the fruit at harvest is when central umbels are about to attain yellow colour. Cumin is harvested in about 100-110 days. Fennel takes 170-175 days to mature and harvesting is done before the fruits are fully ripe, umbel attains a slight greenish yellow colour. A good quality fennel for chewing purposes, commonly known as 'Lucknowi saunf' is produced by harvesting the umbel 30 to 40 days after pollination. In fenugreek, the harvest time is judged when the colour of leaves and pods turn yellow. The right time of harvesting is usually done in the morning hours to avoid shedding losses.

POST HARVEST IMPROVEMENT PROGRAMMES FOR OTHER SPICES

Component	Objective & Scale of assistance
Seed spice thresher	The harvesting and post-harvest practices followed in seed spices are unhygienic which results in contamination of the products with foreign matters like stalks, dirt, sand, stem bits etc. The seeds are separated by beating the harvested and dried plants with bamboo sticks or rubbing the plants manually by hand or trampling under the feet of the cattle.In order to educate the farmers and to avoid contamination in the final product, Board will introduce use of threshers which are operated manually as well as with power for separating the seeds from the dried plants.The cost of such a power and manual thresher are estimated as Rs.1.20 lakhs and Rs.40,000/- respectively. It is proposed to provide 50% of the cost aas subsidy subject to a maximum of Rs.60,000/- for a power thresher and Rs.20,000/- for a manually operated thresher.
Pepper thresher	Traditionally pepper berries were separated from the spikes by trampling under the feet which is un-hygienic and labour intensive. The pepper threshers are mechanically operated, labour saving and separates berries in a hygienic way. Subsidy is offered @ 50% of the cost of the equipment subject to a maximum of Rs.15,000/- per thresher.
Bamboo mats	In order to enable the growers to dry pepper under hygienic condition, the Board has been supplying bamboo mats of size 12' x 6' at 90% subsidy to tribal growers and 50% subsidy to others. The Board will arrange for centralized purchase of mats and supply to growers after collecting the non-subsidy portion.
Turmeric boilers	Fresh turmeric is cured for obtaining dry turmeric. Curing involves boiling of fresh rhizomes in water and drying in the sun. Over cooking spoils the colour of the final product while undercooking renders the dried product brittle. In conventional curing where the rhizomes are boiled in copper or galvanised iron or earthen vessels there are chances for overcooking or undercooking.Improved scientific cooking involves using boilers and perforated trough made of GI of MS sheet extended with parallel handle. This ensures optimum cooking of turmeric, which provides better colour and quality to the final produce. Hence it is proposed to popularize the use of turmeric boilers among turmeric growers for production of quality turmeric suitable for exports. The cost of one such boiler is estimated at Rs.3,00,000/- for community use. Individual growers or groups are eligible to benefit under the scheme. 50% of the cost turmeric boiling unit subject to a maximum of Rs.1,50,000/- is offered as susbidy.

Component	Objective & Scale of assistance
Turmeric polisher	Dried turmeric has a poor appearance and rough dull colour outside the surface with scales and root bits. The appearance is improved by smoothening and polishing the outer surface by manual or mechanical rubbing. Manual polishing is done by rubbing the dried turmeric fingers on a hard surface or trampling them under feet by wrapping in gunny bags. Manual polishing is an unhygienic practice, which affects quality of the product.The improved method of polishing turmeric is by using hand operated / power operated polishers. This ensures hygienic and effective polishing which gives better appearance to the product. Hence it is proposed to popularize the practice of using improved polishers for polishing turmeric. The average cost of such a polishing machine is estimated as Rs.40,000/- for a capacity of 250-500 kg. 50 % of the cost of turmeric polisher subject to a maximum of Rs. 20000/- is provided as subsidy.
Supply of IPM kits	Chilli is one of the major items in our export basket of spices. Indiscriminate use of pesticides has resulted in pesticide residues and detention of Indian chilli consignments in Australia, Spain and some other countries. The presences of pesticides have caused serious trade disruptions. Preventive measures are to be taken to keep pesticide residue within the permitted level to increase chilli exports from India. Under the programme IPM kits containing pheromone traps, bio agents like Trichoderma, Trichogramma, neem pesticides and seed worms of vermi-compost were supplied at an estimated cost of Rs.5000/- per ha to substitute the hazardous chemicals. 50% of the cost of the IPM kits or Rs.2500/- per ha is offered as subsidy
Supply of polythene/ silpauline sheets	Most of the importing countries have prescribed stringent quality standards for agro products including spices. These standards are becoming more and more stringent and are becoming effective trade barriers impacting spice exports. It is therefore necessary to educate, motivate and support the growers to adopt hygienic post-harvest practices to improve the quality of spices and to create a perception all over the world that India is a source of good quality spicesIn order to enable small and marginal farmers to dry spices under hygienic conditions, Board is supplying polythene sheets / silpauline sheets (of size 8 x 6 m, 250 GSM & 120 GSM respectively) at 50% subsidy to tribal growers and at 33.33% to other growers. Board will arrange for centralized purchase of polythene sheets and supply it to growers. The non-subsidy portion will be collected from the growers.
Mint distillation unit	Distillation of mint oil is usually done as a on farm process. At present majority of the farmers are using field distillation units which are poor in fuel efficiency, take long time for distillation and give low recovery of oil with poor quality. Hence there is a need to popularise efficient and improved field distillation units. In order to encourage installation of improved field distillation units, it is proposed to provide 32.5% of the actual cost of improved distillation unit having a capacity range from 500 - 600 kg herbage per shift subject to a maximum of Rs.1.18 lakhs as subsidy.

Component	Objective & Scale of assistance
Pepper/clove ladders	Pepper grows to a height up to 30 feet and traditionally, harvesting has been done manually by using single bamboo pole. This is risky and laborious. At present, there is acute labour shortage in the pepper growing areas for harvesting pepper and also there is severe shortage in the availability of bamboo. Besides the bamboo poles are easily perishable. In view of the above, it is proposed to distribute ladders for harvesting pepper and clove. 50% of cost of ladder subject to a maximum fo Rs 5000/- is provided as subsidy.
Spice cleaners/ graders	The objective of the scheme is to popularize the mechanization of pepper
for pepper	cleaning and grading to improve the quality. Board will provide pepper cleaning/grading machines @ 50% assistance subject to a maximum of Rs 35,000/-.
Spice slicing machine	The objective of the scheme is to motivate the growers to adopt slicing of
for ginger	ginger / turmeric before drying using simple slicing machines to improve the quality. Board will provide 50 % cost of the slicing machine as subsidy subject to a maximum of Rs.7000/- to the growers.
Tamarind dehuller /	The objective of the scheme is to popularize nutmeg desheller / tamarind
nutmeg desheller	dehuller machines among the nutmeg / Tamarind farmers to reduce the labour cost as well as to improve the quality of the produce. Board will provide 50% of the cost of the equipment subject to a maximum of Rs.42500/- is provided as subsidy.
Nutmeg driers	The main objective of the scheme is to popularize the drying of nutmeg / hygienically using dryers among the nutmeg farmers to produce quality nutmeg for exports. Board will provide 50% of cost of the drier subject to a maximum of Rs.30000/- as subsidy .
Packing and storage	The objective of the scheme is to promote packing and storage facilities
unit for garlic	for garlic and improve quality of the produce for export. Board will provide 50% of the cost of equipments required for setting up of primary processing facilities and accessories for storage of garlic subject to a maximum of Rs. 5 lakhs as subsidy. No assistance will be given for construction of building purpose.
Herbal spice extractors	The objective of the scheme is to promote primary processing facilities
& Deydration units	viz. extraction / dehydration of the herbal spices in the herbal spice growing areas. It is proposed to provide 50% of cost of herbal spice extraction unit or dehydration unit subject to a maximum of Rs. 1.20 lakh per unit as subsidy.
Dry/fresh ginger peeling	The objective of the scheme is to improve the quality of ginger by promoting
& storage unit	setting up of primary processing facilities for ginger in the growing areas. Board will provide 50% of the cost of equipments required for setting up of primary processing facilities and accessories for value addition and storage subject to a maximum of Rs. 5 lakhs per unit as subsidy. No assistance will be given for construction of building purpose.
Seed spice cleaning &	The objective of the scheme is to set up primary processing facilities viz.

Component	Objective & Scale of assistance
storage unit	threshing, cleaning, grading and storage of seed spices for export. Board will provide 50% of the cost of equipments required for setting up of primary processing facilities and accessories for storage subject to a maximum of Rs. 5 lakhs per unit as subsidy. No assistance will be given for construction of building purpose.
Spice Washing	The objective of the scheme is to motivate the growers to adopt washing
Equipment	of spices after harvest by using washing equipments to improve quality of the produce for export. 50% cost of the equipment subject to a maximum of Rs.1.10 lakhs per unit is offered as subsidy.

HARVESTING AND POST-HARVEST PROCESSING OF MEDICINAL PLANTS

Medicinal plants constitute the basis of primary health care for a majority of the population and are a critical source of income for many rural people particularly in area near forests. They are a source of primary health care for more than 80% of the population in developing countries who are dependent on traditional systems of medicine. Moreover, herbal medicines are becoming more popular in recent years with their increasing acceptability in both developing and developed countries. The resurgence of herbal drugs is mainly due to realization of harmful side effects of many modern drugs. The ever increasing demand of herbal drugs led to a spurt of large scale commercial collection from the wild. With the ever increasing demand of medicinal plants the supply line is adversely affected, leading to destructive harvesting and adulteration for genuine drugs.

Identification

The fact that most medicinal plants used by the traditional practitioners and manufacturers of Ayurvedic products are collected from the wild have problems for identification and obtaining material of uniform quality due to anumber of reasons. Sometimes a plant name may refer to more than one species: the name 'Brahmi' can refer to either *Bacopa monneiri* or *Centella asiatica*, which have entirely different phyto-chemical compositions. Shankhpushpi is one of the Rasayana drugs in Ayurveda and several plant species are being reported as Shankhpushpi, viz. *Convolvulus pluricaulis*, *Clitoria ternatea*, *Evolvulus alsinoides* and *Tephrosia purpurea* in different regions in India but in Ayurvedic Pharmacopoeia of India *Convolvulus pluricaulisis* is the official drug. Also the effect of agro-climatic conditions on the chemical composition and therapeutic properties of a medicinal plant species is well-known; seven varieties of Terminalia chebula originating in different parts of India are known to have different therapeutic properties.

Collection/Harvesting of Medicinal Plants

Medicinal plants should be collected/harvested appropriate season or time period to ensure the best possible quality of both source materials and finished products. It is well known that the quantitative concentration of biologically active phyto constituents varies with the stage of plant growth and development. The best time for collection/harvest (quality peak season/time of day) should be determined according to the quality and quantity of biologically active constituents rather than the total vegetative yield of the targeted medicinal plant parts during harvest. Care should be taken to ensure that no foreign matter, weeds or toxic plants are mixed with the harvested medicinal plant materials.

The time of harvest depends on the plant part to be harvested. Detailed information concerning the appropriate timing of harvest is often available in national pharmacopoeias, published standards, official monographs and major reference books. Following are the general guidelines need to be practiced while harvesting/collection of medicinal plants:

- Medicinal plants should be harvested by following sustainable harvesting practices.
- Medicinal plants should be harvested under the best possible conditions, avoiding dew, rain or exceptionally high humidity. If harvesting occurs in wet conditions, the harvested material should be transported immediately to an indoor drying facility to expedite drying so as to prevent any possible deleterious effects due to increased moisture levels, which promote microbial fermentation and mould.
- In general, the collected/harvested plant materials should not come into direct contact with the soil. If underground parts (such as the roots or rhizomes/bulbs) are harvested, any adhering soil should be removed from the plants as soon as they are harvested/collected. Contact with soil should be avoided to the extent possible so as to minimize the microbial load of harvested medicinal plant materials where necessary, large drop cloths, preferably made of clean muslin, may be used as an interface between the harvested plants and the soil.
- Collected/harvested plant material should be placed in clean baskets, mesh bags, other well aerated containers or cloths that are free from foreign matter, including plant remnants from previous collecting activities. After collection, the plant materials may be subjected to appropriate preliminary processing, including elimination of undesirable materials and contaminants (by hand picking), washing (to remove excess soil), sorting and cutting.
- The harvested raw medicinal plant materials should be transported promptly in clean, dry conditions. They may be placed in clean baskets, dry sacks, trailers, hoppers or other well-aerated containers and carried to a central point for transport to the processing facility. If the collection site is located some distance from processing facilities, it may be necessary to air or sun-dry the raw medicinal plant materials prior to transport.
- If more than one part of the medicinal plant are to be collected, the different plant materials should be gathered separately and transported in separate containers. Crosscontamination should be avoided at all times. Any mechanical damage or compacting of the raw medicinal plant materials, as a consequence, for example, of overfilling or stacking of sacks or bags that may result in composting or otherwise diminish quality should be avoided. Decomposed medicinal plant

materials should be identified and discarded during harvest in order to avoid microbial contamination and loss of product quality.

- All containers used for harvest should be kept clean and free from contamination by previously harvested medicinal plants and other foreign matter. If plastic any possible retention of moisture that could lead to the growth of mould. When containers are not in use, they should be kept in dry conditions, in an area that is protected from insects, rodents, birds and other pests, and inaccessible to livestock and domestic animals.
- The collected medicinal plant materials should be protected from insects, rodents, birds and other pests as well as from livestock and domestic animals.
- Collecting/harvesting implements (tools), such as machetes, shears and mechanical tools, should be kept clean and maintained in proper condition. Those parts that come into direct contact with the collected medicinal plant materials should be free from excess oil/grease and other contamination. Cutting devices, harvesters, and other machines should be kept clean and adjusted to reduce damage and contamination from soil and other materials. They should be stored in an uncontaminated, dry place or facility free from insects, rodents, birds and other pests, and inaccessible to livestock and domestic animals.

Primary Processing

Primary processing practices viz. drying, storage and packaging conditions are very important in order to maintain the quality of herb after harvest. Temperature and relative humidity are the main parameters affecting quality attributes of fresh herb during storage.

Harvested or collected raw plant materials should be promptly unloaded and transported at the processing facility. Different plant parts should be transported separately. Immediately it should be unpacked on arrival at processing centre. Prior to processing, the plant materials should be protected from rain, moisture, heat and any other conditions that might cause deterioration. Medicinal plant materials should be exposed to direct sunlight only where there is a specific need for this mode of drying.

Medicinal plant materials that are to be used in the fresh state should be harvested/collected and delivered as quickly as possible to the processing facility in order to prevent microbial fermentation and thermal degradation. If storage is required, it should be stored under refrigeration, in jars, in sandboxes, or using enzymatic or other appropriate conservation measures immediately following harvest/collection, and transported to the end-user in the most expeditious manner possible. The use of preservatives should be avoided, if used they should conform to national and/or regional regulatory requirements in both the source country and the end-user country.

Clean cemented floor or good tarpaulin sheet should be used for laying out the harvested crop. Remove all weeds and other foreign matter.

All medicinal plant materials should be inspected during the primary-processing stages of production, and any substandard products or foreign matter should be eliminated mechanically or by hand. For example, dried medicinal plant materials should be inspected, sieved or winnowed to remove discoloured, mouldy or damaged materials, as well as soil, stones and other foreign matter. Mechanical devices such as sieves should be regularly cleaned and maintained. All processed medicinal plant materials should be protected from contamination and decomposition as well as from insects, rodents, birds and other pests, and from livestock and domestic animals.

Washing

The harvested material should be washed with clean water. Soil should be removed properly from rhizomes and roots. Do not use contaminated water for washing. Do not wash seeds and delicate parts of the medicinal plants. All excess water should be drained from the herb before drying.

Drying

Drying is the most common way to preserve quality of medicinal plants. The physical and chemical properties of medicinal plants are determined by their moisture content. The first step in many postharvest operations is removal of water that is, drying. Drying is basically defined as the decreasing of plant moisture content, aimed at preventing enzymatic and microbial activity, and consequently preserving the product to extend shelf life. For this purpose, adequate dryers are needed, using temperature, velocity and humidity values for drying air that provides a rapid reduction in the moisture content without affecting the quality of the active ingredients of medicinal plants.

Drying of medicinal plants must meet the following requirements:

1. Moisture content has to be brought down to be at an equilibrium level that is defined for certain relative air humidity and temperature which is defined as storage condition;
2. minimum quality reduction in terms of active ingredients, colour, flavor and aroma; and
3. microbial count must be below the prescribed limits.

In India, generally herbs are dried either in sun or shade depending upon the requirements. However, traditional drying methods, such as drying in the shade or in the sun, have many drawbacks due to the inability to handle the large capacity of mechanical harvesters and to achieve the high quality standards required for medicinal plants. High ambient air temperature and relative air humidity during the harvesting season promote insect and mould development in harvested crops.

Furthermore, intensive solar radiation adversely affects quality, causing losses in essential oils or colour changes in dried plants. Thus, traditional natural drying in the sun or in the shade does not meet the exact required standards. To overcome these problems hot air-convective drying is widely used.

When medicinal plant materials are prepared for use in dry form, the moisture content of the material should be kept as low as possible in order to reduce damage from mould and other microbial infestation.

Medicinal plants can be dried in a number of ways:

- In the open air (shaded from direct sunlight);
- Placed in thin layers on drying frames, wire-screened rooms or buildings.
- By direct sunlight, if appropriate (fleshy material).
- In drying ovens/rooms and solar dryers.
- By indirect fire; baking; lyophilization; microwave; or infrared devices.
- Vacuum drying
- Spray dryer: Examples: Papaya latex and pectin's, etc.

If possible, temperature and humidity should be controlled to avoid damage to the active chemical constituents. The method and temperature used for drying may have a considerable impact on the quality of the resulting medicinal plant materials.

For example, shade drying is preferred to maintain or minimize loss of colour of leaves and flowers; and lower temperatures should be employed in the case of medicinal plant materials containing volatile substances. The drying conditions should be recorded. In the case of natural drying in the open air, medicinal plant materials should be spread out in thin layers on drying frames and stirred or turned frequently. In order to secure adequate air circulation, the drying frames should be located at a sufficient height above the ground.

Efforts should be made to achieve uniform drying of medicinal plant materials and so avoid mould formation. Drying medicinal plant material directly on bare ground should be avoided. If a concrete or cement surface is used, medicinal plant materials should be laid on a tarpaulin or other appropriate cloth or sheet. Insects, rodents, birds and other pests, and livestock and domestic animals should be kept away from drying sites.

For indoor drying, the duration of drying, drying temperature, humidity and other conditions should be determined on the basis of the plant part concerned (root, leaf, stem, bark, flower, etc.) and any volatile natural constituents, such as essential oils.

Vacuum Drying

This is conducted in steam- heated ovens with perfect closure, and a pump is used to exhaust the vapour laden air. The low pressure maintained within the oven ensures rapid and complete drying.

Advantages of vacuum drying

1. Rapid drying.
2. Relatively low temperature.
3. Cleanliness and freedom from odour and dust.

4. Independence of climate conditions.
5. Control of temperature.
6. Elimination, of risk of fire.
7. Compactness.

Packaging

- After removal of damaged material and foreign matter, the good dried crop should be packed in clean, dry sacks, bags or boxes, preferably new.
- Packaging materials should be stored in a clean dry place free from pests and inaccessible to animals.
- Reusable packaging materials such as jute sacks, plastic bags, etc., should be well cleaned and dried before re-use.
- The packed crop should be stored in a dry place away from the wall and off the ground and be protected from pests and farm and domestic animals.
- Whenever possible, the packaging materials used should be agreed between the supplier and the buyer.

Storage

1. Packed dried crop should be stored in a dry, well ventilated building, with minimal variation in diurnal temperature and with good air ventilation. When necessary, be equipped with air-conditioning and humidity control equipment as well as facilities to protect against rodents, insects and livestock. Shutter and door openings should be protected by wire screens to keep out pests and farm and domestic animals.
2. The floor should be tidy, without cracks and easy to clean. Plant material should be stored on shelves which keep the material a sufficient distance from the walls; measures should be taken to prevent the occurrence of pest infestation, mould formation, rotting or loss of oil; and inspections should be carried out at regular intervals. It is recommended that packed dried crops should be stored: - in a building with concrete floors; - away from the wall; - well separated from all other crops.
3. Continuous in-process quality control measures should be implemented to eliminate substandard materials, contaminants and foreign matter prior to and during the final stages of packaging. Processed medicinal plant materials should be packaged in clean, dry boxes, sacks, bags or other containers in accordance with standard operating procedures and national and/or regional regulations of the producer and the end-user countries.
4. Materials used for packaging should be non-polluting, clean, dry and in undamaged condition and should conform to the quality requirements for the medicinal plant materials concerned. Fragile medicinal plant materials should be packaged in rigid containers. 5. Dried medicinal plants/herbal

drugs, including essential oils, should be stored in a dry, well-aerated building, in which daily temperature fluctuations are limited and good aeration is ensured.

6. Fresh medicinal plant materials should be stored at appropriate low temperatures, ideally at 2-8°C; frozen products should be stored at less than -20°C. 7. Small quantity of crude drugs could be readily stored in air tight, moisture proof and light proof container such as tin, cans, covered metal tins or amber glass containers. Wooden boxes and paper bags should not be used for storage of crude drugs.

Personnel

1. Personnel handling medicinal plant material should: - maintain a high degree of personal hygiene; - be provided with suitable changing facilities and toilets with hand washing facilities.
2. Personnel should not be permitted to work in the herbal material handling area if they are known to be suffering from, or to be carriers of, a disease likely to be transmitted through medicinal plant materials, including diarrhoea.
3. Personnel with open wounds, sores, and skin infections should be transferred away from herbal materials handling areas until completely recovered.

POST HARVEST TECHNOLOGY OF PLANTATION CROP

Coconut

The main coconut producing countries are SriLanka and the Philippines. Coconut water is a refreshing drink which also hygienic and nutritive. The white flesh is a calories rich extensively used for cookery in South India. From grated wet coconut kernel good quality oil as well as protein and carbohydrate can be extracted to prepare coconut honey. Undesirable flavour changes in desiccated coconut is due to oxidative rancidity of the oil. The residual cake after the oil has been extracted is used as cattle food.

The dried coconut meat is known as copra and is the source of coconut oil, which is used in enormous quantities for making fats for baking and confectionery. An important product used in the chocolate and confectionery industry is desiccated coconut. The nut is removed from the hard shell and the thin brown rind pared off. This must be very thorough or the finished coconut will be contaminated by flecks of brown. The pared nuts are then washed free of milk and skin residues and put through shredding machines, which give a milled product of varying degree of fineness. After shredding, the ground meat is subjected to a steam or hot water treatment to destroy *salmonella*and lipases and to reduce bacteria counts to low levels. The pasteurized meat is then dried on wire mesh belts by means of hot air, which reduces the moisture content to less than 4%. The critical moisture level of desiccated coconut is 5% and the product should be well below this to ensure good shelf life.

After drying, the desiccated coconut is put through a multiple sieving operation to produce the four main grades. Extra fine, macaroon (fine), medium and coarse. A variety of other coconut products is available to the chocolate and confectionery trade such as, moist white sweetened coconut, toasted coconut, creamed coconut and tenderized coconut.

Top quality desiccated coconut should be white in colour, free from foreign matter, and uniform in granulation against standard, and the expressed oil should have a free fatty acid content of less than 0.1%.

Harvesting Methods

- Climbing
- Power tiller operated ladder
- Climbing cycle / equipment

Method of harvest

- Harvesting of coconuts is commonly done by climbing the tree with the help of a rope ring round the feet or ankles of the climber or by using a ladder.
- On reaching the top, the climber taps the nut in the lowermost bunch with its harvesting knife to test its maturity.
- If he is satisfied, he cuts the bunch at the base of the stalk when it drops down to the ground.
- If the ground is very hard or if tender nuts are to be harvested, the bunches are lowered by using a rope. The climber also cleans the crown and removes the dry leaves, sheaths and spathes.
- In the West Coast and certain other tracts where coconut leaves are required for thatching houses, one or two lowermost leaves are also cut down at the time of harvest.
- The cutting down of green leaves is considered undesirable as it affects the yield of trees to some extent.
- In some places where the trees are not tall, harvesting is done by cutting the bunches with a knife, attached to a long bamboo pole.

Nuts which are to be stored for making ball copra are not harvested till they are completely ripe and dry.

Yield

- Average yield : 80-100 nuts/palm/year depending on the variety.
- Dwarf varieties : 70-80 nuts/palm/year
- Tall varieties : 80-100 nuts/palm/year
- Hybrids : 100-130 nuts/palm/year

Post Harvest

About Post Harvest

- Fully matured nuts should be harvested
- For collection of seed nuts/tender coconuts the bunches should be harvested and brought down by using ropes to prevent the damage of nuts.
- Copra to be dried to 6% moisture by sun drying or by using copra driers
- The storage period of copra can be increased up to 6 months by storing the copra in polythene tar coated gunny bags.
- For household storage the nuts may be kept in vertical position.

Dehusking: Manual dehusking with the help of an iron rod driven to the ground is strenuous and skill oriented. Presently mechanical devices are used for dehusking.

Copra Processing: The optimum moisture content in copra is 5-6 percent. Sun drying, smoke drying, kiln drying and indirect hot air drying are commonly used drying methods.

(i) ***Sun Drying:*** Traditional system of copra drying is by spreading the cups (Split open coconut) on any open surface for sun drying. It takes about 8 days for sun drying. The deposition of dirt and dust on wet meat during sun drying results in deterioration of copra quality. Further, cloudy weather and low atmospheric temperature also reduce the quality of copra.

(ii) ***Solar Dryer:*** Use of a closed type solar dryer avoids the quality deterioration of copra due to deposition of dirt. Drying time is reduced to 3-4 days. A batch type of solar cabinet dryer with a capacity of 100 nuts developed at CPCRI takes only 3 days for drying.

(iii) Indirect Drying:

1. Small Holder Copra Dryer: An indirect copra dryer of 400 nuts per batch capacity (using agricultural waste as fuel) developed at CPCRI is gaining popularity among coconut growers. The dryer requires only 3 sq.m for housing and could be carried by 2-3 persons. The drying time required per batch is 36 hours spread over 4 days. Kerala Agro Industries Corporation (KAICU) is manufacturing this type of dryer.
2. Large Holders Copra Dryer: Large size copra dryer, the capacity of 3500-4000 nuts is developed at CPCRI. The unit is suitable for large holding and copra processing societies.
3. Smoke Free Copra Dryer for Medium Holding: CPCRI has developed this with a capacity of 1000 nuts per batch. This can dry coconut in 24 hours. It has got unique furnace where in the fuel used is only shell.

4. Electrical Copra Dryer: CPCRI has developed an electrically operated dryer with forced hot air circulation. Its capacity 1000 nuts per batch with a drying time of 28 hrs.

(iv) Ball Copra: Ball copra is of superfine quality and commands a premium price in the market. It is prepared by storing fully mature nuts for 10-12 months, when kernal will get detached from the shell. CPCRI has developed dryer to prepare ball copra in shorter time by giving different heat treatments.

(v) Copra Grading: The copra is graded in the order of its market value. The grading is mainly based on moisture content, foreign matter and black copra. The maximum limits for them are 10 per cent, 2 per cent and 5 per cent respectively. However, the good quality copra should have the following requirements:

THE GOOD QUALITY COPRA

Particulars	Percentage (%)
Moisture	6
Oil content	71
Acid value	2.5
Foreign matter	0.5
Mouldy cups	5
Wrinkled cups	5 (free)
Black copra	1 (free)

MILLING COPRA

Particulars	
Foreign matter (per cent by weight – maximum)	1.0
Mouldy and black kernels (per cent by count –maximum)	10.0
Wrinkled kernels (per cent by count – maximum	10.0
Chips (per cent by weight – maximum)	10.0
Moisture content (per cent by weight – maximum)	6.0

EDIBLE BALL COPRA

Particulars	
Size (diameter) minimum in mm	75.0
Foreign matter (per cent by weight – maximum)	0.2
Mouldy and black kernels (per cent by count –maximum)	2.0
Wrinkled kernels (per cent by count – maximum	10.0
Chips (per cent by weight – maximum)	1.0
Moisture content (per cent by weight – maximum)	1.0

Safe storage of copra

Copra obtained from commonly cultivated varieties / cultivars is attacked by various insect pests in store. Among these ham beetle, *Necrobia rufipes* and saw toothed grain beetle, *Oryzaphilus surinamensis* are of major importance, which can cause more than 15% loss to copra when stored for more than six months.

Following precautions are to be taken for the safe storage of copra for more than three months:

- Dry the produce to four per cent moisture content.
- Avoid heap storage, which causes maximum damage.
- Store copra in netted polythene bags or gunny bags.

Pests of Copra Source

- Insects cause serious damage to copra in storage.
- Apart from loss in tonnage, insect attack adversely affects the quality.
- In severe cases, the financial loss on this account may even exceed that arising from loss by weight.
- Following are the pests of copra.

Carpophilus dimidiatus (Corn Sap Beetle)

- It is the most common insect in copra godowns.
- In India also it is commonly present in copra stores.
- The beetle is small, flat, broad, and about 3 mm in length and 1.5 mm in width.
- It can be readily distinguished from the other beetles by its shorter wing covers, which do not completely cover the abdomen.
- The beetle completes its life cycle in about five weeks.

Necrobia rufipes (Copra beetle or Red – legged harm beetle)

- Necrobia beetle has a world-wide distribution and is known as' Harm Beetle' in America and' Copra Bug' in the Pacific.
- It is generally metallic blue in colour but sometimes with a greenish lustre.
- Its length varies from 3.5 mm to 5.5 mm. The first five joints of the antennae are brown, the rest black, and the terminal three are expanded.
- It completes its life cycle in about 40 days.

Ahasverus advena (Foreign grain beetle)

- It is known as "Foreign Grain beetle," and is present throughout the world.
- It has been recorded feeding upon a variety of stored products.

- It is abundant in copra stores, but is practically mycetophagous.
- The adult beetle is very small, measuring about 2 mm in length and 0.75 mm in width, and is chestnut brown in colour.
- The life cycle of the beetle is completed in about thirty days.

Oryzaephilus surinamensis (Saw-toothed Grain Beetle)

- This Cucujid beetle is commonly known as "Saw Toothed Grain beetle," and it is a serious pest of stored grains; copra and other stored food products throughout the world.
- The adult beetle is small, flat, elongate, dark brown, about 3 mm in length and 0.75 mm in width.
- The pro-thorax has three longitudinal ridges with two depressed areas and its lateral margins are each armed with six prominent tooth-like projections.
- The beetle breeds throughout the year in the moulds and crevices of degenerated copra, and completes its life cycle in about thirty days.

Tribolium castaneum (Red flour beetle)

- It is commonly known as "Red Flour beetle ", and is primarily a pest of cereal products like wheat flour, ' suji ' and' maida " but causes damage to quite a number of other stored products including copra.
- The beetle is small, flat, elongated, reddish brown, about 3.5 mm in length and 1.25 mm in width.
- The first seven joints of the antennae show a gradual increase in size and the last three are thickened into a club.
- In India, its life cycle varies from six to twelve weeks.

Trogoderma granaria (Khapra beetle)

- This pest is known as' Khapra ' in the towns and villages of India, and its principal food is wheat.
- The adults are harmless, and only the grubs are destructive. The beetle is about 3.5 mm in length and 1 mm in width, and has a characteristic oval shape.
- The head is disproportionately small and is concealed from above by the pronotum.
- The life cycle in humid places is completed in about two months.

Corcyra cephalonica (Rice moth)

- This pest is commonly known as the Rice Moth and is a serious pest of stored paddy, rice, cereals, copra and other stored food products.
- The head and thorax of the moth are brownish in colour.
- The fore-wings are brown with a pinkish tinge, speckled with darker scales running longitudinally.

- The hind wings are shiny and silver grey.
- The wing expanse is about 25 mm.

Ephestia cautella (Fig moth or Almond)

- This is frequently referred to as the' Almond' or the' Fig Moth' and is one of the most important insects infesting dried fruits, cocoa, spices, cashew nut, copra and groundnut.
- The moth is about 6 mm in length with a wing expanse of 15 mm.
- Its colour varies with food and environments, but mostly the fore- wings are greyish-brown, some what mottled with a distinctly darker coloured transverse band.
- The average life cycle on mouldy copra is six weeks.

Tea

Tea is a beverage prepared by pouring boiling water over dry processed leaves. Different factors affect the quality of tea like soil in which plant is grown, climate, processing techniques used etc.

PERCENTAGE COMPOSITION OF TEA LEAVES

Constituents	%	Constituents	%
Polyphenols	22.2	Ash	5.6
Protein	17.2	Reducing sugars	3.5
Caffeine	4.3	Pectins	6.5
Crude fibre	27.0	Ether extract	2.0
Starch	0.5		

Types of Tea

Black tea

Various processing steps involved in the manufacture of black tea are withering, rolling, fermentation, drying, grading and packing.

Green Tea

Here withering and fermentation are completely omitted. It is made by first steaming the leaves to prevent the leaves from changing colour and to inactivate the enzymes. Steaming is then followed by rolling and drying. The leaves retains much of its original green colour especially the finer leaves.

Oolong Tea

Oolong tea is partially fermented. The fermentation period is too short to change the colour of the leaf completely. It is only partially blackened. The beverage is intermediate between those produced from green and black teas.

Characteristics of Tea Beverage

- High quality tea beverage is clear and bright. A flavour full cup of tea has a quality known as briskness but a subtle aroma.
- The strength of the beverage determines whether the tea is mild or robust.
- Green tea is a pale straw colour whereas black tea is deep amber.
- Some astringency is inherent in tea but green tea beverage is more astringent than black.

Processing

The fresh tea leaves are placed in withering troughs set on wire mesh. Large fans blow air through the leaves to reduce the moisture content. The leaves must be evenly spread so that uniform withering occurs. The leaves must be spread by hand and all lumps or piles removed so that maximum air flow reaches all the leaves and no heat is generated in the tea. Moisture content is lowered by about 70% or to a predetermined level depending on the condition of the leaf.

For proper withering to occur, the air must be at the proper temperature. If the leaf is wet, the air is heated slightly for a short time to help remove excess moisture.

After withering, the leaves are rolled by large rolling machines to provide a twist to the leaves and to break the cell walls to expose the inside to oxygen to begin the oxidation process. The leaves are fed down onto the machines directly from the withering troughs which are overhead on the second floor.

The sifter separates particles by size. Small leaves which have been rolled successfully the first time fall to the bottom and then are moved to the next process. The larger particles which fall off the end go back to be rolled again.

Green teas skip the fermentation stage and go directly to the drying room. Leaves for black and oolong teas however, are spread evenly on special raised flat beds to allow fermentation to the desired degree. Although the process has been traditionally called "fermentation" it is actually an oxidation process that depends on tea enzymes reacting to oxygen. Temperature control is critical. The cooling ducts at the top of the room help maintain the desired temperature.

The chemical reactions taking place in the leaves directly affect the cup quality and determine the aroma and briskness of the brew.

Fermentation must be strictly regulated to produce good tea.

Next, the tea leaves travel up a conveyor belt to the dryer. They pass through the dryer at controlled temperature for about 24 minutes or longer if necessary. The goal is to stop the fermentation process and transform the tea into a stable product.

After drying the tea must cool down before the sorting and grading process. If the sorting takes place while the tea is still warm it can lose its character and bloom.

The first step in the process is passing the bulk tea through machine sorters which do most of the size grading - whole leaf, broken leaf, fannings. They are then hand graded to further fineness. The tea all comes from the same plant but the grading separates the leaves by size. Broken leaves and fannings work well in teabags while whole leaves are reserved mostly for loose tea brewing.

Post Harvest Technology- Importance and Role

Post harvest loss reduction technology encompasses the usage of optimum harvest factors, reduction of losses in handling, packaging, transportation and storage with modern infrastructure machinery, processing into a wide variety of products, home scale preservation with low cost technology. Use of thermal processing, low temperature, drying, chemical and biological reactions coupled with other preservation techniques are applied to enhance the storability. Containers and packaging materials confer portability as well as extend the shelf-life. Adoption of these techniques could make available a large quantity of food by avoiding losses and provide better quality food and nutrition, more raw materials for processing, thus ensuring better returns to the farmers.

Importance of Post-harvest technology lies in the fact that it has the capability to meet food requirement of growing population by eliminating losses making more nutritive food items from raw commodities by proper processing and fortification.

Post-harvest technology has potential to create rural industries. India, where 80 percent people live in the villages and 70 percent of them depend on agriculture has experienced that the process of industrialization has shifted the food, feed and fibre industries to urban areas. This process has resulted in capital drain from rural to urban areas, decreased employment opportunities in the rural areas balance trade in favour of urban sector and mismatched growth in economy and standard of living between rural and urban people. It is possible to evolve appropriate technologies which can establish agricultural based rural industries. The farmer whose role has been reduced to producer can be transformed into producer cum processor and thus getting more dividends for hard labour, input, kind of risk taken and generating resource for socio-economic advancement keeping pace with the modern times.

Drying

The drying process arrests fermentation and reduces the moisture content in the fermented tea from 55-60% to 2-3%.

Key Points for Value-adding to Quality, during drying:

a. Maintaining exact inlet and exhaust temperature is important to get uniform firing.

b. Proper loading of drier is also essential to get uniform firing:

    - Over loading causes back pressure and results in uneven drying.

c. Maintaining drying time and temperature is very important.

- Longer the drying time - detrimental to quality & flavour.
- Shorter the time - results in thin liquor.
- Higher the temperature - results in bitter tea.

d. In the ECP drier, the optimum inlet temperature for CTC process is 100±5°C, while exhaust temperature is maintained at 55°C.

e. In the FBD drier, the desirable inlet temperature ranges from 120-125°C, while the exhaust temperature is maintained at 60-65°C.

f. Magnet can be fixed in the drier mouth, which will help in removing iron particles from made tea.

Sorting and Grading

Sorting is to remove fibre from the bulk tea and separate the tea particles into various shapes and sizes relevant to trade requirements.

a. Sorting enhances the appearances and quality of liquor.
b. Flakes of coarse leaf in the primary grade can be separated to avoid harshness.
c. Use of stainless steel meshes is desirable for good sorting.
d. Meshes should always be tight; loose meshes cause loss of bloom and uneven grading.

Storage and Packing

Black tea, being a highly hygroscopic material picks up moisture from the atmosphere, if packing is not proper.

a. Three per cent (3%) moisture level is advisable when packed – if the moisture content of tea reaches 6% the keeping quality is impaired.
b. Tea should not be stored in open condition.
c. Use of bins for storage of the finished product is very essential.
d. Tea should be cooled to ambient condition, before it is taken to bins.
e. Tea can be passed under powerful magnet grills to ensure iron free tea while packing.

Coffee

Coffee is an important beverage used all over the world.

Coffee Harvesting and Its Impact on Coffee Quality

Production of speciality coffee needs a proper plan for harvesting the coffee cherries as it gives good economic returns for producers. The time of harvest varies in different places. According to the processing method to be implemented, harvesting the coffee cherries without causing damage to the tree is an important task. In most of the coffee-producing countries, coffee is harvested once per year. As the coffee cherries mature, the coffee fruit contains suitable chemical compositions which lead the fruit to the best quality.

The coffee fruit also contains volatile compounds that are responsible for the aroma and flavour properties of the coffee. These compounds are present at a very low amount at the early stage of the coffee cherry, but later on, it increases as the coffee transformed to the maturity stage. There are two strategies (strip and Coffee processing methods) for harvesting the coffee cherries, which are widely used.

Strip Picking

This strategy is usually done by machinery or by hand. The whole coffee cherries are harvested at one time. The harvested coffee may not achieve the desired quality due to the mixture of underripen or overripen coffee cherries. In order to use machinery for harvesting the coffee cherry, the following factors are critically important such as the topography, inclination, spacing, alignment, and the height of the plant. Considering these factors, the producer should choose the right harvesting system that suits their crop.

Selective Picking

The major challenges in the coffee sector are obtaining uniform maturity, and at the same time, it is the procedure to provide consistent quality coffee. It is important to note that in most flowering plants the climatic condition during the growing period can change the number of flowering plants which affect the maturation uniformity. In this harvesting system, only the ripened coffee cherries are harvested selectively by hand from the whole tree or branches. The harvested coffee meets the standards due to the uniformity of the cherries. This strategy needs several picking rounds and is labour-intensive. Considering the advantage and disadvantages of both harvesting strategies is the key for coffee growers. The coffee cherry has different layers that surround the beans, such as skin, pulp, mucilage, parchment, silver skin, and finally, the coffee beans (Fig. 1).

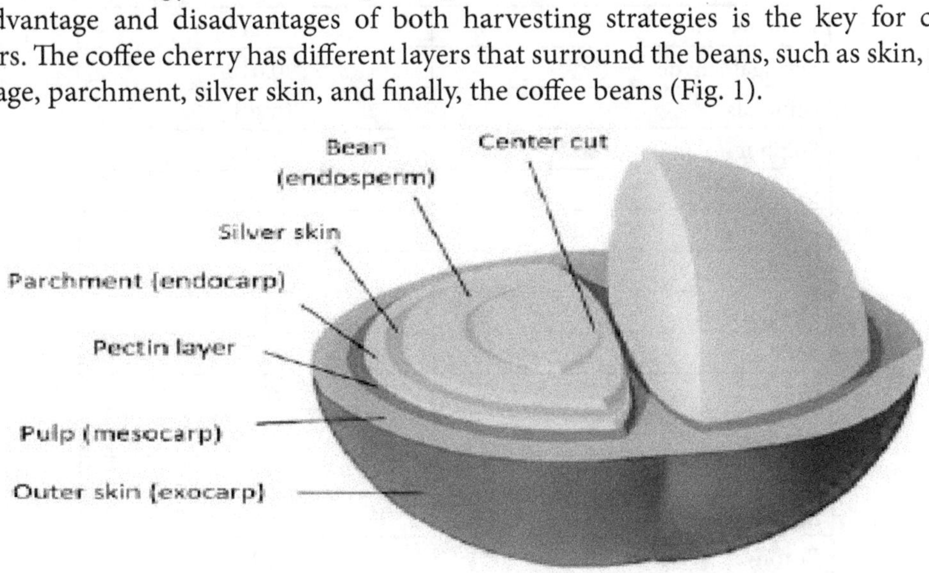

Fig. 1. The coffee cherry anatomy.

The pulp and mucilage are rich in nutrients, and its chemical compositions are presented in Tables 1 and 2. These days the coffee pulps are being used for making a coffee wine by fermenting the coffee pulp. So far, there are two patents already registered in Korea. The ripe coffee cherries have colour ranges from bright red to deep red and yellow, depending on the plant variety and for ripe cherries.

TABLE 1. COMPOSITION OF COFFEE PULP.

Components	%
Ether extract	0.48
Crude extract	21.4
Crude protein	10.1
Ash	1.5
Nitrogen-free extract	31.3
Tannins	7.8
Pectin substance	6.5
Non-reducing sugars	2.0
Reducing sugars	12.4
Chlorogenic acid	2.6
Caffeine	2.3
Total caffeic acid	1.6

TABLE 2. CHEMICAL COMPOSITION OF COFFEE MUCILAGE.

Components	Concentration (g/L)
Glucose	35.65
Galactose	37.67
Lactose	1.06
Proteins	0.119
Syringaldehyde	0.610
Minerals	(mg/L)
Sulphur	30.19
Calcium	37.08
Potassium	239.8
Magnesium	10.05
Phosphorus	41.55
Sodium	7.18
Iron	0.65
Copper	2.45
Zinc	0.14
Manganese	0.07
Boron	0.16
Barium	0.02

Components	Concentration (g/L)
Arsenic	0.47
Lithium	0.01
Silicon	1.58
Strontium	0.07

Parchment Coffee

1. Pulping This method requires equipment and adequate supply of clean water. Fruits should be pulped on the same day to avoid fermentation before pulping. Fruits may be fed to the pulper through siphon arrangement to ensure uniform feeding and to separate lights and floats from sound fruits. The pulped parchment should be sieved to eliminate any un-pulped fruits and fruits skin. The skins are separated by pulping should be let away from the vats into collection pits so that microbial decomposition of the skin will not affect the bean quality when it gets mixed up with the bean.

2. Demucilaging and washing

The mucilage on the parchment skin can be removed by

A) Natural fermentation: The mucilage breaks down in the process of fermentation and it takes 24- 36 hours for arabica and 72 hours for robusta. Cool weather delays the process of fermentation. Under fermented or over fermented beans affect quality. When correctly fermented the mucilage comes off easily and the parchment does not stick to the hand after washing and the beans feel rough and gritty when squeezed by hand. When the mucilage breakdown is complete, clean water is let in and the parchment washed pebble clean with three to four changes of water.

B) Treatment with alkali: Removal of mucilage by treatment with alkali takes about one hour for arabica and one and a half to two hours for robusta. The beans obtained after pulping are drained off excess water and spread out in the wax uniformly and furrowed with wooden ladles with long handles. A 10% solution of caustic soda (NaOH) is evenly applied into the furrows using a water can.10 litres of alkali is sufficient to treat 25-30 forlits (1 forlit = 40 litres) of parchment. The parchment is agitated thoroughly by the ladles so as to make the alkali to come into contact with the parchment and trampled by feet for about half an hour. When the parchment is no longer slimy and makes a rattling noise, clean water is let in and the parchment washed clean with 3 or 4 changes of water.

C) Removal of mucilage by friction: There are machines, which pulp and demucilage the beans in one operation. However, a number of naked and bruised beans may result in the parchment. It is, therefore, necessary to adjust the machines carefully to obtain uniform pulping and demucilaging. Cup-test results have indicated that there is no difference in cup quality coffee processed by different method.

Drying

The next stage is drying the parchment in the sun until the moisture content is sufficiently reduced to permit storage of beans till they are dispatched to curing works. Proper drying contributes to the healthy colour of the bean and other quality factors. Under dried parchment turns mouldy and gets bleached storage and subsequent curing operations. The parchment is spread on clean tiled or concrete drying floor to be dried slowly by spreading to a thickness of about 7 to 10 cm. stirring and turning over coffee, at least once an hour, is necessary to facilitate uniform drying. The parchment should be heaped up and covered in the evening until next morning. Sun drying may take about 7 to 10 days under bright weather conditions. At the right stage of dryness the parchment becomes crumbly and the beans split clean without a white fracture when bitten between the teeth. Drying is complete when a sample forlit of coffee records the same weight for two days consecutively. At this stage, coffee is shifted to the stores and bagged in clean, new gunnies. When coffee is being a dried, all naked bean, pulper nipped and bruised beans, blacks, greens and other defective beans are sorted out and dispatched to curing works separately.

Cherry Coffee

For preparation of cherry coffee fruits should be picked, as and when they ripe. Green and under-ripe should be sorted out and dried separately. The fruits should be spread evenly to a thickness of about 8 cm on clean drying ground in which the cherries are stirred and ridged at least once every hour. The cherry is dry when a fistful of the drying cherry produces a rattling sound when shaken and a sample forlit records the same weight on two consecutive days. The cherry should be fully dry at the end of 12 to 15 days under bright weather conditions. In India most of the Arabica coffee is processed as parchment coffee while bulk of Robusta coffee is made as cherry coffee. After drying coffee should be packed in gunny bags and stored on raised wooden platform in well ventilated and dry store house. The parchment and cherry coffee should not be stacked together. It is advisable to dispatch the dried coffee to curing factories at the earliest possible.

Subsequent processing of both these types is carried out mechanically at curing factories. At curing houses, the parchment coffee is hulled, graded (based on size standards of beans) and polished. The cured coffee is called green coffee which is traded in the market. Coffee quality is a cumulative index of many characteristics of coffee such as physical appearance in the raw and roast as well as liquor qualities like aroma, body and acidity in the cup.

Treatment of Coffee Processing Effluents

Preparation of parchment coffee involves the usage of high amount of water and it is estimated that for wet processing of one tone of clean coffee by using conventional pulper and washer, the water requirement is 80,000 litres for Arabica and 93,000 litres for Robusta.

The resultant waste water cause pollution problems as these effluents contain high amount of suspended and dissolve organic solids. Their pollution load measured in terms of BOD (biological oxygen demand) ranges from 2.5 to 12g/litre. Hence it is recommended that coffee effluents should be subjected to biological treatment methods involving anaerobic digestion followed by aerobic lagooning in order to treat high strength organic waste.

Post-harvest Operations and Their Impact on Coffee Quality

Coffee Processing Methods

After harvesting of the fruits, green coffee beans are obtained by one of three different methods known as dry, wet, and semi-dry processing. Commonly, there are three different coffee processing methods (Fig. 2). These methods are wet, dry, and semi-dry processing, and recently, digestive bioprocessing is practised on a small scale to produce the world's most expensive coffee (kopi luwak and black ivory coffee). Although all methods aim at removing the fruit flesh of coffee cherry, they do it in different ways. After harvesting, the coffee cherry follows washing with water to separate floaters (overripe coffee cherries, undeveloped coffee cherries, sticks and leaves). Those processing methods are briefly described below.

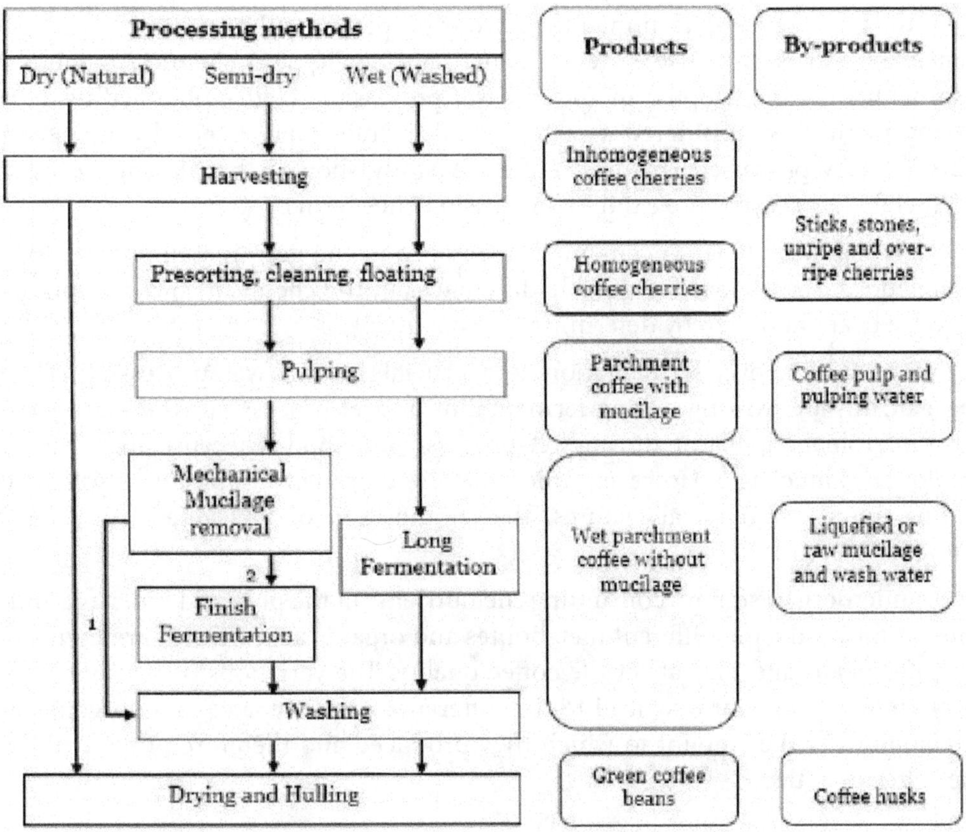

Fig. 2. Coffee processing methods

Wet Processing

This processing method demands the use of some particular facilities and ample amounts of water. When this method adequately implemented, it maintained the inherent quality of coffee beans and produced uniform green coffee beans with minimum defectives. Arabica coffee is widely obtained using wet-processing method except in some countries like Ethiopia, Brazil, and Yemen where they processed in both wet- and dry-processing system, although few percentages of robusta coffee are obtained in this method. Right after harvesting, the coffee cherries are washed and then de-pulped using a machine. The coffee mucilage has pectin substances (polysaccharides), and it has a sticky nature and is challenging to get rid of using water.

After de-pulping, chemicals mostly enzymes used to remove mucilage or the natural fermentation proceeds depend upon the environmental condition and fruit ripeness. Natural plant enzymes are present in the coffee fruit that facilitates the degradation of mucilage; however, it is not enough for a full and sufficient process.

Microorganisms are responsible for degrading the polysaccharide substances from the parchment coffee. Finally, after fermentation the end product is a "parchment" or "washed" coffee. The type of coffee processing methods (wet and dry) that are performed to obtain the green beans determined the flavour properties and created a typical quality difference. Several reports also stated the final cup quality and chemical composition are also defined by the adopted method in wet processing. Coffee obtained using wet-processing method is considered as a high-quality coffee and received a higher price compared to dry-processed (natural) coffee. A study showed that during processing, several metabolic activities are exhibited in green coffee beans.

These metabolic changes occur mainly because of the germination processes, and stress metabolism is a cause for notable differences in the chemical composition of the green coffee beans and thereby determines the quality.

As mentioned above, fermentation is a crucial step in wet processing. Diverse microorganisms are existing during fermentation in wet processing. The reason for an excellent microbial population during wet-processing method is because of the nutrient-rich pulp and mucilage. However, microbial heterogeneity subjects to distinction depending upon geographical features, the composition of the coffee fruit, and the fermentation methods.

These microorganisms are consuming the nutrients in the pulp and mucilage and are responsible for producing different metabolites and organic acids, which are then stored in the coffee beans and may affect the coffee quality. The variations of microorganisms diversity and environment may lead to the difference of the type of these organic acids and metabolites or the amount to which they produced and therefore provide different unique coffee qualities.

Semi-dry (Semi-wet) Processing

The semi-dry (pulped natural) processing is an intermediate process between wet and dry processing. Like wet processing, the coffee exocarp and significant portion of mesocarp are separated while de-pulping is operated. However, unlike wet-processing method, the sticky part (mucilage) remains and is allowed to dry on the parchment instead of complete removal of it with further fermentation until 11–12% moisture content is achieved.

After drying the parchment coffee in well-aerated raised beds or cement patios, the coffee bean is then separated from its parchment using machine or any locally available materials like wooden mortar in small scale farms. Despite having less body compared to naturally processed (dry) coffee beans, the cup quality of semi-dry-processed coffee has a bright and clean cup and is somewhat similar to wet-processed beans. A comparison of the primary and secondary metabolites of coffee, which processed by the wet and semi-dry (semi-wet) method was comprehensively investigated.

These metabolites are supposed to be very important, because during roasting their degradation results in the development of those nonvolatile or volatile compounds essential for the formation of coffee properties, pigmentation, bitterness, body, astringency, sweetness, aroma, and so on. A very similar result between semi-wet- and wet-processed coffees was found regarding the distributions of nine chlorogenic acid classes (3-CQA, 4-CQA, 5-CQA, 3-FQA (feruloylquinic acid), 4-FQA, 5-FQA, 3,4-di-CQA, 3,5-di-CQA, and 4,5-di-CQA), but the total CGA content was statistically lower in semi-wet-processed coffee except for di-CQA.

The reason for low CGA content in semi-washed beans might be associated with soaking water for a shorter time which leads to the loss of water-soluble components due to leaching and fermentation. Generally, semi-dry-processing methods are more regularly practised in only Brazil and some parts of Sulawesi and Sumatra.

Dry Processing

Dry processing is the oldest, cheapest, and simple processing method. The coffee beans that are obtained using dry-processing method are usually called unwashed (natural) coffee. In this processing method, to get the green coffee beans, the harvested cherry is dried in the sun or other mechanical drier and followed by separating the dried outer parts. Sun drying is a lengthy process, and 95% of Arabica coffee from Brazil, Ethiopia, Haiti, Indonesia, Paraguay, India, and Ecuador are sun-dried.

Few steps are involved in this processing method compared to wet- and semi-dry-processing methods. It is a time-consuming processing as the whole coffee cherry takes some time to dry compared to drying the parchment coffee in wet processing. Dry processing has a high risk of secondary fermentation because of the mucilage, which is very hygroscopic remains with the coffee cherry. The fruits are spread over thin layer and regularly raked to maintain uniform temperature from the top-bottom layer.

The drying step in anywhere may take from 10 days to 3 weeks. However, the geographical location and seasons affect the drying process. In some big farms, the mechanical drier is used to speed up the drying process. Because of the existing lowland high temperature and condensation effect, it speeds up the drying process of coffee beans more quickly at the surface (cement or bricks) than raised bed, which is made of mesh wire or bamboo mats. Drying is the most critical step in this processing method because it affects the final quality of the coffee. The overdried fruits have brittle characteristics and produce many defective (broken) during hulling; however, on the other side, under-dried fruits are highly exposed to deteriorations because of fungus and bacteria growth.

Researchers reported that the number of imperfect beans with musty, earthy and greenish colour coffee defects frequently found in dry processed coffee. The bean size and roast volume of dry-processed coffee beans were larger, whereas the wet-processed coffee beans have a higher moisture content.

However, concerning the physical parameters (colour, shape, size) dry-processed coffee beans are no longer regarded as better than wet-processed. Coffee beans that are obtained with dry-processing method are heavy in body, smooth, and sweet and have complex characteristics.

Digestive Bioprocessing

The world most expensive coffees are obtained through digestive bioprocessing method. This processing method is rare and applied on a small scale. Digestive bioprocessing is a way of passing the coffee cherry through the animal intestine or digestive tract, which is the coffee beans exposed for acids, enzymes, and fermentation treatments. The coffee cherries are eaten by the civet cat and passed through the digestive system to produce civet coffee (*kopi luwak*) in Indonesia.

Similarly, elephant dung coffee (black ivory coffee), produced after the coffee cherries passed through the elephant's digestive tract, which is commonly produced in Thailand. During civet digestion, the breaking down of proteins creates a unique flavour and aroma of civet coffee.

The annual production of kopi luwak is less than 500 pounds, and the price is 600 dollars (Canadian) per pound; it leads the indisputable status of being the most expensive and rarest coffee in the world. Currently, we are working on analysing the elephant dung coffee, and it is compared with other coffees that are processed with commonly known processing methods (dry and wet processing). The experiments were conducted in Nepal.

The coffee cherries were harvested and fed by the elephants. Then the parchment coffee (the skin and pulp completely digested) and coffee cherry (the pulp is not digested) were collected from the elephant dung, and washing, drying, and other processes to obtain the green beans are continued. The black ivory coffee has a price tag of 1000 USD per kilogramme bean. This price makes it one of the world's expensive coffees.

Drying

Drying is the most critical steps in coffee processing methods. Drying is performed to reduce the moisture contents of coffee bean to the required level (10–12%) and to separate the parchment from the coffee beans easily. Sufficient drying is a crucial step to avoid the developments of moulds, which leads to significant losses and affects coffee quality. The drying methods were significantly affected the amount of low molecular weight (LMW) carbohydrates that present in the green coffee beans.

Several researchers reported the responses of coffee to various drying processes. The coffee cherry kinetics and drying characteristics under different drying temperature (40, 50, and 60°C) were investigated. The effect of drying and storage conditions on the quality of washed and natural coffee was evaluated. This research was focused on the relevance of suitable drying along with sufficient storage conditions to maintain the coffee quality. There are two ways for drying the coffee, such as natural drying using sunlight and mechanical drier.

Sun (natural) Drying

This drying method is the most common, widely practised in many countries and the cheapest way of drying the parchment coffee or the whole coffee cherry. This activity is highly dependent on climatic conditions and seasons. The removal of the mucilage and pulps at an early stage in wet-processing reduces the time needed for drying. There are different kinds of sun-drying methods. Coffee is typically dried on large patios made of cement or asphalt concrete with a 0.5–1% slope to drain the water.

Using the natural sunlight, drying of coffee takes 7–15 days for parchment coffee and 12–21 days for coffee cherries in patios. This duration varies depending on the weather conditions. The parchment coffee needs special attention than cherries to reduce physical damage (cracking).

Raking is required to allow the coffee to dry uniformly, but it should be carried out carefully. In some tropical countries, during the hottest hours of the day, covering sheets used to avoid the cracking of the coffee beans because of overheating.

The physicochemical characteristics of coffee that are dried on various types of drying grounds and the thickness of the layers (thin and thick) were evaluated by researchers. Their result showed the wet-processed parchment coffee dried on thin layer which resulted in the best beverage quality.

Mechanical Driers

During the mechanical drying, the beans are heated by hot air that passes through the machine and dries the coffee. Also, it carries away the moistures from the drying coffee. Controlling the drying temperature is the most critical part, and it should not exceed 45°C for whole cherries and 40°C for parchment coffee. Commonly there are two kinds of mechanical drier, such as static and revolving driers. Revolving driers include column driers, vertical driers, rotary driers, cascade driers, and flex driers, and there are tray driers with a stirrer.

All these driers can work with fuel oils, gasses, woods, and other energy sources. A report indicated that the best quality coffee was gained when the drying processes were done at two stages, an initial period the drying was performed at low temperature (20°C) and the second stage followed by higher temperature (60°C).

Hulling and Sorting

Hulling is the next step after drying. The dried coffee cherries (dry-processed) or parchment coffee (wet-processed) hulled to remove the covers and get the coffee beans out of it. Once the coffee bean reaches this stage, it means all the essential quality, such as colour and moisture content, has already been achieved. Hulling is done by using different hulling machines, or locally on a small-scale farm, wooden mortar and pestle used to separate the coffee beans from the parchment or dried coffee cherries.

Commonly there are two kinds of coffee hulling machines. One is which rubbed off the parchment by friction, and this might create heat. The second type is just cut the parchment and stripped off. It is essential to take care of the coffee beans and avoid physical damage even not to heat the beans during hulling because it affects the colour and taste of the coffee. The last layer that encloses the coffee beans is thin silver skin, and this may be removed or may not be removed during the hulling process.

If the coffee beans' silver skin is not separated during hulling; it needs a machine called polishers to separate it and get the green coffee beans. Finally, the green coffee beans are ready to be cleaned and sorted according to colour, size, and density.

The hulled coffee beans then undergo sorting processes, which is done by machine and hand. Hand sorting is most widely used, and it requires intensive labour for sorting the coffee beans based on size, colour, and density. The same-size and larger coffee beans get a premium price in the market, and a high percentage of defect may lead to a lower grade, and the price is also low.

Sorting is a crucial step because it affects the roasting condition. Uniform size coffee beans should be roasted to achieve uniform roasted beans. Sieve machines are used to screen the coffee beans according to their size. The sieving principle is applied in the sorting machine, which is using a big vibrating flatbed. There are different kinds of machines used to sort coffee beans. Electronic devices are also used to separate the coffee beans by their colours. However, this is not always sufficient to detect and separate the good and bad coffee beans. After sorting is properly done, the coffee beans are then packed with the right packaging materials and transferred to the storage house.

Storage

Right after the coffee beans are graded, they have to be kept in a storage house until they shipped and sold in the market. The temperature and relative humidity of the storehouse should be controlled to maintain the coffee quality without losing its intrinsic sensorial characteristics and physical and chemical properties and to store for a longer time.

Researchers reported that as the storage relative humidity is higher for an extended time, it decreases the compositions of reducing sugars in green coffee beans. Several research reports are available about the effects of storage conditions on the sensorial quality of coffee beans. He also mentioned that the 60% relative humidity with longer storage causes to cellular degradation and leads to oil leaking, which also contributes to the chemical compositions of green coffee beans. When the storage duration is prolonged, the oil becomes more acidic, and it reduced the quality of the product.

Roasting, Grinding and Drewing

The most critical factor in the coffee value chain is roasting, where the physicochemical changes lead to the fulfilment of the roasted coffee characteristics. Roasting is considered the essential steps in the formation of the aroma and flavour properties. The essential reactions during coffee roasting and which are responsible for the colour, volatile compounds, and flavour developments are called Maillard reaction and caramelization. Minerals are vital catalysts in the various biochemical reactions responsible for the formation of different aroma and flavour compounds.

Amino acids have an important role in the formation of nitrogen/sulphur heterocyclic compounds called melanoidins during roasting because of Maillard reaction or caramelization, which are considered as crucial compounds for flavour development. When the roasting temperatures are higher than 200°C, the precursors in green coffee are transformed into roasted coffee constituents, which lead to the development of diverse aroma test and colour.

However, the coffee's intrinsic quality is predetermined in the green bean by its precursor composition, and the roaster only can unlock the full potential by applying the appropriate and optimised roasting conditions. Optimising the appropriate roasting conditions is undoubtedly the most critical ways for achieving the desirable coffee aroma.

Researchers reported that as the roasting time is extended, the following changes occurred: the sugar content is reduced and then raised; caffeine contents showed insignificant changes; protein continuously decreased; free fatty acid improved; and unsaponifiable compounds declined. Another report mentioned that the lipid and organic acid increased, while the trigonelline and caffeine content showed almost unchanged.

During roasting net losses of matters in the form of water vapour, CO_2, and volatile compounds are exhibited. The roasting duration and the final temperature of the coffee beans determine the development of flavour compounds. A different flavour profile may occur because of the time–temperature and roasting conditions, even though the same coffee beans and roaster are used. The physical properties and kinetics of aroma development of coffee showed differences as the coffee is roasted for a shorter time at high temperature compared to beans roasted for a longer time at low temperature.

Fast roasting generates more soluble solids and causes less degradations of CGA, and the loss of volatile was lower. However, the fast roasted coffee is considered affected by lipid oxidation because of high migration of oil from the inner part of the coffee beans to the surface. The determining factors of the final coffee quality are roasting profile, roasting degree, and the technology used for coffee roasting. There are different kinds of roasting conditions. The most common roasting types are discussed below.

Light Roast

The lightly roasted coffee developed light brown colour, and it is preferred to make a coffee with a mild body. It has light fragrant, floral or fruity coffee notes. In this roasting condition, the coffee beans should not develop oil on their surface. It is also characterised with pronounced acidity and high caffeine content. During light roasting, the internal temperature of the coffee bean reached approximately 204°C

Medium Roast

Medium roasted coffee developed medium brown colour, and the surface of the coffee beans should not have oils. It is characterised by balanced aroma flavour and acidity. Traditionally this is the most preferred roasting condition. It has medium caffeine content. To achieve this roasting condition, the internal temperature of the coffee beans reached approximately 215°C.

Medium-Dark Roast

Medium-dark roasted coffee has a dark brown colour, and the surface of roasted beans developed oils. It is characterised by fully bodied deep flavoured and little spicy notes. It has low caffeine content. This roasting condition is attained when the internal temperature of the coffee bean reached nearly 229°C. Overall, roasting coffee in medium to dark conditions causes an increase in ketones, esters/lactones, aldehydes, aliphatic acids, and aromatic acids, but a reduction in caffeine content is observed.

Dark Roast

It has nearly a dark colour and produced oils on the surface of the coffee beans. The darker the coffee beans, the less the acidity. It is also characterised by very low caffeine content with heavy mouthfeel, strong flavour, bitter, and burnt or smoky notes. When the coffee is roasted in the dark condition, the internal temperature of the coffee beans reached approximately 204°C.

Grinding is the next step after roasting. In this step, the coffee beans are crushed and changed into powder at different particle size. It is done using a grinding machine (electrical) or using mortar. However, to achieve a uniform ground size, an electrical grinder is the best choice. Blenders can also be used for grinding coffee beans, and the larger particle size and the broadest distributions can be seen while blending the coffee beans.

The grinding size is usually measured by using sieve analysis. According to the types of coffee to be prepared, the particle size or grinding particle is the most important. There are many types of grinder with the adjuster to regulate the particle size of coffee grounds. During grinding the important volatile compounds are dispersed into the surroundings from the powdered coffee, so it must be executed right before the intended brewing (1–12 min after grinding) to maintain the volatile compounds.

The coffee brewing and extraction methods differ according to the personal preference, the culture, geographical, and social and financial factors. All these brewing and extraction methods of coffee vary depending on the time of extraction, the pressure, the brewing tool, the temperature of the water, and the extraction amounts. The brewing water quality (electrical conductivity) is another important aspect for maintaining the original coffee flavour and test.

The size of the coffee ground determines the rate and the total amount of extraction. If the coffee is too fine (for example super-fine) extraction will not happen as water cannot pass through, and if it is too course the grinds will be under-extracted. It depends on the coffee ground size, the coffee brewing methods also vary.

POST HARVEST TECHNOLOGY OF NUTS

Cashew

Harvesting of Cashew Nuts

Fully mature nuts are to be harvested. Cashewnuts when fully mature look greyish brown. It is better to pick the fallen nuts because there is a possibility of harvesting immature nuts which leads to poor quality kernel during processing. However if one wants to use cashew apple for preparation of cashew apple products one has to pluck the fruits from the tree. One can know the maturity of nuts by just tapping the fruits. If it is fully mature the fruit just drops.

Storage of Cashew Nuts

Raw nuts immediately after harvest are to be separated from cashew apple and sun dried for 2 to 3 days. If the raw nuts are not dried properly, during subsequent storage, they get spoiled due to microbial infestation. Nuts after drying can be stored in gunny bags well protected from rodents and stacked on a platform above the ground level leaving space on all the sides of the room. Well dried raw nuts (moisture content 8 to 9%) could be stored upto one year without any quality deterioration.

Cashew Apple Utilization

Cashew apples are good source of Vitamin C. Besides Vitamin C, it contains sufficient quantity of minerals, water soluble vitamins and tannins. Tannins impart astringency and could be precipitated by the addition of gelatin. Alternatively steeping in salt solution also removes tannin. Juice, syrup, candies, pickles, chutney, jams and jelly, could be prepared from cashew apples. Cashew apples are fermented in Goa to produce an alcoholic beverage called 'Feni'. Attempts are underway at National Research Centre for Cashew for developing cashew apple based products particularly fibre rich cereal / pulses blends.

Nutritional Value of Cashew Nuts

Is it harmful to consume cashew nuts by heart patients? Cashew kernels are quite rich in fat, protein and carbohydrate and does not contain any anti-nutritional factors. Cashew kernel fat is quite rich in unsaturated fatty acids (olecic acid) which are good for health. Saturated fatty acids are less. Further cashew kernel oil does not contain cholesterol.

Groundnuts

Curing and drying groundnuts

- Groundnut should be allowed to wilt for a day to reduce moisture.

- Curing reduces chances of rotting and moulding, preserves oil quantity and quality in the kernels, reduces crop moisture and bulkiness.

Post-harvest Handling

- It is the stage of crop production immediately following harvest, including cooling, cleaning, sorting and packing.
- The moment a crop is removed from the ground, or separated from its parent plant, it begins to deteriorate which affects more its quality than its quantity and can in the long-term result in a definite loss.
- Post-harvest losses are significantly high during harvesting depending on the soil moisture conditions.
- If soil is too dry or produce harvested late, pods can strip at lifting pods susceptible to attack by fungi.

Fungal toxins also called mycotoxins are major constraint to increased productivity and market competitiveness (Myco-fungus. Toxins = poisons). Most common fungal toxins are aflatoxin which are produced by *Aspergillus flavus* and *A. parasiticus*.

Grains especially groundnut are susceptible to contamination through poor production, harvesting, processing and storage.

Effects of Aflatoxins

Aflatoxins are not destroyed by cooking. Cooking or processing grain e.g. peanut butter making does not remove aflatoxin. Livestock must not be fed mouldy grain as they become sick and their meat, milk, eggs become contaminated.

Two main effects of aflatoxins:

Health Effects

- Eating contaminated food results in aflatoxicosis.
- Chronic exposure: leads to liver damage, immune suppression, malnutrition and stunted growth in children
- High-level ingestion (Acute exposure) can lead to death.

Income losses

- Aflatoxins may reduce the market value of peanuts and in serious cases they can be rejected from the market.

Pecan

Cleaning

Harvested nuts are transferred to trailers or large bins and transported to a cleaner. For small jobs, trailer-mounted field cleaners are available. Multiple component, stationary cleaners housed in a building can operate in any weather and can process greater quantities of nuts compared to field units. In a typical cleaning operation, the following tasks are accomplished. First, small sticks are removed. Next, clods larger than the nuts are removed. The nuts then pass through a series of hullers to remove any loose shucks sticking to some nuts. Next, an air stream blows out leaves and light nuts that didn't fill out or were eaten by insects. Finally, the nuts pass over an inspection table where workers remove cracked nuts, sticktights, nuts with weevil holes, irregularly colored nuts or other damaged nuts.

Sizing

Following cleaning, the nuts should be sized to ensure uniform lots. Generally, the larger kernel varieties bring a higher price. Mixing large nuts with small nuts will result in a lower grade for a shipped lot. The more expensive sizers are capable of separating nuts into sizes less than 1/2 inch to over 1 inch in diameter.

Drying

To take advantage of higher prices for early season pecans and reduce loss due to wildlife depredation, some growers choose to harvest earlier with nuts at moisture contents up to 20 percent. This moisture content needs to be lowered to 4 percent for the nuts to develop good eating quality and prevent molding. Peanut drying trailers are commonly used by pecan producers to dry whole pecans.

Heat is added to the air passing through the pecans to lower the relative humidity and speed the drying process. Temperature in drying trailers or bins should not exceed 90 degrees F. With adequate warm air movement through the pecans, it will take approximately 40 hours to reduce nuts with 20 percent moisture to 4 percent moisture. Pecans with 8 percent moisture will take approximately 12 hours to lower moisture percentage to 4 percent.

Grading

Before marketing pecans, they need to be graded. Knowing ahead of time the quality of the nuts will allow you to anticipate the price they will bring, whether they are marketed wholesale or direct to the public. Grading consists of determining the percentage of edible kernel and percentage of kernel grades. For accuracy, it is important to sample the entire lot of pecans, whether it is a few hundred pounds or several thousand pounds.

To take a representative sample, withdraw at least 1/10 of 1 percent by weight of the lot being sampled. Nuts should be drawn equally throughout the lot by probing the top, middle and bottom of the container.

The drawn sample should be thoroughly mixed. From that composite sample, prepare three separate 1 pound samples. Crack each of the three samples carefully by hand and separate the nutmeats. Separate the edible from the inedible kernels. Calculate the percentage of edible kernel for each sample. Divide the weight of the edible kernels by the weight of the sample to obtain percent edible kernel.

The average of the three represents the percent kernel for the lot. However, if extreme variance occurs, three additional 1 pound samples should be cracked. An average of the 6 pounds will then be used to determine the percent kernel. The inedible kernels and pieces should be weighed and reported as a percentage of the lot.

Classifications of Kernels

Grade Classification	Description
No. 1	Kernel bright colored, full bodied, solid (Fancy Product)
No. 2	Kernel lightweight, bright in color; full bodied, slightly off-color (Choice Product)
No. 3	Kernel amber colored, either full bodied or lightweight (Standard Amber, Cutting Stock)

Storage

Growers who choose not to market their pecans during the harvest season will need to place the crop in cold storage. Nutmeats will discolor and turn rancid at warm temperatures due to high oil content. In general, the lower the temperature, the longer the storage life of in-shell and shelled pecans. Freezing does not damage pecans since their moisture content is low.

Food Safety

Food safety concerns are increasing as once unheard of illness-causing microorganisms become more prevalent and as products previously considered safe cause an increasing number of illnesses each year. Produce, recently thought of as a safe product, has been identified as a cause of major foodborne illness outbreaks in recent years.

The three main types of hazards found in food products are chemical, physical and biological. Biological hazards include bacteria, viruses, fungi and parasites.

The most common biological hazards of concern in pecans are molds and bacteria. Both may be controlled by proper sanitation and handling procedures in the orchard, as well as during storage, cleaning, sorting, cracking, shelling and packaging operations.

Contamination may occur before or after harvest, typically as a result of contact with contaminated water, contaminated soil, direct contact with animal manure, contaminated humans and contaminated equipment or facilities.

Controlling food safety hazards is about identifying sources of risk, developing measures to minimize those risks and documenting what is done. Following are some specific suggestions for minimizing food safety hazards in the orchard and during processing and storage. Contaminated spray water can pose a significant risk, especially as harvest approaches. Test all water sources for fecal coliforms prior to using.

The most likely source of contamination while pecans are on the ground is animal manure, though contaminated irrigation water may also pose a risk. Research suggests that the pathogenic bacteria *E. coli* O157:H7 may survive in untreated manure for two months to a year. Therefore, it is prudent to take extra precautions to limit sources of contamination for at least two months prior to harvest. This includes removing cattle from the orchard at least two months before harvest. It is best to also remove cattle from adjacent pastures if runoff from those fields into the orchard is likely to occur. If manure-based compost is applied to the orchard, it should be properly composted. Compost should reach a temperature of 131°F to 149°F for at least three days during composting to ensure safety.

Proper hygiene procedures for orchard workers are important, especially during harvest. Toilets and hand-washing stations must be available and accessible to workers in order to avoid potential contamination. Workers who handle the nuts during cleaning, grading, shelling and packaging operations are prime sources of potential contamination because the crop is concentrated in a relatively small area. Infected workers (whether showing symptoms or not) can easily contaminate the nuts they handle if they do not wash their hands after sneezing or using the restroom. This is particularly true for workers handling shelled nuts. Proper worker hygiene procedures should be established and included in hygiene and health training programs.

Birds, insects, rodents and domestic animals should be kept out of processing and storage areas. If birds roosting or nesting in exposed rafters are a problem, metal screening may be installed to keep them out. A proper pest control plan can prevent infestations. This plan typically includes traps for rodents and spray programs for insects. The plan should be properly documented. It is especially important to be aware of EPA regulations for pesticide use in food processing facilities. Spraying and trapping alone are not enough; proper cleaning and maintenance of food processing and storage areas are vital parts of any pest control program.

A single dirty piece of equipment can contaminate a large number of nuts that pass through it. A dirty storage area may contaminate many or most of the nuts stored in it. Therefore, proper maintenance, cleaning and sanitation of both equipment and facilities are vital to any food safety program.

Walnut Postharvest Technology

Drying Technologies

Current drying methods have drawn increasing concerns due to low efficiency, high energy consumption, large carbon footprint and risks of causing quality and safety deterioration. As the results, the tree nut industries are under the pressure to develop more efficient and sustainable drying technologies that can preserve the quality and safety of dried products. In recent years, research works have been conducted mainly in two aspects: (1) to improve or optimize the operating conditions of the current HA drying practice; (2) to develop novel drying technologies based on thermal and non-thermal methods.

Improve the Hot Air Drying Process

Elevating the drying temperature is a potential approach to reduce the drying time and improve the efficiency of the current HA drying practices. However, keeping the nuts at high temperatures for long times is not desirable, as it may induce significant quality deterioration to nuts, such as lipid oxidation and browning. Enhancing the drying rates by intensifying the heat and moisture transfer during the drying while maintaining the product qualities has become the key approach. Since the hull and shell have much higher IMCs than the kernel, most thermal energy in the heated air is consumed for raising the temperature and evaporating moisture in the hull and shell, particularly in initial drying stages.

As the results, the temperature increase in kernels is significantly slowed down, and the moisture removal from kernel to the environment is greatly restricted due to the reverse moisture gradient from the hull to the kernel. Based on these characteristics, researchers developed a new drying strategy by using high temperature heating in the beginning of drying to quickly heat up and partially dry the hull and shell; at the same time, due to the relatively low thermal conductivity of the kernel and shell, the kernels should not be over-heated; then before kernels reached a temperature that was high enough to cause significant oil quality deterioration, drying temperature was decreased to finish the drying. With the aids of experimental studies and mathematical modeling of the heat and moisture transfer during the drying, the feasibility of this new drying strategy was verified.

Drying in-shell walnuts by HA with step-down temperatures reduced the drying time and energy consumption by up to 40% and 24%, respectively, and obtained similar oil quality and kernel color in the dried products compared to the conventional practice. Similarly, using HA drying with step-down temperature and tempering for in-hull almonds significantly reduced the drying time and did not affect the quality of the dried almonds in terms of oil quality or incidence of concealed damage.

Sorting the nuts into different groups based on IMCs first and then drying them separately is another strategy of improving the drying efficiency, moisture uniformity and product quality. Researchers studied the correlation between the MCs of walnuts and terminal velocities and developed a sorting method called the 'air knife'. The walnuts were separated into low and high MC groups and dried separately, which resulted in 18–28% energy saving compared to without sorting.

Meanwhile, the uniformity of moisture in the dried products was greatly improved. This technology has been commercialized and installed in-line in walnut hulling and drying facilities in California. Similarly, the correlations between terminal velocities of pistachios and almonds with their MCs have also been studied. Researchers have shown that in-hull almonds, in-shell almonds and loose hulls at harvest could be separated based on the thickness and terminal velocities of different groups, and suggested sorting and removing almond hulls prior to drying benefited the improvement of moisture uniformity and energy saving.

Fig. 3. (A) 'air knife' sorter and (B) SIRHA dryer for walnuts.

Novel thermal and non-thermal drying technologies

Infrared (IR) heating has the advantage of high heating intensity compared to convective heating. IR radiation could penetrate 2–5 mm depth into food surface, which matches the typical thickness of the nut hull and shell. Therefore, IR heating was an ideal technology to pre-dry the tree nuts. A commercial-scale sequential infrared and hot air (SIRHA) drying technology was developed by Researchers, which had 14.2 ton/h throughput and achieved 13.6–26.5% drying time reduction and 10–20% energy savings, respectively.

Researchers used SIRHA drying for pistachios and almonds, and achieved 9% and 40%, respectively, compared with HA drying only. Additionally, research has shown that the percentage of shell splitting for pistachios increased with drying rates, and thus spray-rinsing with water and then IR drying could effectively improve the shell splitting.

Heat pump drying utilizes the thermal energy in the air flowing out of the dryer by dehumidifying and condensing the water vapor containing in the outflow air, retrieving the enthalpy of heat evaporation, then circulating the heat back to the dryer inlet. HA heat pump drying was used for walnuts and higher drying efficiency was achieved without affecting the product quality. IR-assisted heat pump drying for walnuts reduced 20% drying time and 10% energy consumption compared to HA drying. Solar heat pump drying reduced the drying time and increased the energy utilization ratio of pistachios.

Dielectric heating such as microwave (MW, wavelength range: 300 MHz–300 GHz) and radiofrequency (RF, wavelength range: 3 kHz–300 MHz) can penetrate the foods. The electromagnetic wave activates the water molecules through dipole rotation and/or ionic polarization, generating heat volumetrically within foods, while non-polar molecules are not affected. HA drying at 50°C assisted with RF heating at 27.12 MHz and 6 kW reduced 58.3% drying time of walnuts compared to HA drying only. Intermittent MW drying could be used to dry pistachios without affecting the nut quality.

Some non-thermal technologies have also been studied as assistance to conventional drying processes. For example, high-power ultrasound treatments improved the drying rate and energy efficiency of pistachios. The drying time of almonds was reduced by 58.33% with 40 min ultrasound treatment.

Such phenomena should be attributed to the enhancement of moisture transfer by the molecule vibrations induced by ultrasonic field. Under vacuum, the boiling point of water and vapor pressure in the drying chamber were significantly lower, and thus increase the driving force of moisture transfer during the drying process. Vacuum drying was able to dry walnuts in a shorter time compared with conventional practice. IR- and MW-assisted vacuum drying improved the drying efficiency of almonds.

Blanching and kernel peeling

Tree nut kernels with bright and white color are appealing to the consumers and are typically required for producing meals or milk, or consuming raw in salads. For this purpose, blanching and peeling of kernels are usually needed. The peel on the kernel surface, also known as 'pellicle', has dark color and usually has high contents of natural antioxidants, such as tannins, phenolics and flavonoids. They protect the kernel from natural oxidations and show good antibacterial activities.

However, due to the abundant antioxidants and fiber, pellicles usually have bitter taste, high chewiness, and low solubility in drinks, which lowers the sensory satisfaction and affect the palatability. Kernel peeling is mainly accomplished by physical or chemical methods.

Physical peeling involves blanching and mechanical abrasive peeling and is the recommended method in the almond industry. However, abrasive peeling is not recommended for walnut kernels since walnut kernel has irregular shape and abrasion may also result in significant loss of nut flesh. During blanching, nut kernels are subjected to either water soaking or steaming at high temperatures, which cause the pellicle to swell and crack.

After blanched, the tree nuts go through soft rubber rollers to mechanically peel off the pellicle by abrasion. Besides, blanching also showed the capability to protect the color of the tree nut kernels, and control the cross-contamination of aflatoxins in almonds.

The disadvantages of blanching include the increase of MCs, softening of the nut flesh and leakage of polyphenols into the water, which brings up the need to further dry the kernels after peeling and loss of antioxidant activities. The large water consumption is another concern for the sustainability, particularly in the drought regions.

Chemical peeling usually refers to the hot lye peeling by NaOH, Na_2CO_3, and $Ca(OH)_2$, etc. In a typical process, nut kernels are soaked in hot lye solutions for several minutes to corrode the pellicles, then rinsed with water.

Factors, such as alkali type, concentration, temperature, and soaking time, are important for the peeling performance. Although lye peeling is efficient and effective, it may cause significant quality deteriorations, such as texture softening, surface browning, loss of crude protein and fat, decrease of antioxidant activities, and increased oxidation. If the rinsing of peeled nuts was not performed adequately, the chemical residues on the nut surface cause additional food safety concerns. More importantly, disposal of the wastewater from lye peeling requires excessive use of chemicals and may cause severe environmental impacts.

For these reasons, novel and safe peeling methods have been studied and developed. IR radiation can penetrate the pellicle and heat up the kernel, which cause moisture evaporation and accumulation of vapor pressure under the pellicle. Meanwhile, IR heating may cause pyrolysis of pectin substances. When the vapor pressure reaches a critical level, the pellicles crack and can be peeled.

Researchers developed a cryogenic peeling system for walnuts, in which the walnuts were held at -160°C by cold gas/liquid N_2 and moved dynamically downwards, and the shrinking pellicles were removed by upflowed air. Studies have shown that walnuts and almonds with their pellicles peeled off had shorter shelf life compared to the unpeeled ones. Therefore, applications of edible coatings containing antioxidant substances, such as Mastic gum, chitosan incorporated with green tea extract, walnut phenolic extracts, and mango kernel starch, etc. are popular research topics in recent years. Meanwhile, the sensory quality of the nuts with edible coatings should not be compromised.

Roasting

Roasting is a commonly used processing to improve the palatability of tree nuts, which is usually done at a temperature higher than 90°C. Maillard reaction between the carbonyl group of reducing sugars and the amino groups of proteins in the nuts is responsible for the nonenzymatic browning and formation of substances with desired 'roasted aroma and flavor' (e.g., pyrazines, furans, and pyrroles). During roasting, nuts are further dried and are subjected to some change in texture properties, which gives rise to the crunchy mouthfeel. The texture properties of tree nuts (hardness, fracture force, firmness, etc.) are significantly affected by the roasting temperatures. Meanwhile, roasting is also an effective measure to reduce the aflatoxin contents in tree nuts.

Conventionally, tree nuts are roasted by HA or oil. HA roasting is usually performed at a temperature ranging from 100 to 180°C for up to 60 min. The main problems with this method are the long processing time, high energy consumption and non-uniform roasting, which should be attributed to the slow convective and conductive heat transfer with HA heating. During oil roasting, tree nuts are immersed and fried in a vegetable oil at a temperature near 180°C for 7–8 min, followed by drying to remove the oil from surface. Although the roasting process do not significantly reduce the contents of macronutrients, it also has some disadvantages. The major concern associated with roasting are the severe oxidation of polyunsaturated fatty acids that give rise to 'off-flavor' and the risk of toxin generation over the smoke point under the high temperature processing. Some main chemical reactions related to the quality change during tree nut roasting are shown in Fig. 4.

In addition, oil roasting results in high oil intake in the products. Therefore, new roasting technologies need to be more efficient and cost effective, while not compromising the product quality and safety. Formation of acrylamide, a group 2a carcinogen by WHO, from the free asparagine and reducing sugars in the nuts arises another food safety concern. Researchers found IR roasting caused the highest acrylamide content in pistachios, and MW roasting led to the lowest. Increasing the roasting temperature and time, and MW power facilitated acrylamide formation, since the formation rate of acrylamide increased with temperature.

Researchers suggested that MW roasting was a promising method to replace the conventional HA roasting for almonds. Adding of salt during roasting can mitigate the acrylamide formation in tree nuts, which should be due to the prevention of intermediate (such as Schiff base) formation under the inhibition of cations. In response, researchers suggested that almonds should be roasted at the lowest possible temperature to minimize the acrylamide formation.

Researchers found that using a step-down temperature heating may reduce the acrylamide contents in baked goods. Researchers suggested that use of IR heating together with HA or MW heating reduced the roasting time while obtaining similar product quality.

Future research in tree nut roasting should focus on the development of novel roasting methods that combine the advantage of different heating methods, optimization of roasting parameters (using stepwise temperature roasting strategy) and addition of appropriate food-safe additives to preserve product quality, reduce acrylamide formation, and improve microbial safety and energy efficiency.

Postharvest processing technologies are essential in preserving the quality and safety of tree nuts. In this chapter, the status, recent advances, and challenges of drying, disinfection, disinfestation, blanching, peeling, and roasting of almonds, walnuts, and pistachios in scientific research, as well as in industrial productions are summarized.

Current processing practices can be improved by optimizing the operating parameters. Novel thermal and non-thermal drying technologies, such as IR, MW, RF, ultrasound, vacuum, etc. and combined technologies, have shown great potential to improve the drying efficiency and safety of nuts without compromising the nut quality. It was noted that the postharvest processing methods of different tree nuts share some similarities but are also different due to the differences in physical properties and chemical compositions.

Each processing practice has significant impacts on the quality and safety of the final products, such as lipid oxidation, loss of nutrients and formation of acrylamide. Therefore, suitable processing technologies and operating conditions must be carefully selected. Additionally, processing wastes of tree nuts, such as hulls and shells, have potential to be utilized as high value bioresources for producing value-added products. Being important economic crops with large market size, improvements in the postharvest processing technologies of tree nuts could translate into significant global influence in reducing the energy consumption and environmental impacts for improved sustainability, economic values, and competitiveness of tree nut industry.

9

MATURITY INDICES, HARVESTING AND POST-HARVEST HANDLING OF FRUITS AND VEGETABLES

INTRODUCTION

Postharvest losses refer to measurable quantity and quality loss of food crops at harvest, storage, transportation, processing, marketing and preparation before consumption. It occur throughout the value chain, as a result of technical and managerial setbacks during harvest, handling, transportation, processing, packaging, marketing and distribution. Postharvest losses impact on environment and climate following unnecessary emissions of greenhouse gases produced during production, processing and transportation of fruits and vegetable (FV) which ultimately end into loss.

Agriculture intensification projects aim to improve income and nutrition status of households, however, increase in production without corresponding increase in postharvest management skills, storage facilities, and processing may heighten postharvest losses. On the other hand, investment efforts made to save food after harvest usually cost less and are less harmful to the environment. A minimum postharvest losses reduction can potentially reduce production cost.

FRUITS AND VEGETABLE NUTRITIVE VALUES

FV are highly valued in human diet mainly for their minerals and vitamins content. They are good sources of vitamin C, folate, β-carotene (pro-vitamin A), potassium, iron, zinc and calcium. Also, they are substantially high in antioxidants, bio-active compounds and fibers. Examples of antioxidants found in FV are carotenoids (β-carotene, lycopene), ascorbic acid, phenolic, polyphenols (terpenoids and flavonoids) and anthocyanins. FV provides anti-cancer, anti-oxidative, anti-microbial and immune modulating effects apart from nutritional benefits to human health. Consequently, consumption of FV reduces incidence of cardiovascular diseases, cancer, and type-two diabetes in human. On that regard, Food and Agriculture Organization (FAO) and the World Health Organization

(WHO) of the United Nations (UN) have set average FV intake at 400 g/day per person. Also, due to their bound antioxidants, pectin and carotenoids; FV are used as functional ingredients, bio-pigments and colorants in processed foods.

FV QUALITY AND SAFETY ASPECTS

Quality is the combination of characteristics and properties that gives the commodity value for food and fulfills consumer's requirements. Quality focuses on the nutritional value, safety, sensory (color, taste), physical appearance (size, shape, absence of defects) and shelf life of FV. FV quality are determined by pre-postharvest production factors such as; location, climate, soil type, water quality, plant nutrition, use of pesticides and plant growth regulators; Postharvest handling; storage and processing. To producers quality is considered as high yield, resistance to diseases, ease to harvest, good appearance and few defects as well as good shipping quality, while marketers and distributors consider quality as good appearance, firmness and shelf life. On the other hand, consumers consider quality as good appearance, firmness, size, good flavor, convenience, nutritive value and good edible quality that defines their buying behavior.

Increased consumer awareness on food safety has led to public health concern on freshly consumed food like FV, related to food-borne illness, safe use of pesticides and ripening chemicals. FV are colonized by a range of spoilage and pathogenic microorganisms. They get contaminated by dust, soil, and water during harvesting, handling, processing, distribution and preparation. Consumption of contaminated FV and their products has been linked to disease outbreaks caused by pathogens such as, *Bacillus species, Salmonella species, klebsiella species, Escherichia coli* and *Listeria monocytogens*.

Safe use of pesticides and ripening chemical, detection and assessment of food adulteration are among important areas of food safety. Washing and sanitizing steps before or during processing reduces the risk of chemical residues, pathogens and other contaminants on FV. To ensure adherence to food safety Researchers recommended that, FV producers and all involved in the value chain should be equipped with skills and knowledge on food safety which includes; clean handling practices, personal hygiene, process hygiene and control of cross-contamination. For that reason, the differences between developed and developing countries in production, handling, processing, storage and distribution should not be an excuse to compromise FV quality and safety standards.

MATURITY

It is the stage of fully development of tissue of fruit and vegetables only after which it will ripen normally. During the process of maturation the fruit receives a regular supply of food material from the plant. When mature, the abscission or corky layer which forms at the stern end stops this inflow. Afterwards, the fruit depend on its own reserves, carbohydrates are dehydrated and sugars accumulate until the sugar acid ratio form. In addition to this, typical flavour and characteristic colour also develop.

It has been determined that the stage of maturity at the time of picking influence the storage life and quality of fruit, when picked immature like mango develop white patches or air pockets during ripening and lacking in normal brix acid ratio or sugar acid ratio, taste and flavour on the other hand if the fruits are harvested over mature or full ripe they are easy susceptible to microbial and physiological spoilage and their storage life is considerably reduce. Such fruits persist numerous problems during handling, storage and transportation. Therefore, it is necessary or essential to pick up the fruits or vegetables at correct stage of maturity to facilitate proper ripening, distant transportation and maximum storage life.

Horticultural maturity

It is a developmental stage of the fruit on the tree, which will result in a satisfactory product after harvest.

Physiological maturity

It refers to the stage in the development of the fruits and vegetables when maximum growth and maturation has occurred. It is usually associated with full ripening in the fruits. The physiological mature stage is followed by senescence.

Commercial maturity

It is the state of plant organ required by a market. It commonly bears little relation to physiological maturity and may occur at any stage during development stage.

Harvest Maturity

It may be defined in terms of Physiological maturity and horticultural maturity, it is a stage, which will allow fruits / vegetables at its peak condition when it reaches to the consumers and develop acceptable flavour or appearance and having adequate shelf life.

Vegetables are harvested at harvest maturity stage, which will allow it to be at its peak condition when it reaches the consumer, it should be at a maturity that allows the produce to develop an acceptable flavour or appearance, it should be at a size required by the market, and should have an adequate shelf life.

Time taken from pollination to horticultural maturity under warm condition, skin colour, shape, size and flavour and abscission and firmness are used to assess the maturity of the produce.

TABLE 1: CRITERIA OF MATURITY FOR HARVESTING FRUITS AND VEGETABLES

Fruit	Physical	Chemical
Mango	Olive green colour with clear lenticels, shoulder development size sp. gravity, days from fruit set.	Starch content, flesh colour
Banana	Skin colour, drying of leaves of the plant,	Pulp/peel ratio, starch content
	brittleness of floral ends, angularity of the fruit, and days from emergence of inflorescence.	

Fruit	Physical	Chemical
Citrus	Colour break of the skin from green to orange, size	Sugar/acid ratio, TSS
Grapes	Peel colour, easy separation of berries, characteristic aroma	TSS 18-12 Thompson seedless, 12-14 for Bangalore Blue, 14-16 for Anab-e-shahi
Apple	Colour size	Firmness as measured by pressure tester
Papaya	Yellow patch or streaks.	Jelliness of the seed, seed colour

TABLE 2: TIME TAKEN FROM POLLINATION TO HORTICULTURAL MATURITY

Sl. No	Vegetables	Time to harvest Maturity (days)
1.	Ridge gourd	5 -6
2.	Squash	7 - 8
3.	Brinjal	25 - 40
4.	Okra	4 - 6
5.	Pepper (green stage)	45 - 55
6.	Pepper (red stage)	60 -70
7.	Pumpkin (mature)	65 - 70
8.	Tomato (mature green)	35 - 45
9.	Tomato (red ripe stage)	45 - 60
10.	Peas	30 - 35

COMPUTATIONAL METHODS

Methods for Judging maturity

Computational methods

Calendar date

It is one of the commonly used indices of maturity and is reasonably accurate provided flowering and weather during growing season is normal. But standardization requires study for many seasons for given variety, location, rootstock etc., e.g. Mango harvesting period – April to July.

DFFB(Days From Full Bloom)

It is reliable but varies greatly from year – to - year and location –to- location. In such case the optimum date of harvest can be predicted by doing night temperature correction for 15 days fallowing full bloom. For every 10F variation from an average night temperature, a correction of one day is made in the standard figure from full bloom, e.g. Mango 110 -125 day (Var. Alphonso and Pairi), Banana 99 - 107 days in dwarf Cavendish.

Heat units/Day degree

Optimum maturity is computed by the sum of mean daily temperature, above base temperature (10°C/°50F for apple)for a period concerned. The number of degree-days to maturity is determined over a period of several years. °10C /°50F is the temperature at which growth occurs for apple and base temperature varies with crop.

The degree day is based on a growth-temperature relation. However this heat units work only within limited temperature. Heat units are not useful for photoperiod sensitive species.

A Heat unit is calculated by - (daily mean temp – base temp) X No. of Days (flowering to harvest)

Base temperature for tomato, spinach and pumpkin is 150°, 20° and 130°C respectively.

Skin colour

Loss of green colour in citrus and red colour in tomato.

Shape, size and flavour

Sweet corn is harvested at immature stage, smaller cobs marketed as baby corn. Okra and cow pea are harvested at mature stage (pre fiber stage). In chilli, bottle gourd, bitter gourd, cluster beans maturity is related to their size. Cabbage head and cauliflower curd are harvested before un pleasant flavour.

Abscission and firmness

Musk melon should be harvested at the formation of abscission layer. In cabbage and lettuce should be harvested at firmness stage.

FACTORS AFFECTING MATURITY

1. Temperature: Higher temperature gives early maturity. e.g. Gulabi (Pink) grapes mature in 100 days in Western India but only 82 days are enough in the warmer Northern India.

   Lemon and guava takes less time to mature in summer than in winter. Sun-scorched portions of fruits are characterized by chlorophyll loss, yellowing, disappearance of starch and other alcohol insoluble material, increase in TSS content, decrease in acidity and softening.

2. Soil: Soil on which the fruit tree is grown affects the time of maturity. e.g. Grapes are harvested earlier on light sandy soils than on heavy clays.

3. Size of planting material: This factor in propagated fruits affects fruit maturity. e.g. In pineapple, the number of days taken from flowering to fruit maturity was more by planting large suckers and slips than by smaller ones.

4. Closer spacing: Close spacing of hill bananas hastened maturity.

5. Pruning intensity: It enhanced the maturity of Flordasun and sharbati Peaches.

6. Girdling: Process of constricting the periphery of a stem which blocks the downward translocation of CHO, hormones, etc. Beyond the constriction which rather accumulates above it. In Grape vines it hastens maturity, reduces the green berries in unevenly maturity cultivar and lowers the number of short berries. It is ineffective when done close to harvest. CPA has an additive effect with girdling

MATURITY INDEX

The factors for determining the harvesting of fruits, vegetables and plantation crops according to consumer's purpose, type of commodity, etc and can be judged by visual means (colour, size, shape), physical means (firmness, softness), chemical analysis (sugar content, acid content), computation (heat unit and bloom to harvest period), physiological method(respiration). These are indications by which the maturity is judged. Various index are as follows:

Visual indices

It is most convenient index. Certain signals on the plant or on the fruit can be used as pointers. E.g. drying of top leaves in banana, yellowing of last leaf of Peduncle in jackfruit. Flow of sap from cut fruit stalk of mango slows down if the harvest is done after maturity but in immature fruits, exudation is more and comes with force in a jet form. in papaya, the latex becomes almost watery. The flow gets reduced on maturity in Sapota.

In fruits like banana and Sapota, floral ends become more brittle and shed with a gentle touch or even on their own. In Sapota, the brown scurf on the fruit skin starts propping. In mango, lenticels become more prominent and the waxy bloom gradually disappears. Grapes develop translucent bloom. Other changes like angularity in banana, development of creamy wide space between custard apple segments and the flattening of the eyes in pineapple and tubercles in litchi serve as reliable maturity indices.

Seed development

It can also be used as an index of fruit maturity, e.g. endocarp hardening for stone and fiber development for dessert in mango.

Start of bud damage

Occasionally it can be used as an index of fruit maturity in mango.

Calendar date

For perennial fruit crops grown in seasonal climate which are more or less uniform from year to year, calendar date for harvest is a reliable guide to commercial maturity. This approach relies on a reproducible date for the time of the flowering and a relative constant growth period from flowering through to maturity. Time of flowering is largely dependent on temperature, and the variation in number of days from flowering to harvest can be calculated for some commodities by use of the degree-concept.

TECHNOLOGIES AND PRACTICES TO REDUCE POSTHARVEST LOSSES

FV are metabolically active, undergoing ripening and senescence changes, which has to be controlled in order to preserve quality and prolong shelf life. Application of postharvest technologies has proven quantifiable postharvest losses reduction in different part of the world and could be a strategic pathway to reduce poverty, hidden hunger and malnutrition in SSA. They also, have enhanced horticultural industries to meet local and international demand of FV for nutrition and food security. The choice of technology depends on crop type, climatic conditions, affordability and ease of use.

At the same time, proper postharvest handling practices should be adhered to prevent damages and bruises on FV. Postharvest technologies intend to slowdown ripening and senescence changes, thereby minimize crop spoilage and microbial growth. Some of the postharvest technologies involve use of chemical and physical methods which are efficient in reducing microbial contamination. Physical methods include Modified Atmosphere Packaging (MAP), nano-composite package, active and intelligent packaging. Other emerging technologies are cold plasma, irradiation, ultra-sound and combined methods.

Temperature management

Temperature is an important factor that influences the postharvest life of FV. It determines FV postharvest quality and has direct influence on growth of spoilage microorganisms. Therefore, pre-cooling is an important step for fresh produce pre-treatment. For effective temperature management removal of field heat is achieved by several methods such as, hydro-cooling, ice toping, evaporative cooling, forced air cooling and vacuum cooling that precede further processing.

On the other hand, low temperatures result into chilling injury and loss of cellular structure caused by collapsing of FV tissues. Chilling injury affects mainly tropical and subtropical FV due to their sensitivity to temperatures below 12°C. Some of the deleterious effects associated with low temperatures on FV, are failure to ripen, pitting, surface and internal discoloration, off flavors and microbial growth.

However, high temperatures are implicated with heat injury and consequences on water soluble micronutrient losses such as, vitamin C and B group vitamins. For instance, high temperatures due to sunlight can cause localized bleaching, sunburn and sunscald.

Heat treatment

Heat treatment include blanching, hot water dip, saturated water vapor, hot dry air and hot water rinse with brushing. Heat treatments can be of short (up to 1 h) or long (up to 4 days) duration at 37–55°C or less than a minute in hot water of up to a temperature of 63°C and is referred as heat conditioning method. Blanching is the pretreatment method before FV are subjected to further processing and storage. Blanching is used to inactivate deteriorating enzymes; maintain freshness color; stabilize texture and nutrients; and achieve sterilization by killing microorganisms. Usually, blanching involves dipping of fruits or vegetables in boiling or near to boiling water for 1–3 min, depending on their size, shape and quantity.

Postharvest benefits of heat treatment are linked to reduction of chilling injury, delayed ripening through inactivation of enzymes, controlling onset of fungal growth, killing of insects and other microorganisms. Heat treatment can be used; to preserve color of FV; to prevent development of off flavors; to firm or harden tomatoes, carrots and strawberries; and to add to prolonged shelf life of plums, peaches, carrots and grapes. Example; mango fruits treated at 50°C retained overall fruit eating quality and weight as compared to untreated at five days of storage in the study conducted in South Africa.

Low temperature treatment (refrigeration and freezing)

Low temperatures treatment is one of the effective methods to preserve quality and extending postharvest life of FV. Low temperatures (refrigeration and freezing) storage can keep FV quality attributes like texture, color, nutrients, aroma and flavor over a long storage period. Refrigeration temperatures ranges from 1–4°C, while freezing temperatures ranges from -18 to -35°C. For tropical FV of the SSA countries, low temperature conditioning is recommended before subjecting them to refrigeration or freezing temperatures. Low temperature conditioning is an alternative approach to prevent chilling injury in stored FV. The approach use intermediate temperatures prior to low temperatures storage to eliminate possibility of low temperature stress which trigger chilling injury.

10

PRE-HARVEST FACTORS AFFECTING QUALITY ON POST-HARVEST LIFE OF FRUITS AND VEGETABLES

Postharvest management starts with pre-harvest managements. Once the fruits are harvested, the overall quality of fresh fruits can hardly be improved but it can be maintained. The final market value of the produce and acceptance by the consumers depends upon the grower's ability to apply best available pre-harvest technology followed by harvesting and then to apply best available postharvest handling practices. The pre-harvest factors influencing postharvest quality are frequency of irrigation, use of fertilizers, pest control, growth regulators, climatic conditions like wet and windy weather, natural climates such as hailing, high wind velocity, heavy rainfall, and tree conditions (age, training pruning, light penetration, etc), which influences overall fruit quality and suitability for storage by modifying physiology, chemical composition, and morphology of fruits. One such pre-harvest factor is spray of Gibberellic acid (10 ppm), if applied at color break stage, results in delay in color development and maintains firmness. This is important because it in extending harvesting period. Similarly, the use of calcium solution as foliar sprays increases firmness of fruits as well as extends the shelf life

PRE-HARVEST FACTORS

1. ***Genetic / variety*** - Varieties with shorter shelf-lives are generally prone to higher post harvest losses. Varieties with thick peel, high firmness, low respiration rate and low ethylene production rates would usually have longer storage life. The cultivars that have ability to withstand the rigors of marketing and distribution will have lesser losses after harvest. Varieties with resistance to low temperature disorders and/or decay-causing pathogens can be stored well for longer duration with minimum storage losses. Hence, while growing horticultural crops, one must choose such varieties that inherently have got good quality and storage potential in addition to the high yield and pest resistance potential.

2. ***Light*** - light regulates several physiological processes like chlorophyll synthesis, phototropisum, respiration and stomatal opening. The duration, intensity and quality of light affect the quality of fruits and vegetables at harvest. Most of the produce needs high light intensity (3000-8000 f.c.). Absorption of red light (625-700 nm) through pigments, phytochrome, is essential for carbohydrates synthesis which determines the shelf life of the produce. The vase life of the carnation and chrysanthemums is longer under high light intensity than low.

   Citrus and mango fruits produced in full sun generally had a thinner skin, a lower weight, low juice content and lower acidity but a higher TSS. And citrus fruits grown in the shade may be less susceptible to chilling injury when subsequently stored in cold storage.

   In tomatoes, leaf shading of fruits produced a deeper red colour during the ripening than in the case of those exposed to light. The side of the fruit that have been exposed to sun will generally firmer than the non exposed side. In general, the lower the light intensity the lower the ascorbic acid content of plant tissues. In leafy vegetables, leaves are larger and thinner under condition of low light intensity.

3. ***Temperature*** - all type of physiological and biochemical process related to plant growth and yield are influenced by the temperature. The higher temperature during field conditions decreases life and quality of the produce. At high temperature, stored carbohydrates of fruits, vegetables and flowers are quickly depleted during respiration and plant respires at the faster rate. The produce which is having higher amount of stored carbohydrates show longer storage/vase life. For example- high temperature during fruiting season of tomato leads to quick ripening of fruits on and off the plant.

   Orange grown in the tropics tend to have higher sugars and TSS than those grown sub tropics. However, tropical grown oranges tend to be green in colour and peel less easily and it is due to the lower diurnal temperature that occurs in the tropics.

4. ***Humidity*** - High humidity during growing season results in thin rind and increased size in some horticultural produce and this produce is more prone to high incidence of disease during post harvest period. Humid atmosphere may cause the development of fungal and bacterial diseases, which damages produce during storage and transport. Damaged produce remove water very quickly and emit a larger concentration of ethylene than healthy ones. Low humidity may cause browning of leaf edge on plants with thin leaves or leaflets. High humidity can maintain the water - borne pollutants in a condition so that they can be more easily absorbed through the cuticles or stomata's. Reduced transpiration leads to calcium and other elemental deficiency.

5. ***Mineral nutrition*** - balanced application of all nutrient elements is necessary for the maintaining growth and development of the plants. The application

of fertilizers to crops influences their post harvest respiration rate. Excess or deficiency of certain elements can affect crop quality and its post harvest life. Numerous physiological disorders are also associated with mineral deficiencies which ultimately lead to post harvest losses.

Nitrogen - High N fertilization reduces while moderate to high K improves PH life and quality of anthurium, cut flowers and many horticultural produce. Application of K in water melon tend to decrease the PH respiration. High levels on N tend to decrease flavor, TSS, firmness and color of the fruit and in stone fruits it increases physiological disorders and decrease fruit colour.

Generally, crops that have high levels of nitrogen typically have poorer keeping qualities than those with lower levels as. High nitrogen increases fruit respiration, faster tissue deterioration thereby reducing their storage life.

- ***Phosphorous*** - Application of phosphorous minimizes weight loss, sprouting and rotting in bulb crops compared with lesser application. Phosphorous nutrition can alter the post harvest physiology of some produce by affecting membrane lipid chemistry, membrane integrity and respiratory metabolism. The respiration rate of low-phosphorous fruits will be higher than that of high phosphorous fruits during storage.
- ***Potassium*** - potassic fertilizers improves keeping quality, its deficiency can bring about abnormal ripening of fruits and vegetables. Potassium helps in reducing some physiological storage disorders, e.g. superficial rind pitting in oranges.
- ***Calcium***- the storage potential of the fruits is largely dependent on the level of Ca and it is associated with produce texture. The higher level of N, P and Mg and low levels of K and Bo lead to the Ca deficiency in fruits and reduce its storage life. Reduction in calcium uptake causes lateral stem breakage of poinsettia. Calcium treatment delays ripening, senescence, reduces susceptibility to chilling injury, increase firmness and reduces decay subsequent to storage in avocados and improves the quality.

Physiological disorders of storage organs related to low Ca content of the tissue are

- Bitter pit in apples
- Cork spot in pears
- Blossom end rot in tomato
- Tip burn in lettuce and hallow heart in potato etc.
- Red blotch of lemons

Zn is known to act as vehicle for carrying ions across tissue and increase Ca content of the fruit.

- Adequate supply of Bo improves the mobility of Ca in the leaves and the fruits and subsequently increases fruit firmness, TSS, organic acids and reduce the incidence of the drought spot, bitter pit and cracking disorders. And impart diseases resistance.

- The incorporation of 4% Ca into proto pectin of middle lamella form bond with the cellulose of the cell wall and thus delayed softening in fruits.
- Infused Ca inhibits the internal browning, retarded respiration, and reduced the metabolism of endogenous substrates. Post climacteric respiration of apple decreased as peel Ca level increased from 400 to 1300 ppm. Ca may reduce the endogenous substrate catabolism by limiting the diffusion of substrate from vacuole to the respiratory enzymes in the cytoplasm (limited membrane permeability).

STORAGE DISORDER AND STORAGE CHARACTERISTIC OF COX'S ORANGE PIPPIN APPLE IN RELATION TO THEIR MINERAL CONTENT

Disorder	Composition (mg 100 g-1)				
	N	P	Ca	Mg	K/Ca
Bitter pit			< 4.5	>5	>30
Break down		<11	< 5		>30
Lenticel blotch pit			<3.1		
Loss of firmness	>80	<11	< 5		
Loss of texture		<12			

Application of $CaCl_2$ delayed the accumulation of free sugars, decreased inorganic contents, mold development, softening and development of red colour in strawberry. In pears reduced cork spot, increased flesh firmness, total acidity and juiciness and in apple even after 90 day of storage at ambient condition shown acceptable quality.

6. ***Water relation and Irrigation*** - stress due to excessive or inadequate water in the medium reduce the longevity of the produce. Crop like carnation require 850 to 1200 g of water to produce one gram of dry matter. In general, <5 % of water absorbed in the plant system is utilized for the development of different plant components. Moisture stress increases the rate of transpiration over the rate of absorption and irregular irrigation/ moisture regime leads fruits/vegetable cracking (potato and pomegranate cracking). Higher level of moisture stress affects both yield and quality by decreasing cell enlargement.

Crops which have higher moisture content generally have poorer storage characteristics. An example of this is the hybrid onions, which tend to give high yield of bulbs with low dry matter content but which have only a very short storage life. If fully matured banana harvested soon after rainfall or irrigation the fruit can easily split during handling operations, allowing micro organism infection and PH rotting.

If orange is too turgid at harvest (early morning) the flavdeo/oil gland in the skin can be ruptured during harvesting , releasing phenolic compounds and causes Oleocellosis or oil spotting (green spot on the yellow/orange coloured citrus fruit after degreening).

Quailing - 'harvested produce is kept in the basket for few hours in the field before being transported to pack house, this will allow the produce to loose little moisture'. Some growers have practice of harvesting lettuce in the late in the morning/ early afternoon because when they are too turgid the leaves are soft and more susceptible to bruising.

In green leafy vegetables, too much rain or irrigation can results in the leaves becoming harder and brittle, which can make them more susceptible to damage and decay during handling and transport.

Mango hot water treatment is better if there is delay of 48 hr. between harvest and treatment and resulted better efficiency of hot water in disease control.

Generally, crops that have higher moisture content or low dry matter content have poorer storage characteristics. Keeping quality of bulb crops like onion and garlic will be poor if irrigation is not stopped before three weeks of harvesting.

7. Canopy Manipulation

A. Fruit thinning - increases fruit size but reduces total yield. It helps in obtaining better quality produce

B. Fruit position in the tree - Fruits which are exposed to high light environment possesses higher TSS, acidity, fruit size, aroma, and shelf life compared to which lies inside the canopy. Hence better training system should be practiced to circulate optimum light and air, e.g.: Grapes, Mango, peaches, kiwifruits

C. Girdling - increases the fruit size and advance and synchronized fruit maturity in peach and nectarines. Increases fruitfulness in many fruit tree species.

8. ***Rainfall*** - Rainfall affects water supply to the plant and influences the composition of the harvested plant part. This affects its susceptibility to mechanical damage and decay during subsequent harvesting and handling operations. On the other hand, excess water supply to plants results in cracking of fruits such as cherries, plums, and tomatoes. If root and bulb crops are harvested during heavy rainfall, the storage losses will be higher.

9. ***Seasons / Day and day length-*** seasonal fluctuation and time of the day at harvest will greatly affects the postharvest quality of the produce. Synthesis of higher amount of carbohydrates during the day time and its utilization through translocation and respiration in the night is responsible for the variation in the longevity of the cut flowers. Roses and tuberose have been found to show longer keeping quality in the winter season under ambient condition than in the summer seasons.

Generally produce harvested early in the morning or in the evening hours exhibits longer PH life than produce harvested during hot time of the day.

Day length - If long days Onion (temperate) grown during short day (tropics) condition it leads to very poor storage quality.

10. ***Carbon dioxide*** - quality planting material, early flowering, more flowering, increased yield and rapid crop growth and development at higher level of

CO_2. Production of chrysanthemum under green house at 1000 - 2000 ppm of CO_2 showed an increase in stem length, fresh weight, leaf no. and longevity of cut flowers.

11. ***Use of Agro chemicals*** - Pre-harvest application of chemicals such as BA, IAA, GA3, growth retardants like B-9, CCC, A-Rest and Phosphon-D have bee reported to improve quality and longevity of flowers crops. Application of GA3 @ 50-100 ppm improves PH quality of roses by anthocyanin development. And it stimulate the accumulation of N, K, Mg and S. Pre-harvest spray with Alar(1500ppm), MH(500ppm), and Cycocel(500ppm) increased vase life of Aster. Beneficial effect of leaf manure, K and GA3 is found to enhance the longevity of tuberose flowers.

    Use of chemicals on the plants to prevent the pathogen will have direct impact on extending the postharvest life. Generally, if produce has suffered an infection during development its storage or marketable life may be adversely affected. Banana which suffers a severe infection with diseases such as leaf spot may ripen pre maturely or abnormally after harvest and in mango it is rapid postharvest loss. Pre harvest application chemicals like MH on onion filed prevent them sprouting during storage.

12. ***Pest and Diseases*** - infection by fungi, bacteria, mites and insects reduces the longevity as well as consumer acceptability. Tissue damage caused by them show wilting and produce ethylene leads to early senescence. Vascular diseases/stem rot /root rot of floral corps hinder the transport, affects the post harvest life and quality. The potato tuber moth may infest tubers during growth if they are exposed above the soil and subsequently in the storage.

QUALITY OF POST HARVEST PRODUCT

Post harvest quality represents market quality, edible quality, transport quality, table quality, nutritional quality, internal quality and appearance quality. Quality means a combination of characteristics, attributes and properties that gives the values to human and enjoyments. Consumers consider good quality in relation to colour, flavour and nutrition. Quality of the produce is the final manifestation of inter-relation between the commodity and its environment. The genetic characteristics and physiological status of the commodity determine the typical post-harvest behavior and quality of the produce and these two are the major bases for the interaction. Pre-harvest factors viz, environmental factors such as temperature, relative humidity, water potential, light, cultural practices and pest management techniques determined the inherent quality of the produce.

11

POSTHARVEST DISEASES AND PHYSIOLOGICAL DISORDERS

Losses due to postharvest disease may occur at any time during postharvest handling, from harvest to consumption. When estimating postharvest disease losses, it is important to consider reductions in fruit quantity and quality, as some diseases may not render produce unsaleable yet still reduce product value. For example, blemished fruit may not be sold as fresh fruit but may still be suitable for processing, in which case, it brings a lower price. It is also important to take into account costs such as harvesting, packaging and transport when determining the value of produce lost as a result of postharvest wastage. Aside from direct economic considerations, diseased produce poses a potential health risk. A number of fungal genera such as *Penicillium*, *Alternaria* and *Fusarium* are known to produce mycotoxins under certain conditions. Generally speaking, the greatest risk of mycotoxin contamination occurs when diseased produce is used in the production of processed food or animal feed. In most cases, fresh produce which is obviously diseased would not be consumed.

Losses due to postharvest disease are affected by a great number of factors including:

- commodity type
  - cultivar susceptibility to postharvest disease
  - the postharvest environment (temperature, relative humidity, atmosphere composition, etc.)
- produce maturity and ripeness stage
  - treatments used for disease control
- produce handling methods
  - postharvest hygiene.

CAUSES OF POSTHARVEST DISEASE

Correct identification of the pathogen causing postharvest disease is central to the selection of an appropriate disease control strategy. Many of the fungi which cause postharvest disease belong to the phylum Ascomycota and the associated Fungi Anamorphici (Fungi Imperfecti). In the case of the Ascomycota, the asexual stage of fungus (the anamorph) is usually encountered more frequently in postharvest diseases than the sexual stage of the fungus (the teleomorph). Important genera of anamorphic postharvest pathogens include *Penicillium*, *Aspergillus*, *Geotrichum*, *Botrgtis*, *Fusarium*, *Alternaria*, *Colletotrichum*, *Dothtorella*, *Lastodiplodia* and *Phomopsis*. Some of these fungi also form ascomycete sexual stages.

In the phylum Oomycota, the genera *Phytophthora* and *Pythium* are important postharvest pathogens, causing a number of diseases such as brown rot in citrus (*Phytophthora citrophthora* and *P. parasitica*) and cottony leak of cucurbits (*Pythium* spp.). *Rhizopus* and *Mucor* are important genera of postharvest pathogens in the phylum *Zygornycota*. *R. stolonifer* is a common wound pathogen of a very wide range of fruit and vegetables, causing a rapidly spreading watery soft rot. Genera within the phylum *Basidiomycota* are generally not important causal agents of postharvest disease, although fungi such as *Sclerotium rolfsii* and *Rhizoctonia solani*, which have basidiomycete sexual stages, can cause significant postharvest losses of vegetable crops such as tomato and potato. While diseases caused by these pathogens are primarily field diseases, the development of symptoms often accelerates after harvest.

The major causal agents of bacterial soft rots are various species of *Ertuinia*, *Pseudomonas*, *Bacittus*, and *Xanthomonas*. Bacterial soft rots are very important postharvest diseases of many vegetables, although they are generally of less importance in most fruit. This is because most fruit have a low pH which is inhibitory to the majority of bacterial plant pathogens.

ECONOMICAL AND HEALTH IMPACT OF POSTHARVEST DISEASES

Postharvest diseases cause qualitative and quantitative losses of vegetables and make them unfit for human consumption due to potential health risks. A large number of postharvest diseases are caused by black, white, and yellow fungi-derived carcinogenic mycotoxins and mutagenic secondary metabolites. Losses due to postharvest disease may occur during the handling of produce from harvest to consumption. Primary and secondary agricultural practices are also important and costs such as harvesting, packaging, and transport must be taken into account when estimating the value of the produce lost as a result of postharvest wastage.

Fresh vegetables are highly perishable, and they have relatively short shelf lives. Fresh vegetables are living, respiring tissues that start senescing immediately after harvest. They are mostly comprised of water, with most having 90–95% moisture content. Because of the perishable nature of vegetables, special skills are required for postharvest handling. *Aspergillus flavus* is a saprophytic soil inhabitant fungus that infects postharvest vegetables and produces carcinogenic secondary metabolite aflatoxin in tropical, subtropical, and temperate geographic regions of the world.

It also causes animal and human diseases (causing aflatoxicosis and/or liver cancer) due to consumption of contaminated food and feed and through invasive growth (causing aspergillosis), which is often fatal to humans who are immunocompromised. A holistic approach is needed for regulating aflatoxins under the trade/export market with biosecurity including bio-warfare, biodiversity, and biosafety for liberalized trade under the World Trade Organization (WTO).

CHALLENGES OF POSTHARVEST LOSSES IN CROPS

Application of good postharvest management practices which are supported by good technologies and also improving postharvest systems will maintain the quality of vegetables and reduce quantitative losses. Losses in vegetables are the result of

(i) Poor knowledge about the right harvesting index; thus, a large proportion of the harvested beans are usually over-mature

(ii) Poor handling practices, such as the use of plastic sacks for bulk packaging and transportation which results in mechanical damage that serves as entry points for disease-causing organisms leading to rotting of the pods

(iii) Poor transport practices such as the use of trucks that have no cover, thus exposing the produce to direct sunlight and high temperature

(iv) The absence of low-temperature storage facilities and transport systems, and

(v) Rough handling practices during distribution in retail markets.

DETECTION OF POSTHARVEST PATHOGENS

Pathogens were isolated on agar medium and identified on the basis of macroscopic and microscopic analysis of colony and conidia/spore morphology by Microscopy, Sero-diagnostics (ELISA, Dot-blot assays), and nucleic acid (PCR) based methods.

Why do we need, want, or should detect emerging postharvest pathogens (diseases) in vegetable crops?

- Determine presence and quantity of the pathogen (s) for quarantine legislation.
- Assess the effectiveness of Integrated Disease Management (IDM) modules.
- Issuing of Sanitary and Phytosanitary (SPS) certificate vegetable produce for safe export/transboundary movement under trade.
- Quantify spatial and temporal pathogen populations in a specific location.
- Quantify pathogen populations in relation with regional and seasonal yield losses.

PHYSIOLOGICAL DISORDERS

Apple: Bitter Pit

Symptoms

Small brown lesions of 2-10 mm in diameter (depending on the cultivar) develop in the flesh of the fruit. The tissue below the skin becomes dark and corky. At harvest or after a period of cold storage the skin develops depressed spots on the surface. These most often start to appear as water soaked spots on the skin near the calyx. These spots generally turn darker and become more sunken than the surrounding skin and are fully developed after one to two months in storage.

Fruit located on vigorous, leafy, upright growing branches have a greater potential to develop bitter pit than does fruit that develops from spurs or on horizontal wood near the tree's main frame. Young trees that are just coming into bearing are the most susceptible. Immature fruit are more susceptible to bitter pit than fruit harvested at the proper harvest maturity. Older trees, which are less vigorous and produce larger crop loads, reduce their susceptibility to bitter pit except in the very sensitive cultivars.

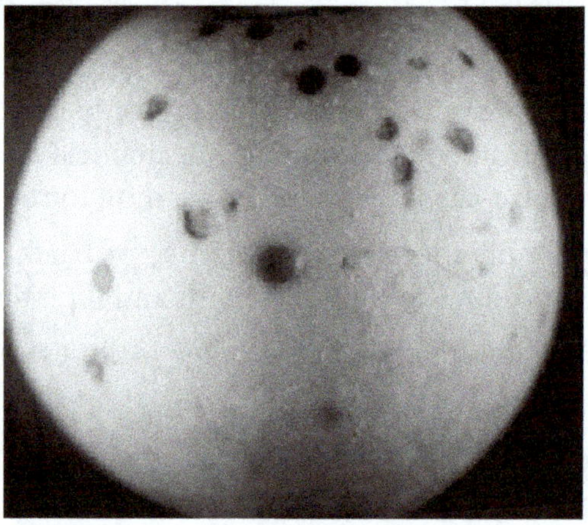

Fig. 1. Bitter Pit. Apple

Highly Susceptible Cultivars

Granny Smith, Yellow Newtown, Jonathan, Golden Delicious (California), Gravenstein, Red Delicious (California), Baldwin, Boskoop, Bromley's Seedling, Cleopatra, Cox's Orange Pippin, Grimes Golden, Merton, Rhode Island Greening, Worcester, Stayman (Canada), Sturmer, White Winter Pearmaine, Prima Starking, Starkrimson, Marigold, Northern Spy, and York Imperial.

Less Susceptible Cultivars

Rome Beauty, Golden Delicious (Canada), Macintosh, Spartan, Haralson, Lobo, Winesap (California), Red Gravenstein, Stonetosh, Gala, and Fuji.

Physiology

Initiation of symptoms may begin four to six weeks after petal fall when affected tissues have a higher rate of respiration and ethylene production. This is a period of greater protein and pectin synthesis with greater migration of organic ions into the affected areas. Affected areas retain starch grains not seen in healthy tissue. A mineral imbalance in the apple flesh develops with low levels of calcium and relatively high concentrations of potassium and magnesium. Low levels of calcium impair the selective permeability of cell membranes leading to cell injury and necrosis. Other explanations for the cause of bitter pit include the dissolution of the middle lamellae by oxalic and succinic acid, and changes in proton secretion and potassium permeability.

Control

Cultural practices that reduce the incidence of bitter pit are annual bearing, moderate tree vigor, smaller fruit size, calcium sprays, summer pruning and harvesting mature fruit. Avoid nitrogen and magnesium summer sprays, fertilizer treatments that result in lowering the soil pH or induce excess vigor, and fluctuating soil moisture. Early thinning and over thinning can increase bitter pit. Do not over dormant prune, which would result in a light crop and large fruit.

Summer sprays of calcium chloride($CaCl_2$), calcium nitrate [$Ca(NO_3)_2$] and/or a postharvest dip in a calcium solution are recommended. Both $CaCl_2$ and $Ca(NO_3)_2$ summer sprays are effective at reducing the incidence of bitter but the response varies with the cultivar and both may cause russeting and leaf burn under certain conditions. In California, apply at least three sprays at one month intervals beginning in mid-June. Use 20 pounds of $Ca(NO_3)_2$ per acre in 400 gallons of water or 12 pounds of $CaCl_2$ per acre in 400 gallons of water.

Apply sprays under conditions of rapid drying to prevent russeting. Orchards with a history of severe bitter pit should receive four to six sprays or more at two week intervals. Fruit coverage is mandatory. Postharvest dips before storage may reduce bitter pit symptoms. Use 2-3% by weight $CaCl_2$ in the drench water. Calcium dips can cause fruit injury, follow the label recommendations carefully.

Preharvest sampling for bitter pit is possible. Two weeks before harvest select large fruit from upright limbs of light cropped, vigorous trees. Dip the fruit in a solution of 2,000 ppm ethephon in water (about 1½ teaspoons of ethephon to one gallon of water) to hasten the ripening process. Hold the fruit for two weeks at room temperature. If bitter pit develops, delay the harvest as long as possible. Cool the harvested fruit as soon as possible and delay the packing for at least four weeks. The delay will allow the bitter pit to fully develop. The affected fruit is then removed during the packing process.

Apple: Internal Browning

Symptoms

Brown discoloration in the flesh, firm but moist, usually originating in or near the core. Brown areas have well defined margins and may include dry cavities resulting from desiccation. Symptoms can range from a small spot of brown flesh to nearly the entire flesh being affected in severe cases. When the entire apple is affected, a margin of healthy, white flesh usually remains just below the skin. Symptoms develop early in storage and may increase in severity with extended storage time.

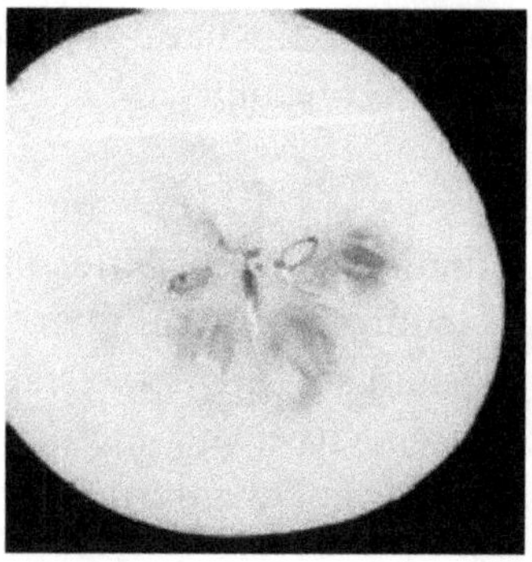

Fig. 2. Internal Browning. Apple

Physiology

Evidence to date indicates that internal browning occurs as a result of CO_2 injury to the apple. Injury incidence and severity increase with increasing concentrations of CO_2 in the storage atmosphere. Variability in susceptibility of apple varieties and in apples of different maturities may be due to anatomical differences (cell size, size of intercellular spaces) rather than biochemical differences. Later harvested fruit generally have greater susceptibility to internal browning and it has been shown in many fruit that diffusivity of CO_2 decreases as fruit mature and ripen. A reduction in the ability of CO_2 to diffuse out of the apple fruit would result in higher internal CO_2 concentrations and thus more chance of injury. Internal browning can be more severe in fruit when watercore is also present. The accumulation of sorbitol in the intercellular spaces would result in reduced gas diffusion in the affected tissue. Some reports suggest that internal browning injury is more severe at 0°C (32°F) than at 2.2°C (36°F). CO_2 solubility is greater at lower temperatures; however, the influence of temperature requires additional study before warmer storage temperatures could be safely recommended. Apple fruit can accumulate acetaldehyde and ethanol as well as various organic acids during exposure to high CO_2 concentrations; however, it is not known if these are causal agents of fruit injury.

Control

Avoid harvesting overmature fruit. Harvest at the optimum maturity, especially for controlled atmosphere storage. Maintain CO_2 concentrations below %1 in controlled atmosphere storage and air storage. CO_2 concentrations of less than %0.5 are recommended for Fuji apples in California. Assure good air circulation in storage rooms to prevent pockets with higher CO_2 concentrations. CO_2 can develop to damaging levels in air storage or marine containers if fruit temperature is °3C or warmer and ventilation is not good. Proper temperature management and good ventilation will prevent buildup of CO_2. Avoid heavy wax coatings and thoroughly and rapidly cool fruit after waxing and packaging.

Apple: Scald

Symptoms

Irregular brown patches of dead skin which can become rough when severe, developing within 3 to 7 days upon warming of the fruit following cold storage. The warm temperatures do not cause the scald but allow symptoms to develop from previous injury which occurred during cold storage. Symptoms may be visible in cold storage when injury is severe. In this case, the symptoms intensify upon warming the fruit. Scald is usually not evident until after 3 months of storage. Storage scald can involve only the skin or extend 6 mm (¼-inch) into the flesh depending on the variety. Scald can be more severe on the greener side of the fruit.

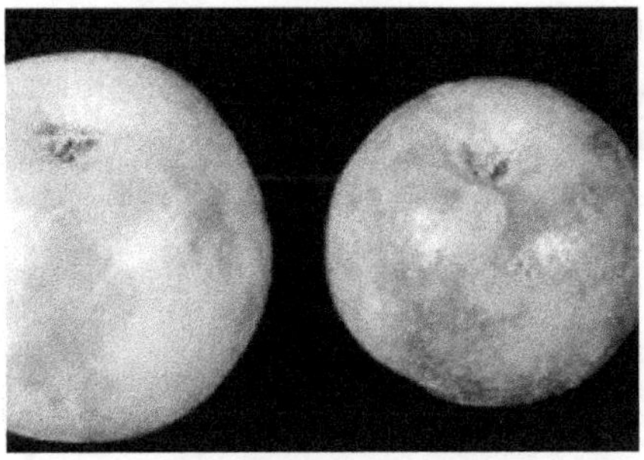

Fig. 3. Scald - Apple

Physiology

There continues to be much debate about the cause of storage scald in apples, but most agree that storage scald is a type of chilling injury. The general theory is that alpha-farnesene, a naturally occurring volatile terpene in the apple fruit, is oxidized to a variety of products (conjugated trienes). These oxidation products result in injury to the cell membranes which eventually result in cell death in the outermost cell layers of the fruit. Ethylene promotes the formation of alpha-farnesene and oxygen is required to oxidize alpha-farnesene to conjugated trienes. Both ethylene scrubbing and low-oxygen storage reduce the incidence of storage scald. The reasons for variability in susceptibility to storage scald among apple varieties is not known; however, the concentration of naturally occurring antioxidants in the apple skin has been investigated. These differences between apple varieties may hold the key to the cause of storage scald.

Control

For many years, proper harvest maturity, ventilation in cold storage and 15% mineral oil wraps were the only methods available for scald control and control was incomplete. Mineral oil wraps were shown to absorb the volatile alpha-farnesene and ventilation also prevented accumulation of alpha-farnesene in the storage room air. Today, the most common method used to control scald is application of an antioxidant immediately after harvest. Diphenylamine (DPA) is commonly used. Ethoxyquin is also effective for some varieties, but can cause damage to some apple varieties (check labels for recommendations).

Antioxidants should be applied within one week of harvest for maximum control. Wait 16 hours after DPA application before cooling of fruit (unless bitter pit is of great concern as cooling delays can increase bitter pit incidence). Low oxygen controlled atmosphere storage can provide a non-chemical control method in some cases. Oxygen concentrations between 0.5 and 1.0% can significantly delay the development of scald, perhaps up to 10 months. Fruit tolerance to such low oxygen levels as well as their effectiveness for scald control must be tested prior to use.

Apple: Sunburn

Symptoms

Initial symptoms are white, tan or yellowed patches found on the sun exposed side of the fruit. With severe skin damage, injured areas can turn dark brown on the tree. White, tan or yellow patches often turn brown within a few weeks in cold storage (sometimes called sunburn scald but not controlled by antioxidants, such as diphenylamine, as is storage scald).

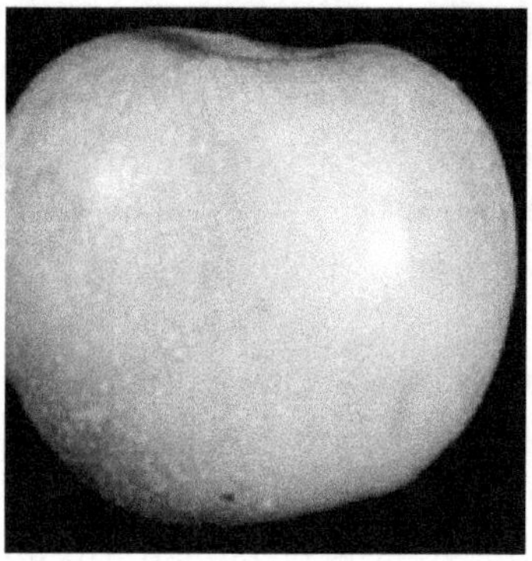

Fig. 4. Sunburn

During unusually hot weather following cool or moderate conditions, and especially when accompanied by water stress, injury to the skin and flesh can occur. Injured cortex tissue is brown and firm and may become spongy and sunken. Fruit exposed to the sun after removal from the tree, either on the orchard floor or in field bins, can develop severe sunburn.

Physiology

Sunburn results from heat stress to the fruit leading to injury of the affected cells. The temperature of the surface of the fruit can be as much as 18°C (32°F) above air temperatures when the fruit is exposed to solar radiation and 8 to 9°C (14 to 16°F) warmer than the shaded side of the fruit. The transpiration or evaporation of water from the apple fruit helps to cool the fruit while it is attached to the tree. Upon harvest from the tree or during periods of water stress on the tree, fruit surfaces and flesh can increase to much higher temperatures resulting in greater sunburn.

The greatest problem occurs when fruit are suddenly exposed to high temperatures and intense sunlight such as when the weather rapidly shifts from being cloudy and cool for many days to sunny and warm. The shifting of branches during the season due to heavy fruit loads can suddenly expose fruit to the sun as can harvesting fruit into field bins which are left in the sun. A gradual increase in temperature and solar radiation could allow the fruit to adapt whereas a sudden change does not allow for adaptation and often results in injury.

Control

The best method of control is to avoid sudden exposure of fruit to intense heat and solar radiation. Proper tree training and pruning are critical. Summer pruning must be carefully accomplished to avoid excessive sunburn. Overhead sprinkling and white washing of trees have both been tested with mixed success in reducing sunburn. While some reports indicate that overhead sprinkling reduces sunburn, the potential benefit must be weighed with potential increases in apple scab and fireblight.

White washing showed no benefits in tests in California. Avoid water stress in cropped orchard trees to reduce heat stress. Once injury has occurred, there is no treatment to prevent browning in storage. The browning which occurs in storage as a result of sunburn cannot be prevented with diphenylamine (DPA) drenches. Careful sorting to remove affected fruit upon packing is the only solution once the injury has occurred. Injury will be easier to spot after a few weeks of storage because of the change in color from yellow to brown.

Apple: Watercore

Symptoms

A preharvest disorder resulting in water soaked regions in the flesh, hard and glassy in appearance, only visible externally when very severe. Water soaked areas are found near the core and around the primary vascular bundles but may occur in any part of the apple or involve the entire apple. Symptoms often increase rapidly as fruit become overmature but does not increase postharvest. If symptoms are mild to moderate, they may disappear completely in storage. However, when watercore is severe, internal breakdown can develop.

Physiology

The water soaked appearance of watercore affected fruit results from the accumulation of sorbitol-rich solutions in the intercellular spaces. Sorbitol is the carbohydrate source translocated into the fruit from the tree. Sorbitol must be converted to fructose by the apple fruit. The reason for accumulation in the intercellular spaces is not known. It is speculated that sorbitol may be translocated to the fruit faster than it can be assimilated perhaps due to an inability of the apple tissue to convert sorbitol to fructose. Comparative studies have shown differences in vascular tissues between susceptible and resistant varieties. The browning and breakdown that results from severe watercore is likely due to reduced gas diffusion in the affected tissue and may involve an accumulation of ethanol and acetaldehyde.

Control

The most effective way to reduce the incidence of water core is to avoid delayed harvests. As fruit approach harvest maturity, samples of fruit from the southwest quadrant of the tree should be cut to look for water core. Fruit should be harvested before water core develops extensively. A subsample of fruit at harvest should be cut prior to storage. Lots with moderate to severe water core should not be placed in CA storage but should be marketed quickly.

Asian Pear: Flesh Spot Decay

Symptoms

Partial browning of spots and/or development of cavities in Asian pear flesh. It appears along and around the vascular bundles when the symptoms are severe, but there is no external indication of the disorder. Generally, Flesh Spot Decay is more pronounced above the equator of the fruit (towards the stem end), but it can also be observed all the way down to the calyx. Cavities are usually dry and surrounded by apparently healthy tissue. This disorder can occur in fruit while still on the tree. It is more obvious, during cold storage.

Causes

The cause of flesh spot decay is still unknown. However, climatic factors, such as a fluctuating hot and cool summer, or high rainfall right before harvest increased incidence of this disorder.

Control

Since definite causes have not been identified; there is no effective way to control Flesh Spot Decay. The inability to predict or diagnose Flesh Spot Decay without cutting the fruit makes it difficult to study the problem. Further research needs to be carried out to determine the causes and variety susceptibility, and to understand preharvest and/or postharvest practices that will reduce Flesh Spot Decay losses. Meanwhile, AVOID whenever possible the following conditions that might induce Flesh Spot Decay: low crop load and therefore large fruit; advanced maturity; sunburn; erratic irrigation (amount and timing) near harvest; harvesting fruit during warm temperatures and cooling the fruit within 24 hours.

Asian Pear: Internal Browning

Symptoms

Development of brown to dark brown water-soaked areas in the core and/or flesh occur during storage of Chinese pears. A high incidence of internal browning symptoms have been reported after 2-3 months of cold storage of 'Hosui' and 'Shinko'. Unfortunately, there is no visible external indication of internal browning.

Causes

The cause of internal browning is still unknown. Ripening–senescence predisposes the fruit to the disorder while it is in cold storage.

Control

To avoid internal breakdown development during cold storage and marketing, pick fruit grown under dry, hot, and long summer conditions of the San Joaquin Valley conditions before 180 days (3000 degree-days) after full bloom.

Pick fruit when most of the pears on the tree are still green, although a few at the top may begin to develop some light-yellow spots and smooth surface.

Fruit picked when the skin is completely yellow and smooth will develop internal breakdown after one or two months in cold storage.

We recommend cooling immediately at harvest, as cooling delays yellowing of the pears which results in increased incidence of internal browning in Chinese pears.

We suggest that growers who are having internal breakdown problems in Chinese pears keep records of the number of days after full bloom (degree-days) needed for their fruit to start turning light yellow to help them decide when to pick in subsequent years.

Cherry: Surface Pitting & Bruising

Symptoms

Pits are small sunken areas on the fruit surface, larger flattened areas are called bruises.

Physiology

Symptoms are primarily caused by a mechanical impact or compression. Although any other mechanism that damages fruit tissue may cause similar symptoms. Damage becomes apparent after several days at room temperature or longer at lower temperatures. Pitting is associated with damage to cells near the epidermis which collapse over time. Bruising appears to be associated with injury to cells much below the epidermis. Impact damage increases cherry respiration, ethylene production, and susceptibility to decay and damaged fruit loose titratable acidity at a faster rate compared to undamaged fruit.

Control

Damage can occur during picking, handling, transport to the packinghouse, packing, or transport to market. Fruit factors such as high soluble solids concentration, warm temperature, preharvest use of gibberillic acid, and high fruit weight have been shown to reduce fruit susceptibility to damage. Postharvest pitting damage can be reduced by:

1. Keeping drop heights lower than 30 cm (12") for fruit landing on a smooth surface,
2. Preventing fruit from falling on rough surfaces,
3. Installing padding,

4. Keeping water drop heights less than 20 cm (8") in shower type hydrocoolers, and
5. Using the slowest possible fruit speeds on cluster cutter machinery.

Bruising is mainly caused by pickers holding fruit when they remove it from the tree. They should be trained to touch only fruit stems. Packinghouse design features that reduce pitting also minimize bruising.

Grape: Berry Shrivel

Symptoms

After ripening begins, Emperor berries become flaccid and sunken. Some affected berries develop color; some remain white. The flaccid berries are usually interspersed with normal ones, but occasionally several berries at the tip of a lateral or at the cluster apex are affected.

With Calmeria, whole clusters may be involved. Sometimes the disorder is difficult to recognize. The clusters may be a slightly duller green but the berries must be touched to confirm the flaccid condition. The amount of fruit involved usually increases up to harvest.

Physiology

The cause is unknown. With Emperor, 16 grams per acre of gibberellin applied about two weeks after fruit set when the berry diameters average 10-15 mm will reduce the shrivel by 50 to 70% compared to the untreated. Similar treatments are not tolerated by the Calmeria variety.

Grape: Shatter

Symptoms

Shatter refers to the loose berries, those that have detached from the stem.

Physiology

For some reason, seedless varieties are usually less well attached to the capstem than seeded varieties. This is defined with less of a "brush" than in seeded varieties. When a seeded variety is pulled from the capstem, a substantial amount of flesh remains attached to the stem. Conversely, a seedless variety may be removed with little or no flesh. The force required to remove a seedless berry from its capstem is much below that required for a seeded berry.

In general, shatter increases in severity with increasing maturity and/or the longer the fruit remains on the vine. It is also believed to be associated with any stress to which the fruit is subjected such as heat, water insufficiency, weakly growing vines, etc. Shatter varies considerably from season to season, and among vineyards.

Girdling at fruit set, as done for berry weight increase, also increases the strength of the berry attachment to the capstem and so reduces the shatter potential. However gibberellin, also applied at fruit set, weakens this attachment and is dosage dependent. As the rate of gibberellin is increased, the berry weight benefit decreases with each additional increment, but the shatter potential continues to increase.

Shed packing, with some delay between harvesting and packing, decreases the shatter somewhat as compared to field packing where the fruit is packed soon after harvest. Much of the shatter occurs during the packing operation itself with additional shatter all the way to the final retail sale. Bagging the clusters reduces the shatter during packing. Also the loose berries are inside the bag and are sold along with the cluster. With bagging less fruit is packed per box.

Grape: Waterberry

Symptoms

The affected berries become watery, soft, and flabby during ripening due to the interrupted flow of sugar and other constituents into the berries. The affected berries tend to have a more metallic, opaque, or dull green appearance; berry color will be lighter and more variable in red and black varieties. This results from the breakdown and necrosis of the cluster stem structure supporting cluster parts as well as individual berries. Ultimately, water movement may be affected, causing berry shriveling and drying. The earliest symptom is the development of small (1-2 mm) dark spots on the cap stems (pedicles) and/or other parts of the cluster framework. These spots become necrotic, slightly sunken, and expand to affect more areas. The cluster tips, shoulders, and upper laterals are most affected; the entire framework may be involved in severely affected clusters. The nectroic cluster stem tissue will have a dark brown to purplish-black or black color.

Physiology

Waterberry is thought to be a physiological disorder which appears more in certain cultivars, years, and individual vineyards. It is associated with fruit ripening and most often begins to develop shortly after veraison (berry softening).

Studies in California have shown it to be associated with vineyard areas with a high nitrogen status; symptoms can be induced by foliar spraying with nitrogen compounds or fertilizing a low nitrogen status vineyard. Clusters with symptoms have higher total nitrogen and ammonium levels in the stem structure. Vine canopy shading and/or cool weather during veraison (berry softening) and fruit ripening can increase waterberry, further suggesting nitrogen metabolism involvement in this disorder. Other mineral nutrients such as calcium, magnesium, and potassium have not shown a direct relationship to waterberry; applications of these elements have not influenced the incidence.

Control

There is no known control. However, it is recommended to avoid over fertilization with nitrogen due to waterberry's association with a high nitrogen status. Folier nutrient sprays of nitrogen should be avoided in waterberry-prone vineyards. Likewise, ammonium or urea fertilizers should not be applied via drip irrigation near the beginning of or during fruit ripening. Trellising and canopy management practices such as shoot thinning that minimize shading within the canopy may be beneficial. Supplementing other nutrients such as calcium, magnesium, and potassium has not been effective. Crop load differences or gibberellic acid treatments have not been shown to affect waterberry. Ethephon may possibly hasten its development but not increase the overall incidence by harvest. It has been well demonstrated that no pathogen is involved; fungicidal spray treatments are ineffective. Trimming off affected berries during harvest and packing is a common practice, although labor intensive. A delay between picking and packing may enhance the ability to see affected berries for trimming. This is because waterberries will tend to welt first on a cluster. However, the advantage of this practice must be weighed against the effects of water loss from the frtui to be packed.

Olive: Chilling Injury

Symptoms

Internal browning begins in the flesh around the pit and radiates outward toward the skin as time progresses. Skin browning indicates an advanced stage and/or greater severity of CI.

Physiology

CI stimulates respiration and ethylene production rates of fresh olives. Exposure to CO_2 levels above %5 aggravates CI, while %2 O_2 is beneficial in maintaining flesh firmness and green color of the skin in olives kept at 5°C (41°F) or higher temperatures.

Control

Avoid exposure of fresh olives to temperatures below 5°C (41°F). Ideal storage conditions are 5 to 7.5°C (41 to 45°F) and 90-95% relative humidity.

Peach and Nectarine: Field Inking

Symptoms

Field inking symptoms appear as brown and/or black spots or stripes that are restricted to the skin. Symptoms become evident within 48 hours after harvest. However, inking symptoms are triggered during harvest and during transportation to the packinghouse. Field inking usually begins in the field, although symptoms may take more time to become apparent.

Causes

Abrasion damage in combination with heavy metal contamination is required for field inking development. The skin cells, where the anthocyanin/phenolic pigments are located, collapse and their contents react with heavy metals turning their color dark brown/black. Iron, copper and aluminum are the most deleterious heavy metal compounds that can combine with anthocyanin/phenolic pigments (metallo-polyphenol complex). Only 5-10 ppm of iron (Fe) is enough to induce inking at the physiological fruit pH (~ 3.5). This contamination can occur in the field within 15-20 days before harvest or during harvesting operations. When applied close to harvest, pre-harvest foliar nutrient, fungicide and insecticide sprays that contain the above-mentioned metals have the capacity to induce field inking on peach and nectarine fruit when followed by abrasion damage.

Fig. 6. Field Inking

Control of Field Inking:

- Reduce fruit abrasion damage by treating fruit gently, use air-ride suspension on trailers, and avoid long hauling.
- Reduce fruit contamination by keeping picking containers dirt free and clean; avoid dust contamination on fruits.
- Check your water quality for contamination with heavy metals {Fe, copper (Cu) & aluminum (Al)}.
- Test your pesticides for presence of heavy metals (Fe, Cu & Al) early in the season (5-10 ppm Fe is the critical level).
- Do not spray foliar nutrients or preharvest fungicides containing Fe, Cu, or Al within 21 days of predicted harvest.
- Chemical manufacturing companies should attempt to identify and remove from their. Products any potential sources of contaminants that may contribute to inking formation, and to develop safe pre-harvest spray intervals (PHI) for foliar nutrients, fungicides, miticides, and insecticides.
- Growers need to know the composition of the chemicals commonly used in their tree fruit pre-harvest and postharvest operations and understand how they may affect inking incidence.
- In orchards where inking is a problem, delay packing for ~48 hours so you will be able to remove fruit with field inking before placing fruit in the box.
- Fine tune your post-harvest fungicide application to assure that your residues are above the effective minimum recommended, but well below the maximum residue limit MRL) or tolerance.

- As a long-term solution, it is suggested that chemical manufacturers attempt to identify and remove the possible sources of contamination from their products that may cause skin discoloration before distributing them.

Peach and Nectarine: Packinghouse Skin Burning

Symptoms and Causes

Skin burning symptoms appear as brown and/or black areas that are restricted to the skin. In contrast to field inking, these symptoms are mainly triggered during packing operations, principally at the brushing-washing point, although abrasion that occurred prior to packing will also contribute to its development. Specifically, the damage is triggered by exposure to high pH and/or dehydration caused by high-velocity, forced-air cooling during packing. Symptoms can be observed very soon after packing, but the symptoms rapidly increase during cold storage due to dehydration. In fact, it has been observed that most of the intense skin damage in packed fruit occurred on the exposed part of the fruit above the tray receptacle and no damage occurred under the price-look-up (PLU) sticker. Our studies demonstrated that fruit skin phenolics in susceptible cultivars and breeding lines are not stable at high pH leading to phenolic transformations. These transformations at different pH values result from *copigmentation* of phenolics with anthocyanins resulting in a change in color of the anthocyanin compound and therefore discoloration of the skin.

Fig. 7. Packing house Skin Burning

Different susceptibilities to skin burning have been observed among peach and nectarine cultivars, depending mainly on the specific phenolics in their skin tissues.

Control of skin burning

- Minimize physical damage or abrasion on the fruit surface during pre- and/or post-harvest operations. Handle fruit gently, use air-ride suspension on trailers, avoid long hauling distances and keep harvest containers free of dirt.
- In a standard packing operation, washing water pH in the brushing-washing or hydrocooling operation should be continuously maintained around 6.5-7.0. The installation of automated systems using oxidation-reduction potential (ORP) to monitor and/or adjust active/effective chlorine and pH levels is critical to increase disease control effectiveness and decrease potential skin burning development.

- Based on our results, we recommend dry packing (without brushing or a chlorine rinse) for highly susceptible peach or nectarine cultivars.
- Avoid high air velocities during forced-air cooling for skin burning susceptible peach or nectarine cultivars. For these susceptible cultivars, we suggest cooling the fruit by room cooling, without forced-air.
- As a long-term solution, we suggest screening peach and nectarine breeding parents for their susceptibility to copigmentation.

Pear: Core Breakdown

Symptoms

A senescent disorder resulting from storing fruit beyond its postharvest life. A brown, soft breakdown of the core and surrounding tissues. May develop in storage or soon after transfer to warm temperatures. In Bosc, symptoms may be preceded by vascular browning as the fruit ripens. In Bartlett, the watery, brown tissue separates easily from the healthy tissue.

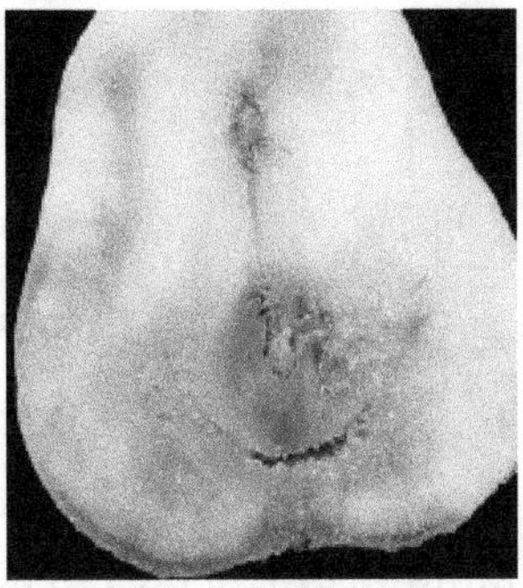

Fig. 8. Senescent Scald and Core Breakdown

Physiology

A part of the normal sequence of ripening and senescence where fruit become ripe, then overripe and then core breakdown occurs. In freshly harvested pears, the time between the ripe stage and the occurrence of breakdown is approximately one week. The longer fruit are in storage, the shorter this time becomes. In fruit which are overstored these changes occur so quickly that at no time is the pear fit to eat, and the breakdown may occur before the fruit becomes ripe. Yellowing in storage is a good indication of susceptilibity to core breakdown. Higher concentrations of CO_2 in the core area prior to core breakdown incidence has been reported and acetaldehyde may accumulate in overstored fruit and may contribute to core breakdown.

Control

Avoid storing pears beyond their postharvest life. To extend postharvest life, fruit should be rapidly cooled and stored at the lowest non-freezing temperature (generally -2 to -1°C, 28 to 30°F). Controlled atmosphere storage can extend the postharvest life as much as two-fold. Monitor fruit color as it approaches its typical storage life. Market any fruit that begins to yellow.

Pear: Senescent Scald

Symptoms

Brown to black discoloration of the skin associated with fruit which have become yellow in storage and lost their capacity to ripen normally. Browning is initially restricted to the skin but progresses rapidly into the flesh, particularly upon transfer of the fruit to warm temperatures. Browned skin becomes weak and is easily sloughed off in later stages of senescence. Often associated with core breakdown. Decay may follow secondarily. Taste and odor of the fruit are very disagreeable even before discoloration begins.

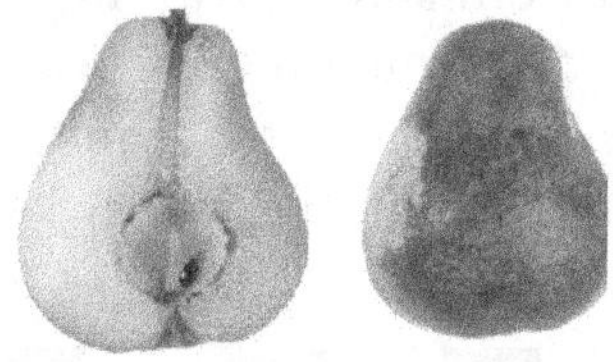

Fig. 9. Senescent Scald

Physiology

Fruit have reached the end of their postharvest life. Tissue begins to die and fermentative metabolism occurs resulting in alcoholic odor and off-flavors.

Control

Avoid storing the fruit beyond its postharvest life. To extend the postharvest life maximally, harvest fruit at optimum maturity for the length of storage, cool fruit rapidly and thoroughly after harvest, and store at the lowest, non-freezing temperature (often -1 to -2°C, 28 to 30°F). Controlled atmosphere storage can extend the postharvest life of the fruit up to twice that achievable in air storage. Watch for change in fruit color from green to yellow. Yellowing is an indication that the fruit's storage life is nearing the end.

Pear: Watery Breakdown

Symptoms

Watery breakdown, as the name implies, is a soft, watery deterioration of affected tissue. When the affected tissue is cut or punctured, juice flows out of the fruit. Typically, the involved tissue is in the outer portions of the flesh, but in severe cases can move into the core tissue. The affected tissue is not discolored during early stages, but becomes brown with time.

It is thought that the random pattern of symptoms development on individual fruits is related to other stresses that occur during handling (in tests, impacted fruits developed watery breakdown around the bruised area).

Causes

The soft, watery breakdown results from softening of the affected tissue. Fruit that are cooled slowly and/or held at too warm a temperature are most severely affected by this disorder. Fruit of any maturity can be affected, but high maturity fruit appear to be most susceptible.

Control

Cool fruit as soon as possible following harvest, and hold them at as low a temperature as possible without danger of freezing. To minimize the incidence of watery breakdown, fruit should be cooled to near storage temperature within 1 day of harvest, and held at 0°C (32°F) or below for up to one-month storage, or -1°C (30°F) for longer storage.

Check the soluble solids concentration in the core area (lowest soluble solids content) to determine the lowest safe storage temperature for each lot of pears.

Persimmon: Chilling Injury

Symptoms

Chilling injury can be a major cause of deterioration of 'Fuyu' persimmons during marketing after exposure to temperatures below 15°C (59°F). Symptom development is fastest at 5-7°C (41-45°F) and slowest at 0°C (32°F), which is the recommended storage and transport temperature for persimmons.

Physiology

'Fuyu' persimmons exhibit chilling injury symptoms if kept at temperatures between 2°C (36°F) and 15°C (59°F). Upon transfer to higher temperatures the severity of the symptoms (flesh softening, browning, and water-soaked appearance) increases and renders the fruits unmarketable. Respiration and ethylene production rates of chilled 'Fuyu' persimmons are higher than those of non-chilled fruits. Exposure to ethylene at 1 ppm or higher aggravates chilling symptoms of 'Fuyu' persimmons, while controlled atmospheres ameliorate these symptoms.

Control

Avoid exposure of 'Fuyu' persimmons to temperatures between 2°C (36°F) and 15°C (59°F). Optimum storage and transport temperature is 0°C (32°F).

Avoid exposure to ethylene above 1 ppm throughout postharvest handling of 'Fuyu' persimmons.

Use of controlled atmosphere of 3-5% O_2 + %8-5 CO_2 at temperatures below 5°C (41°F) reduced chilling injury.

Pomegranate: Chilling Injury

Symptoms

External symptoms include brown discoloration (scald) of the skin, pitting, and increased susceptibility to decay. Internal symptoms include brown discoloration of the white segments separating the arils and pale color (loss of red color) of the arils.

Physiology

Pomegranate are susceptible to chilling injury if stored longer than one month at temperatures between their freezing point (-3°C or 26.6°F) and 5°C (41°F). Upon transfer to higher temperatures, respiration and ethylene production rates increase and other chilling injury symptoms appear; their severity increases with lower temperatures and longer durations of chilling exposure. External and internal browning is related to oxidation of phenolics by polypherol oxidase. Storage in a 2% oxygen atmosphere at temperatures below 5°C (41°F) reduces severity of chilling injury symptoms.

Control

Avoid exposure of pomegranates to temperatures below 5°C (41°F).

Pomegranate: Husk Scald

Symptoms

Scald is a brown superficial discoloration restricted to the husk. There are no internal changes of the arils or the white segments as occurs with chilling injury. At advanced stages, the scalded areas became moldy. The scald symptoms became evident after 8 weeks storage at 2°C(36°F) on the cultivar 'Wonderful'. Symptoms will appear earlier at higher storage temperatures.

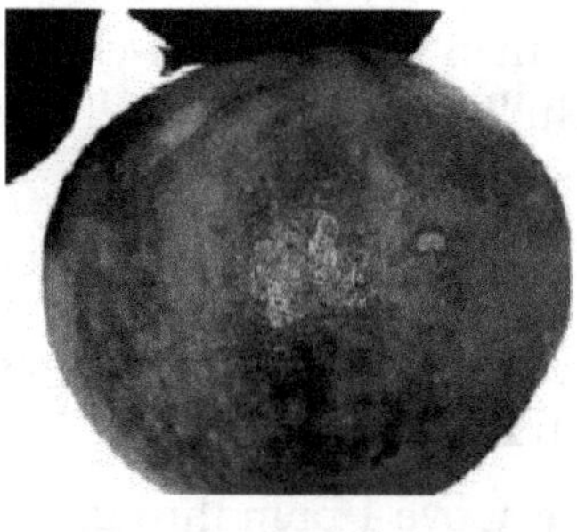

Fig. 10. Pomegranate: Husk Scald

Causes

Husk scald development of pomegranate may be due to phenolic oxidation.

Control

Late harvested fruit is less susceptible than earlier harvested fruit. Scald development may be delayed up to 6 weeks on 'Wonderful' stored at 2°C(36°F). Other potential treatments must be evaluated under California conditions before commercial recommendations can be made.

Stonefruit: Internal Breakdown (Chilling Injury)

Symptoms

Loss of flavor, 'off flavor', lack of juiciness due to leathery texture or flesh mealiness, black pit cavity, and flesh browning (FB). In plum and pluots, in addition to the previous symptoms, flesh translucency and red pigment accumulation (bleeding) are typical.

Causes

CI is triggered when fruit are stored below 8°C. Onset of symptoms is more rapid and severity of symptoms greater when fruit are exposed to temperatures between 2 and 8°C (36 and 46°F) (the *'killing temperature range'*). CI symptoms normally appear after fruit are moved from cold storage to room temperatures and while some ripening is occurring. For this reason, this problem is usually experienced by consumers, not the grower and/or packer. Stone fruit cultivars vary greatly in susceptibility to CI and some of them do not develop symptoms. In most cases, early-season peach and nectarine cultivars are least susceptible and late-season cultivars are most susceptible. Among plum cultivars, there is no seasonal pattern of susceptibility. Recently developed cultivars in the market from different breeding programs are showing less susceptibility to CI.

Control:

- The long-term solution is to breed cultivars using genomic tools that are not susceptible to CI. This work is in process, pursued by a multistate effort (RosBREED).

In the meantime, follow these guidelines:

- Market susceptible cultivars according to their potential postharvest life.
- Use the controlled-delayed-cooling (*'preconditioning'*) treatment to delay onset and reduce intensity of CI symptoms. When this *'preconditioning'* protocol is executed properly, fruit postharvest life almost doubles.
- Storing below 0°C (32°F), but above the freezing point, and avoiding the 'killing temperature range' 2 to 8°C (36 to 46°F) are beneficial to delay CI symptoms and extend market life.
- Enforce proper postharvest practices during handling, transportation and retail handling. Keep fruit near 0°C (32°F) during storage and transportation, and avoid fruit exposure to temperatures between 2 and 8°C (36 and 46°F) ('killing temperature range') during transportation and retail handling.
- Educate fruit handlers (warehouse, retail handlers and consumers) on how to reduce CI incidence and intensity.

12

POST-HARVEST OPERATIONS: GRADING, RIPENING, DE-GREENING, CURING ETC.

SORTING, SIZING AND GRADING

Sorting

Sorting is done by hand to remove the fruits and vegetables which are unsuitable to market or storage due to damage by mechanical injuries, insects, diseases, immature, over-mature, misshapen etc. This is usually carried out manually and done before washing. By removing damaged produce from the healthy ones, it reduces losses by preventing secondary contamination. Sorting is done either at farm level or in the pack-houses. In sorting, only sensory quality parameters are taken into consideration.

Sizing

Before or after sorting, sizing is done either by hand or machine. Machine sizers work on two basic principles; weight and diameter. Sizing on the basis of fruit shape and size are most effective for spherical (oranges, tomato, certain apple cultivars) and elongated (Delicious apples and European pears are of non-uniform shape) commodities, respectively.

Mechanisms/Types of Sizing

A. *Diverging belts/rope grader* - the different speed of belts makes produce rotate besides moving forward to a point where produce diameter equals belt/rope separation, e.g. cucumbers, gherkins, pineapples and large root vegetables.

B. *Sizing rollers* - with increased spaces between rollers, e.g. Citrus

C. *Hand held template*-Sizing can be performed manually using rings of known diameter.

D. *Sizing by weight* - sorting by weight is carried out in many crops with weight sensitive trays. These automatically move fruit into another belt aggregating all units of the same mass. Individual trays deposit fruit on the corresponding conveyor belt, e.g. Citrus, apples and pear and irregular fruits E. *Mesh screens* - e.g., potato, onion, anola etc.

GRADING

The produce is separated into two or more grades on the basis of the surface colour, shape, size, weight, soundness, firmness, cleanliness, maturity & free from foreign matter /diseases insect damage /mechanical injury, e.g. Apple I. Extra Fancy II. Fancy III. Standard IV. Cull (for processing).

Grading may be done manually or mechanically. It consists of sorting product in grades or categories based on weight/size.

Systems of grading : Static and Dynamic.

A. ***Static systems*** - are common in tender and/or high value crops. Here the product is placed on an inspection table where sorters remove units which do not meet the requirements for the grade or quality category.

B. ***The dynamic system*** - here product moves along a belt in front of the sorters who remove units with defects. Main flow is the highest quality grade. Often second and third grade quality units are removed and placed onto other belts. It is much more efficient in terms of volume sorted per unit of time. However, personnel should be well trained. This is because every unit remains only a few seconds in the worker's area of vision, e.g. Onion grading

There are two types of common mistakes: removing good quality units from the main flow and more frequently, not removing produce of doubtful quality.

New Innovation in grading systems

i. Computerized weight grader – operate on the basis of tipping buckets that drops to release the pre weighed item at a particular position. – Apples, citrus

ii. Video image capture & analysis–used for size, colour &external defect grading-coffee bean, apple

iii. NIR Spectrometers – to assess the TSS non – destructively in apple and stone fruits

iv. X-ray imaging and Computer aided tomography

v. MRI - Magnetic Resonances Imaging

vi. Spectroscopy

vii. Acoustic methods

viii. Volatile emission analysis

RIPENING

Fresh fruit and vegetables (FFV) are a major source of essential vitamins and minerals, such as vitamin A, vitamin C and potassium, needed for human wellbeing. They are, however, perishable living products that require coordinated activity by growers, storage operators, processors and retailers to maintain quality and reduce food loss and waste. The extent of coordination can vary greatly from loose in the case of local food supplies to complex for global supply chains. The Food and Agriculture Organization estimated that 32% (weight basis) of all food produced in the world was lost or wasted in 2009. When converted into calories, global losses represent approximately 24% of all food produced. Reducing the loss and waste of FFVs is important because these foods provide essential nutrients and represent sources of domestic and international revenue.

Fresh produce attributes (appearance, texture, flavour and nutritional value) have been traditional quality criteria, but increasingly safety (chemical, toxicological and microbial) and traceability are important for all the role players along the supply chain, from the farm to consumers. Fresh produce is often eaten raw or after minimal processing and food pathogen contamination can present risk of outbreaks of foodborne illnesses.

Listeria monocytogenes, *Salmonella enteritidis* phage, and *Escherichia coli* O157:H7 and O104:H4 are major pathogens contributing to outbreaks of foodborne illness with fresh produce as vectors for these pathogens. Owing to multiple uncertainties along the supply chain, microbial contamination leading to spoilage and postharvest losses can occur at any of the stages in the continuum from farm to consumer. Therefore, postharvest treatments are essential to minimize microbial spoilage and reduce the risk of pathogen contamination for FFV.

Various postharvest physical, chemical and gaseous treatments may be applied to maintain fresh-like quality with high nutritional value and meet safety standards of fresh produce. These postharvest treatments are typically combined with appropriate management of storage temperatures. This article reviews the current status of postharvest treatments and emerging technologies that can be used to maintain quality and reduce wastage of fresh produce.

Physical treatments

Heat treatment

Heat treatment has been studied as an alternative to chemical treatments for harvested FFV. Treatments include hot water dip (HWD), saturated water vapour heat, hot dry air and hot water rinse (HWR) with brushing. Beneficial effects of these heat treatments are linked (i) through changes in physiological processes such as a reduction in chilling injury and delay of ripening processes by heat inactivation of degradative enzymes, (ii) by killing of critical insect contaminations, and (iii) by controlling the onset of fungal decay. Heat treatments can be of short- (up to 1 h) or long-term duration (up to 4 days). Heat treatments have been applied to firm potatoes, tomatoes, carrots and strawberries; to preserve the colour of asparagus, broccoli, green beans, kiwi fruits, celery and lettuce;

to prevent development of overripe flavours in cantaloupe and other melons; and to generally add to the longevity of grapes, plums, bean sprouts and peaches, among others.

It has been demonstrated that heat shock by using hot water washing at temperatures ranging from 37 to 55°C for a duration of 30 s to 3 min can improve the postharvest quality of spinach, rocket leaves, apples and mandarin fruit. A clear mode of action of any water treatment is to wash-off the spores from the fruit surface. Hot water is a better vector of energy than air and has provided comparable reductions in fungal decay. Blue mould on grapefruit caused by *Penicillium* sp. has been controlled by dipping fruit in hot water for 2 min at 50°C.

Improvements in the quality of bell pepper, apples, melons, sweet corn, kumquat and grapefruit have been reported with cold water cleaning in combination with brushing and a short HWR. Hot water treatments also influence the structure and composition of epicuticular waxes. Covering of cracks and wounds and the formation of anti-fungal substances in the wax after heating are thought to be possible modes of action.

Researchers suggested that the combination of *Bacillus amyloliquefaciens* HF-01, sodium bicarbonate and hot water could be a promising method for the control of postharvest decay on citrus while maintaining fruit quality after harvest. To date, commercial applications of heat treatments are limited. Heat treatment provides an alternative to fungicide applications, and in Germany HWD has been used in the storage of organic apples. Treatment of fruit after a few days of cold storage or immediately after the opening of a long-term controlled atmosphere (CA) storage room provides new options for prolonging their subsequent storage life, although acceptance of this technology by fruit growers has been hampered by high-energy costs and also the need for added labour at the peak work period during harvest time.

Edible Coating

Edible coatings are thin layers of external coatings applied to the surface of fresh produce to enhance the waxy cuticle or as replacements for natural barriers where the produce cuticle has been removed. The application of edible coatings on fresh produce provides a partial barrier to the movement of moisture on the surface of fresh produce, thereby minimizing moisture loss during postharvest storage; a gas barrier, thereby establishing a modified atmosphere around the product, which slows down respiration, senescence and enzymatic oxidation and preserves colour and texture; helps to retain volatile compounds contributing to produce a natural aroma and restrict foreign odours; maintains fresh produce structural integrity, and protects against mechanical damages; and serves as carriers of functional or active compounds, such as nutraceuticals, flavouring and colouring agents, antioxidants and antimicrobials, that will maintain/improve product quality and safety. Edible coatings are composed of hydrophobic groups, such as lipid-based waxes; hydrocolloid/hydrophilic groups, such as polysaccharide or protein-based materials; or an integration of both groups in order to improve the functionality of the coating.

Several edible coatings including chitosan, *Aloe vera*, polyvinyl acetate, mineral oils, cellulose and protein based have shown desirable attributes on fresh produce with good barrier properties, without residual odour or taste and efficient antimicrobial activity. However, more research is required to enhance moisture barrier properties of hydrophilic edible coatings, improve coating adhesion and durability during storage. To maximize the benefits of edible coatings for fresh produce, it is important to understand the effect of storage conditions on the desired functions and the adverse effect on fresh produce quality.

The main limitation for the application of edible coating at the industrial level is the cost of scaling up research concepts or investment for new installation of film production and coating equipment, the lack of edible materials with desired physical and functional properties as well as the challenges of regulatory status for the different coating materials. Furthermore, process parameters, such as the method of coating and the amount of additives, can affect the film barrier properties and overall quality of the food product.

One of the commercial coating products is Natureseal, which maintains colour, texture and shelf life of a number of fresh-cut fruits e.g. apples, pears, carrots, celery, etc., has recorded a good success. However, further research development is required to investigate the influence of edible coatings on individual cultivars of fresh-cuts in order to understand the variation in shelf life.

Irradiation

Irradiation exposes food to radiant energy from γ rays and e-beam (high-energy electrons) that penetrate objects and break molecular bonds, including the DNA of living organisms. Ionizing radiation from cobalt-60 or caesium-137, or machine generated electron beams are used as a source of irradiation for extending shelf life of fresh produce. By inhibiting cellular reproduction, irradiation can neutralize pest and food safety problems. The effect depends on the doses, measured in kilograys (kGy).

Low doses of irradiation (less than 1 kGy) only disrupt cellular activity enough to inhibit sprouting of tubers, bulbs and roots and delay senescence. Medium doses (1–10 kGy) reduce microbial loads while high doses (more than 10 kGy) kill a broad spectrum of fungi and bacteria spp. and pests. Most medium- and high-level doses are not appropriate for fresh produce because they can cause sensory defects (visual, texture and flavour) and/or accelerated senescence due to the irreparable damage to DNA and proteins. Irradiation presents an effective postharvest treatment for destroying bacteria, moulds and yeasts, which cause food spoilage, and also control insect and parasite infestation resulting in reduced storage losses, extended shelf life and improved parasitological and microbiological safety of foods. Irradiation has been commercialized for control of potato and onion sprouting, and strawberry decay. Low-dose *ã*-irradiation on mango (0.3–0.7 kGy) resulted in delay in ripening and extension of shelf life by a minimum of 3–4 days.

Recently, researchers reported an irradiation dose of 1 kGy to be the only effective dose in which enhanced shelf life was achieved without any deterioration of various quality attributes of litchi fruit. While much of the focus of irradiation use on FFV has been for extending shelf life and reducing decay, it has been known for many decades that irradiation is effective at killing, sterilizing or preventing further development of a wide variety of insect pests of quarantine importance on perishable FFV.

Despite some misconceptions, exposing food to irradiation does not make the food itself radioactive. The irradiation process produces very little chemical change in food and does not change the nutritional value of food. Extensive research and testing has demonstrated that irradiated food is safe and wholesome.

Chemical Treatments

Antimicrobial and Anti-browning Agents

Over the past decade, the increasing number of reported outbreaks of foodborne illnesses has heightened the concern of regulatory agencies, producers and the consumers about the microbial safety of FFV. Outbreaks have been associated with vegetables such as cabbage, celery, cucumber, leeks, watercress, lettuce and sprouts. Antimicrobial and anti-browning agents offer the possibility to maintain safety and can be grouped into chemical- and natural/bio-based agents. Chemical-based agents include chlorine-based solutions, peroxyacetic acid (PAA), organic acids, hydrogen peroxide (H_2O_2) and electrolysed water. A chlorine-based solution such as NaClO has been one of the commonly used disinfectants for fresh produce, owing to its very potent oxidizing properties and cost effectiveness.

However, its efficacy as an antimicrobial agent is dependent on the levels of chlorine and at high levels may cause taste and odour defects on treated products. Additionally, chlorine-based compounds have been reported to have limited effectiveness in the reductions of microbial load on fresh produce. Surfactants, detergents and solvents, alone or coupled with physical manipulation such as brushing, may be used to reduce hydrophobic nature of the waxy cuticle or remove part of the wax to increase exposure of microorganisms to chlorine. However, chlorine has been associated with the possible formation of carcinogenic chlorinated compounds and this may lead to new regulatory restrictions in the EU.

PAA is a very strong oxidizing agent, with no harmful by-products. PAA has been reported to be effective in controlling *E. coli* O157:H7 and *L. monocytogenes* on apples, strawberries, lettuce and cantaloupe. A 5 log reduction in *Enterobacter sakazakii* was reported for lettuce when treated with PAA. Researchers reported that decontamination treatment of fresh-cut carrot with PAA reduced the initial load of aerobic mesophilic bacteria by about 4 log units and yeasts and moulds by 3.5 log units and no further microbial growth was observed during storage.

H_2O_2 possesses a bactericidal, sporicidal and inhibitory ability, owing to its property as an oxidant and being able to generate other cytotoxic oxidizing species, such as hydroxyl radicals. Treatment with H_2O_2 can extend the shelf life and reduce natural and pathogenic microbial populations in melons, oranges, apples, prunes, tomatoes, whole grapes and fresh-cut produce.

However, H_2O_2 treatment requires a long duration of application and can cause injury on some produce. Also, it is accepted as a generally recognized as safe for some food applications but not yet approved as an antimicrobial agent. However, a recent study found that the newly developed H_2O_2-based sanitizers provoked a significant increase in the respiration rate and the electrolyte leakage of fresh-cut iceberg lettuce compared with tap water washing.

Organic acid, ascorbic acid and calcium-based solutions have been applied largely to slow down enzymatic and non-enzymatic browning, deterioration of texture and microbial growth on fresh produce. Treatment of fresh-cut melon dipped in 0.52 mM citric acid for 30 s prior to modified atmosphere packaging (MAP) maintained microbial safety and prevented translucency and discoloration. Inhibitory effects of organic acids (acetic, lactic and malic acids) combined with MAP on foodborne pathogens, including *E. coli* O157:H7, *S. Typhimurium* and *L. monocytogenes*, on cabbage was reported. However, there are factors limiting the efficacy of antimicrobial and anti-browning agents such as internalization of bacteria and inaccessible sites within fresh produce such as the calyx. These limitations highlight the need for novel means of applying of antimicrobial and anti-browning agents.

Nitric oxide

Nitric oxide (NO) is a highly reactive free radical gas and acts as a multifunctional signalling molecule in various plant physiological processes, such as fruit ripening and senescence of FFV. Endogenous NO concentrations decrease with maturation and senescence in FFV, thereby offering an opportunity for modulation of their levels with exogenous application to exert the opposite effects. Optimum NO levels delay the climacteric phase of many tropical fruits, prolong the postharvest storage life by impeding ripening and senescence, suppress biosynthesis of ethylene, reduce ethylene production and, consequently, delay in fruit ripening. NO gas is applied as a fumigant or released from compounds such as sodium nitroprusside, S-nitrosothiols and also diazeniumdiolates used as a dipping treatment. Reduced ethylene production during ripening in NO-fumigated fruit has been claimed owing to binding of NO with 1-aminocyclopropane-1-carboxylic acid (ACC) and ACC oxidase to form a stable ternary complex, thus limiting ethylene production. Other mechanism of NO action include the inhibition of ethylene biosynthesis, cross-communication with other phytohormones, regulation of gene expression and amelioration of oxidative postharvest stress.

Successful application of NO has been reported for apple, banana, kiwifruit, mango, peach, pear, plum, strawberry, tomato, papaya, loquat, Chinese winter jujube fruit and Chinese bayberry. NO treatments reduce the rate of respiration, water loss and inhibit browning, as well as reducing the incidence of postharvest diseases. NO fumigation in combination with cold storage has a synergistic effect in extending storage life of fruit such as plum and mango. Similarly, NO combined with modified atmosphere conditions extended the postharvest life of green beans, broccoli and bok choy. Commercial application of NO in FFV depends upon the development of a smart carrier/controlled release system for NO.

Sulfur Dioxide

Sulfur dioxide (SO_2) is widely used on table grapes to prevent decay during storage, by either initial fumigation of fruit from the field followed by weekly fumigation of storage rooms or slow release from in-package pads containing sodium metabisulfite. SO_2 technology has also been tested for control of postharvest decay on other fruits such as litchi, fig, banana, lemon or apple. Researchers reported that SO_2 fumigation followed by CA storage (3%O_2+6 or 12%CO_2) is a promising postharvest strategy for fresh blueberries to reduce decay, extend market life and maintain high nutritional value.

Researchers demonstrated that SO_2 is an effective and practical technology for reducing the risk of blueberry grey mould decay during storage and it could be used for the export market. There are disadvantages to SO_2 use; the SO_2 concentration necessary to inhibit fungal growth may induce injuries in grape fruits and stems, and sulfite residues pose a health risk for some individuals, as well as firming of the texture of some fruit species (pomaces), incomplete de-sulphiting and incomplete re-colouring of red fruits. Nevertheless, SO_2 treatment is a widespread process because of its advantages of universal antiseptic action and economic application.

Gaseous Treatments

Ozone

Recent research and commercial applications have verified that ozone can replace traditional sanitizing agents. Ozone is a very pungent, naturally occurring gas with strong highly reactive oxidizing properties. Ozone is reported to have 1.5 times the oxidizing potential of chlorine and 3000 times the potential of hypochlorous acid. Contact times for antimicrobial action are typically four to five times less than that for chlorine. Ozone rapidly attacks bacterial cell walls and is more effective than chlorine against the thick-walled spores of plant pathogens and animal parasites, at practical and safe concentrations.

Researchers reported that the fruit exposed to 2.5 ppm ozone had higher levels of total soluble solids, ascorbic acid content, β-carotene content, lycopene content, and antioxidant activity and also reduced weight loss at day 10 compared with untreated fruit. The sensory attributes of papaya of ozone-treated fruit was also superior in sweetness and overall acceptability endorsing ozone as a non-thermal and safe food preservation technique for FFV. Ozone can be employed in cold storage, washing system or process water sterilization. Researchers reported that irradiation and washing with ozonated water slightly reduced respiration in white asparagus spears, but increased spear tissue toughness. However, neither washing the asparagus spears with ozonated water (3 or 4.5 ppm) nor treating them with radiation (1 kJ m^{-2}) systematically and significantly affected their microbial loads during storage.

Some commercial use has occurred with commodities such as apples, cherries, carrots, garlic, kiwi, onions, peaches, plums, potatoes and table grapes. However, ozone does not penetrate natural openings or wounds efficiently. Additional research is needed to define the potential and limits of the effective use of ozone for postharvest treatments for the quality and safety of FFV.

Ethylene

Endogenous ethylene production and its exogenous application exhibit both beneficial and deleterious effects on horticultural fresh produce. Beneficial effects of exogenously applied ethylene includes triggering ripening, improving fruit colour and quality in some crops, such as bananas and avocados, kiwifruit, persimmon, tomato, mangoes, de-greening of citrus fruit. The deleterious effects of ethylene in postharvest phase horticultural commodities has also been documented, such as shorter storage life, promotion of senescence, fruit softening, discoloration (browning) and russet spotting in lettuce, yellowing of leafy vegetables and cucumbers and increased susceptibility of FFV to decay.

Therefore, ethylene management plays a pivotal role in maintaining postharvest life and quality of climacteric and non-climacteric horticultural produce. Most commercial strategies for maintaining horticultural commodities involve storing at low temperatures, blocking ethylene biosynthesis and its action, minimizing exposure produce to ethylene during ripening, harvest, storage and transport by controlling temperature and atmospheric gas composition. Newly developed ethylene measurement devices will enable to detect critical concentrations during storage and transportation.

Beneficial effects of ethylene biosynthesis inhibitors such as amino-ethoxyvinylglycine alone on postharvest quality have been demonstrated in apples and stone fruits and in combination with controlled atmosphere (CA) storage. Treating FFV with inhibitors of ethylene action, such as 1-methylcyclopropene (1-MCP) or NO alone or in combination with MAP or CA storage, also impedes ethylene production and action consequently extends storage life and maintains quality of FFV.

A recent development has been palladium-promoted zeolite materials, which can be effective ethylene scavengers to prolong the shelf life of climacteric fresh produce, such as bananas and avocados. The material has the potential to be used commercially, as an alternative and/or supplemental treatment to 1-MCP. Researchers developed a carbon-heat hybrid ethylene scrubber for fresh horticultural produce storage purposes. The device comprised a cartridge heater tightly joined to the activated carbon–1% palladium. Application of heat pulses leads to an increase in ethylene oxidation and to auto-regeneration of the activated carbon.

1-Methylcyclopropene

The discovery and patenting of cyclopropenes as inhibitors of ethylene perception represents a major breakthrough in controlling ethylene responses of horticultural products. The process of discovery of the effects of cyclopropenes, and their proposed method of action, has been described. Of the cyclopropenes, 1-MCP proved to be extremely active, but unstable in the liquid phase. However, 1-MCP can be complexed with α-cyclodextrin to maintain its stability; this development represented a major step towards its commercialization as it was then possible to release 1-MCP from the complex to expose to the horticultural products.

Regulatory approval for use of 1-MCP has been obtained in more than 50 countries, and approval for use of the technology continues to occur around the world. 1-MCP is registered for use on a wide variety of FFV including apple, avocado, banana, broccoli, cucumber, date, kiwifruit, mango, melon, nectarine, papaya, peach, pear, pepper, persimmon, pineapple, plantain, plum, squash and tomato. 1-MCP affects many ripening and senescence processes, including pigment changes, softening and cell wall metabolism, flavour and aroma, and nutritional properties, but to varying degrees in both non-climacteric and climacteric products. While aqueous 1-MCP shows similar responses as those treated with gaseous 1-MCP, ripening actors such as activity of cell wall-associated enzymes, e.g. lycopene, antioxidant and volatiles of avocado, are delayed but recover to reach levels similar to those of untreated fruit. The range of responses reflects the enormous diversity of these crops in terms of both inherent diversity and morphological derivation. Several generalizations can be made about responses of crops to 1-MCP:

- Genotype, cultivar and maturity effects can be highly variable;
- Climacteric fruit are affected by 1-MCP treatment, but the capacity to interrupt ripening once initiated varies by fruit and attributes studied;
- Non-climacteric fruit also respond to 1-MCP, providing insights into ethylene-dependent and -independent events during ripening;
- Treated fruits are firmer, slower to soften, slower to change peel colour, and develop aroma and flavour. If 1-MCP concentrations and exposure periods for each fruit must be appropriate to allow ripening to occur, so the final quality of the treated fruit is similar to that of the untreated product; and
- The effect of 1-MCP on physiological disorders is dependent on the role of ethylene. 1-MCP decreases senescent-related and ethylene-induced disorders (senescent breakdown of apples and water soaking of watermelons). 1-MCP can increase susceptibility of apple fruit to CA storage-related disorders (CO_2 injury). Some chilling-related disorders (woolliness and internal breakdown of peaches and nectarines, chilling injury of citrus and bananas) can be increased by inhibition of ethylene production. Other chilling-related disorders (superficial scald of apples and pears and internal flesh browning of avocados and pineapples) are decreased by inhibition of ethylene production.

Registration of 1-MCP for FFV has focused on major or specialty products important to specific countries. The apple has been an excellent crop for use of 1-MCP, and the technology is used extensively around the world to maintain quality through the whole marketing chain from storage to consumer. Success of 1-MCP technology for apples is largely associated with a fruit where maintenance of 'at harvest' quality and only moderate softening to a crisp fracturable texture is desirable.

Researchers found that rapid treatment of fruit with 1-MCP after harvest can afford storage operators more freedom to delay CA storage application, but attention to cultivar, fruit maturity and susceptibility of fruit to storage disorders must be considered. Challenges exist for the effective use of 1-MCP for fruit that ripen uniformly to a melting texture and/

or have major colour, flavour and aroma changes that are expected by the consumer. Failure to ripen normally has been shown in avocado, banana, pear and tomato, where fruits were treated at an early ripening stage or where applied 1-MCP concentrations were too high. Despite the challenges, successful commercialization of SmartFresh for treatment of avocados, bananas, melons, persimmons and tomatoes has resulted from careful attenuation of 1-MCP concentrations and/or ripening stage at harvest.

Recently, 1-MCP formulations have been approved by the Environmental Protection Agency and other regulatory authorities for preharvest applications, and these are marketed as Harvista for FFV. Semi-commercial trials have been carried out in several US locations, Argentina, Brazil, Canada, Chile, New Zealand and South Africa. Harvista has useful effects on delaying fruit drop, slowing fruit maturation and ripening, and in maintaining postharvest quality, including improving the effects of SmartFresh.

Controlled Atmosphere Storage

CA storage refers to the monitoring and adjustment of the CO_2 and O_2 levels within gas tight stores at optimum storage temperature. Thus, the atmosphere is controlled rather than established passively as in the case of MAP outlined in §4e, though the effects of altered atmospheres on metabolism of FFV are essentially the same. In most cases, the concentrations of CO_2 are higher and those of O_2 are lower, optimum concentrations depending on the specific product and the purpose of the CA storage conditions. Reduced O_2 and elevated CO_2 levels affect both primary (glycolysis, fermentation and aerobic respiration) and secondary (e.g. processes involved in ethylene production and action, pigments, phenolics and volatiles) metabolism.

Each FFV has an optimal range of O_2 and CO_2 for maintaining quality and extending shelf life, and these can differ for whole and fresh-cut products of the same fruit or vegetable. Benefits of CA as well as detrimental effects resulting from exposure of FFV to atmospheres outside of the safe range were summarized by researchers. The beneficial effects of CA include:

- Retardation of senescence and associated biochemical and physiological changes, e.g. slowing down rates of respiration, ethylene production and softening;
- Reduction in sensitivity to ethylene action at O_2 levels less than 8% and/or CO_2 levels more than 1%;
- Alleviation of certain physiological disorders, e.g. chilling injury of avocado and superficial scald of apples and pears;
- Direct or indirect effect on postharvest pathogens (bacteria and fungi) and consequently decay incidence and severity e.g. CO_2 at 10–15% inhibits development of *Botrytis* rot on strawberries and cherries; and
- Low O_2 (less than 1%) and/or elevated CO_2 (40–60%) can be a useful tool for insect control in some fresh and dried fruits, flowers and vegetables; and dried nuts and grains.

The detrimental effects of CA include:

- Initiation and/or aggravation of certain physiological disorders, such as internal browning in apples and pears, brown stain of lettuce and chilling injury of some commodities;
- Irregular ripening of fruits, such as banana, mango, pear and tomato, can result from exposure to O_2 levels below 2% and/or CO_2 levels above 5% for more than one month;
- Development of off-flavours and off-odours at very low O_2 concentrations (as a result of anaerobic respiration) and very high CO_2 levels (as a result of fermentative metabolism); and
- Increased susceptibility to decay when the fruit is physiologically injured by insufficient O_2 or too-high CO_2 concentrations.

Use of CA technology is limited to relatively few FFV, the major crop being apples. Other FFV include cabbages, sweet onions, kiwifruits, avocados, persimmons, pomegranates, nuts and dried fruits, and vegetables. Atmospheric modification during long-distance transport is used on apples, asparagus, avocados, bananas, broccoli, cranberries, cherries, figs, kiwifruits, mangos, melons, nectarines, peaches, pears, plums and strawberries.

Limited use of CA for many FFV is related to the high levels of capital investment required for high-quality storage rooms and the maintenance and monitoring of atmospheres; therefore, fruit volumes must be high enough to fill rooms, and extended storage periods are needed to make investments economical. For shipping containers, limitations include maintaining container identity for those with equipment.

Specific CA storage technologies that are used are a function of the FFV, growing region and size of the industry. Where used, static systems are still used for the majority of FFV CA storage, ranging from simple controlled ventilation systems to conventional CA (more than 2 kPa O_2). The use of low O_2 (LO) and ultra low O_2 (ULO) CA storage, where O_2 levels are as low as 1.5–2.0 and 0.8–1.2 kPa, respectively, is increasingly becoming common.

Researchers proposed a hollow fibre membrane module for generating nitrogen-enriched air stream that controls the concentration of both O_2 and CO_2 in the CA storage. Other developments include dynamic CA (DCA), initial low O_2 stress (ILOS) and hypobaric storage. The principle of DCA is that responses of fruit to decreasing O_2 in the storage atmosphere are monitored. As O_2 levels reach the anaerobic compensation point, stress responses by the fruit can be detected and the O_2 increased to slightly higher levels. The most common commercial DCA system is based on chlorophyll fluorescence (DCA-DF), but others include measurement of ethanol in the fruit (dynamic controlled systems, DCS) and the respiration quotient (DCS-RQ). Most research has been carried out with apples and pears, a major impetus for research being the loss of diphenylamine, the antioxidant used widely for control of the physiological disorder superficial scald in Europe.

However, DCA has been tested with avocado and should work with any chlorophyll-containing product. ILOS involves the use of low partial pressures of O_2, e.g. 0.4 kPa, that result in ethanol accumulation for short periods of time that result in delay of fruit ripening.

Hypobaric or low-pressure storage results in very low O_2 around the product and delay ripening and senescence but has yet to find commercial acceptance because of factors such as cost and safety. These issues may be addressed in the future.

Modified Atmosphere Packaging

MAP generally involves the packaging of a whole or fresh-cut product in plastic film bags, and can be either passive or active. In passive MAP, the equilibrium concentrations of O_2 and CO_2 are a function of the product weight and its respiration rate, which is affected by temperature and the surface area, perforations, thickness and permeability to gases of films used in packaging. In active MAP, the desired atmosphere is introduced in the package headspace before heat sealing, but the final atmosphere will eventually be a function of the same factors that affect passive MAP.

Correct equilibrium atmosphere can delay respiration, senescence, and slow down the rate of deterioration, thereby extending product storage life. More recently, active MAP also includes technologies to adsorb substances, such as O_2, ethylene, moisture, CO_2, flavours/odours, and release substances such as CO_2, antimicrobial agents, antioxidants and flavours.

An extensive number of models have been developed to predict respiration rates of FFV under MAP conditions. An example of prediction models is that, where the Pack-in-MAP software for optimum packaging solutions for FFV was developed. Behind the software is an extensive database on product respiration rate, optimum temperature, optimum ranges of O_2 and CO_2 and gas permeability of packaging materials commonly used in MAP.

The software is based on a series of mathematical algorithms to simulate the evolution of internal gas composition in the packaging as a result of food respiration and mass transfer through the packaging material and, when used in a reverse manner, to identify the window of gas permeability that satisfy food requirements, size and number of micro-perforations, if needed.

Current MAP design considers the respiration rate of product as the only important parameter for deciding target gas barrier properties required to achieve an equilibrium-modified atmosphere. However, besides in-package gas composition it is also important to take into consideration the in-package level of humidity, in order to avoid condensation and/or mould and bacterial development in MAP systems. It is well known that the in-package humidity is influenced by respiration and transpiration of the fresh produce as well as the water vapour permeability of the packaging material.

However, most polymeric materials (polyethylene, polypropylene or polyvinyl chloride) used in MAP have lower water vapour permeability relative to transpiration rates of fresh produce; therefore, most water molecules evaporated from the produce do not escape through the film and remain within the package, enhancing the water vapour pressure in the package microenvironment. Under these near-saturation conditions, even minor temperature fluctuation may result in condensation inside the package resulting in produce sliminess and enhancement of microbial growth and decay of produce.

Therefore, the major challenge of modified atmosphere and humidity packaging (MAHP) is finding a solution for creating optimal atmosphere and reducing the risk of water condensation in the package while still maintaining produce weight loss as low as possible. Recent developments on high water vapour permeable films such as Xtend, NatureFlex or biodegradable films or use of hygroscopic additives located within the package headspace or directly integrated into the packaging material. However, the hygroscopic additives should not be used for fresh produce, which have high water activity, in order to avoid excessive weight loss of the packed food.

Recently, researchers reported that the humidity absorption of the trays with NaCl improved the quality and shelf life of fresh mushrooms. In this study, different percentage of NaCl was incorporated in the polymer matrix of a film from which trays were produced. The results indicate that the amount of water vapour absorbed by the tray is directly proportional to the percentage of salt incorporated in the trays, which enhanced the total appearance of the package. The maximal capacity and the rate of moisture absorption by the humidity regulating trays need to be studied further in order to confirm whether the trays have enough moisture absorption capacity to prevent condensation for the selected horticultural crops. Additionally, questions about sustainability and consumer expectations of such humidity regulating trays have to be taken into consideration.

Emerging Technologies

Plasma is an emerging technique for decontaminating FFV. Plasma is composed of ionized gas molecules, which have been dissociated via an energy input. Depending on the mode of particles activation and the excitation energy, they can generate high or low temperatures, referred to as thermal or cold plasma, respectively. Cold plasma at atmospheric pressure can be generated by transforming argon gas into plasma at radio frequency of 27 MHz or by electric discharge between two electrodes separated by dielectric barriers.

Three basic mechanisms have been suggested for the inactivation microbial spores in plasma environments, including the erosion of microbial spore surface atom by atom through adsorption of reactive free radicals 'etching'; direct destruction of DNA via UV irradiation and volatilization of compounds from the spore surface by UV photons through intrinsic photo-desorption. Researchers revealed that at the optimal operating conditions of cold gas plasma treatment about 15 min treatment time was required to achieve 2.72, 1.76 and 0.94 log-reductions in viable cells of *S. enteric* sv. *Typhimurium* on lettuce, strawberry surfaces and potato tissue, respectively.

Recent study by researchers on fresh corn salad leaves showed that the plasma treatment at 20 W for 1 min successfully inactivated *E. coli* by 4 log-cycles. However, more research is required for a complete understanding of the role of microbial cell structure, physiology and stress resistance mechanisms involved in plasma resistance. Also, the effect of plasma treatment on food enzymes and postharvest quality attributes of FFV requires more detailed study. Safety of gases, consumer perception and the translation of laboratory scale to large commercial scale, also requires further investigation.

DEGREENING

Post-harvest treatment of citrus fruits with ethylene under controlled conditions hastens the loss of chlorophyll, a process known as 'degreening'. Degreening consists of chlorophyll degradation to allow the expression of natural pigments masked by the green colour, e.g. Yellowing of citrus fruits. Degreening process can be hastened by applying ethylene and done mainly in non-climacteric fruits like citrus. On a small scale dip treatment in ethrel solution also bring about degreening. The concentration of ethylene required and time of exposure is significantly high in case of degreening when compared to ripening.

The main causes for greening are climatic conditions before harvest. For example, citrus often reaches commercial maturity with traces of green colour on the epidermis (flavedo). Although not different from fruits with colour, consumers sense that they are not ripe enough and have not reached their full flavor. Exposure to low temperature during maturation is necessary for an orange-coloured peel to develop. Hence fruit grown in low altitude tropics fail to degreen completely.

Degreening is Done by 2 Methods

1. ***Exposing to ethylene*** - Degreening is done at 25-300C and 85-95% RH with ethylene gas trickled into room to achieve 20-30 PPM or 10 μL L-1 for 24-72 hr. with regular ventilation of the chamber to prevent CO2 build up and injury. In batch process it is at 20-200 μL ethylene L-1. Trickle process is faster than batch process; since degreening condition accelerate deterioration and decay of citrus fruits. This is most popular methods .

   Citrus fruits are exposed from 1-3 days (depending on degree of greening) to an atmosphere containing ethylene (5-10 ppm) under controlled ventilation, 20-260C and 90-95% RH. Conditions for degreening are specific to the production area.

2. ***Artificial colouring*** - When weather is not conducive for the development for colour in orange; legally permitted dye can be used to colour the peel of the fruits like orange, with Citrus Red No.2 (1-2(2,5-dimethoxy phenylazo)2-napththol) this process is called as 'Colour Add'. It is used on mature fruit which are not intended to processing. Dye is applied to fruit by dip at 49°C for 4 min. for oranges. Rinsed enough to prevent bleeding and residue tolerance is 2ppm / 2 mg kg^{-1} of fruit.

POSTHARVEST HANDLING SYSTEMS: UNDERGROUND VEGETABLES (ROOTS, TUBERS AND BULBS)

The edible portions of this group of vegetables develop mostly underground and include several botanical structures.

Roots: beet, carrot, celeriac, radish, horseradish, parsnip, turnip, sweet potato, cassava, Jicama

Tubers: potato, Jerusalem artichoke, yam

Bulbs: onion, garlic, shallot

Others: ginger rhizomes, taro (dasheen) corms

Vegetables within this group can also be divided into two subgroups based on their postharvest temperature requirements:

Temperate-zone underground vegetables: beet, carrot, celeriac, radish, horseradish, parsnip, turnip, potato, onion, garlic, shallot, daikon, salsify, water chestnut

Subtropical and tropical underground vegetables: sweet potato, yam, cassava, ginger, taro, jicama, malanga

The commodities in this grouping have several common characteristics. They are all storage organs, principally of carbohydrates; they generally have low respiration rates (depends on the stage of development); they are considered relatively nonperishable, especially if the tops are removed; they continue growth after harvest (rooting and sprouting); and they all can be stored for relatively long periods.

HARVESTING

Maturity indices vary with commodity. Many of these products may be harvested and marketed at various stages of development (e.g., "new" or immature potatoes versus mature potatoes, "baby" or immature carrots versus mature carrots). Criteria commonly used to harvest these commodities are as follows:

- Carrot: size, length of root
- Radish: days from planting, size
- Potato: drying of foliage, setting of skins
- Cassava and taro: drying of foliage begins
- Garlic and onion: drying and bending over of tops
- Sweet potato: drying of foliage

Both mechanical and manual harvest are used for this group of vegetables. Most roots and tubers are harvested mechanically and transported in bulk to packinghouses or processing facilities. Garlic and onion for fresh market are mechanically undercut, hand harvested and trimmed, cured in the field, and field-packed or transported to packinghouses. Sweet potato harvesting is still done mostly by hand, although vine cutting and lifting are done mechanically. Vine-killing chemicals may be used on potatoes before mechanical harvesting. Carrots are undercut, lifted by their tops, and detopped during mechanical harvest. Physical damage during the harvesting operations can be extensive and is a major cause of postharvest losses.

POSTHARVEST PROCEDURES

Curing

One of the simplest and most effective ways to reduce water loss and decay during postharvest storage of root, tuber, and bulb crops is curing after harvest. In root and tuber crops, curing refers to the process of wound healing with the development and suberization of new epidermal tissue called wound periderm. The type of wound also affects periderm formation: abrasions result in the formation of deep, irregular periderm, cuts result in a thin periderm, and compressions and impacts may entirely prevent periderm formation.

In bulb crops, curing refers to the process of drying of the neck tissues and of the outer leaves to form dry scales. Some water loss takes place during curing. Removing decayed bulbs before curing and storage ensures a greater percentage of usable product after storage.

When onions and garlic are cured in the field they are undercut, then hand pulled. Sometimes the roots and tops are trimmed and the bulbs then are allowed to dry in field racks or bins from 2 to 7 days or longer (depending on ambient conditions). Sometimes they are pulled and cured before trimming. Curing may be done in windrows with the tops covering the bulbs to prevent sunburn. Where ambient conditions are unfavorable, curing may be done in rooms with warm forced air. Onions develop the best scale color if cured at temperatures of 25° to 32°C (77° to 90°F).

Storage

Onion, garlic, potato, and sweet potato are often stored after curing and before preparation for market (cleaning, grading, sizing, and packing). These products, and other root crops such as carrot and turnip, may be stored from 3 to 10 months in mechanically refrigerated or ventilated storages.

Preparation for Market

The following operations are commonly used to prepare root, tuber, and bulb crops for market:

Cleaning. Dry brush or wash and partially dry, removing excess moisture.

Sorting. Eliminate defective products and plant debris.

Decay control. Postharvest fungicides are used on some of these commodities such as sweet potato; chlorination of wash and flume waters provides sanitation for carrot and potato.

Sizing. Size mechanically or by hand. Mechanical sizers are generally diverging rollers or weight sizers. Modified volumetric sizers are used for potatoes. Carrots present special sizing problems, since they must be sized by diameter (diverging rollers) and by length (manually or with a gravity-length sizer).

Grading. Separate into quality grades.

Packing. Pack into consumer units (bags, trays) and then pack into master shipping containers; or bulk pack into shipping containers (bags, boxes, and bins).

Loading into transit vehicles. Bulk transport to processing plants is sometimes used for onion, potato, and radish. For fresh market, most products are loaded in packed shipping containers (palletized if boxes, manually stacked if bags).

All temperate-zone root crops except potato, onion, and garlic can be hydrocooled. In areas where potatoes are harvested during hot weather, they may also be hydrocooled. Tropical root crops, potato, onion, and garlic are occasionally room-cooled before shipment to market. Potatoes may be cooled to 13° to 16°C, (55° to 60°F) before shipment when they are harvested, packed, and shipped to market during hot weather.

Potatoes and onions destined for storage are cooled during the early phase of the storage period with cool air forced through storage piles or bins. Cooling may be done with cold ambient air or with air cooled by mechanical refrigeration.

Special treatments. For storage, onions and potatoes are generally sprayed with maleic hydrazide (MH) a few weeks before harvest to inhibit sprouting during storage. Aerosol applications of CIPC (3-chloro-isopropyl-N-phenyl carbamate) are often circulated around stored potatoes to further inhibit sprouting. Rodent control is also necessary storages.

Non-refrigerated storage methods. Some growers occasionally store mature potatoes in the ground several weeks before harvest. Ground storage is also used for several of the tropical and subtropical roots, including cassava and jicama. Pits, trenches, and clamps are used for storage of harvested tropical roots and tubers. Pits are occasionally used for short-term, small-scale storage of potatoes in some areas.

Ventilated storage in cellars and warehouses is used for potatoes, sweet potatoes, garlic and onions. Newer facilities with temperature and relative humidity controls provide forced-air circulation through bulk piles of potatoes or onions, or through and around stacks of bulk bins

Recommended Storage Conditions

Temperate-zone Root Vegetables

In California, temperate-zone root vegetables are not usually stored. When they are stored, the following conditions should be maintained: 0°C (32°F), 95 to 98 percent RH, and adequate air circulation to remove vital heat from the product and prevent carbon dioxide accumulation.

Potatoes can be stored up to 10 months under proper conditions. Most long-term potato storage facilities are in the northern U.S. For fresh market, potatoes should be stored under the following conditions: 4° to 7°C (39° to 45°F), 95 to 98 percent RH, enough air circulation to prevent oxygen depletion and carbon dioxide accumulation (about 0.8 cubic feet per minute per 100 pounds of potatoes), and exclusion of light to avoid greening. Greening is due to chlorophyll synthesis and is associated with accumulation of a toxic alkaloid, solanine. For processing (e.g., chipping), the proper conditions are 8° to 12°C (46° to 54°F), 95 to 98 percent RH, adequate ventilation, and exclusion of light. This higher temperature storage retards undesirable sweetening of the potatoes and consequent dark color of the processed products. Seed potatoes are best kept at 0° to 2°C (32° to 36°F), 95 to 98 percent RH, with adequate ventilation.

Garlic should be kept at 0°C (32°F) for long-term storage (6 to 7 months); 28° to 30°C (82° to 86°F) can be used for storage up to 1 month. Ventilation of about 1 cubic meter of air per minute per cubic meter of garlic with 70 percent RH is adequate.

Onions vary in their storage capability. The more pungent types with high soluble solids contents store longer, whereas mild onions with low soluble solids contents are rarely stored for more than 1 month. Storage temperatures should be either 0° to 5°C (32° to 41°F) or 28° to 30°C (82° to 86°F), as intermediate temperatures favor sprouting. Relative humidity should be maintained at 65 to 70 percent, and ventilation rate should be from 0.5 to 1 cubic meter of air per minute for each cubic meter of onions. Avoid light exposure to prevent greening. The storage potential of onions depends on the cultivar.

The use of controlled or modified atmospheres for this group of commodities is negligible. Limited commercial CA storage (3 percent oxygen, 5 percent carbon dioxide) of mild types of onions has been tested recently. Low concentrations of ethylene in the storage environment of carrots and parsnips induce bitterness.

Tropical-zone Root Vegetables

The general storage recommendations for tropical-zone root vegetables, show that most of these products are chilling sensitive. Some of these root crops (e.g., cassava) are successfully field-stored but deteriorate rapidly if harvested and held under ambient conditions.

13

ADVANCES IN POST-HARVEST TECHNOLOGIES TO ENHANCE SHELF LIFE OF CROPS

Fruits and vegetables are highly perishable and have a very short shelf-life. During different handling and marketing operations, there is a huge postharvest loss of agricultural produce. Both qualitative and quantitative losses occur in horticultural commodities between harvest and consumption. Qualitative losses like loss inedibility, nutritional quality, calorific value, and consumer acceptability of fresh produce are much more difficult to assess than are quantitative losses. Quantitative post-harvest losses in India estimated by different committees ranged between 25 and 33% depending upon the crop. The major cause of postharvest loss is the lack of proper infrastructure for processing and packing. These losses can only be minimized to some extent by proper marketing, handling, and processing of agricultural commodities.

According to a national level study conducted under the All India Coordinated Research Project (AICRP) on Postharvest technology of the Indian Council of Agricultural Research (ICAR) the post-harvest losses during different farm handling operations like harvesting, sorting, grading, and packing accounts for about 13%, during farm storage about 6% and during storage at going down, wholesale and retail level about 12% of the produce goes waste. Thus, on average, about one-third of horticulture produce never reaches the ultimate consumer. This results in a considerable gap between gross food production and net availability.

Insufficient knowledge of pre and post-harvest operations and lack of proper facilities for handling like pre-cooling, grading, packaging, transport, storage, processing, and marketing all together compound the post-harvest losses and wastage which in value terms accounts for more than 6,720,000.00 US dollars.

Keeping the huge postharvest losses in mind, there is an urgent need to reduce the postharvest losses of fresh commodities and increase the level of processing as a reduction in post-harvest losses is a complementary means of production.

The important strategies for loss prevention include the development of varieties (genotypes) that have longer postharvest life, use of integrated crop management system, and development of cost-effective adaptable technologies for post-harvest handling, value addition, and by-product waste utilization. The value chain in post-harvest management of horticultural crops mainly comprises pre-harvest factors, harvesting, market preparation (pre-cooling, sorting, grading, packaging, and on-farm storage), transportation, storage, value addition, and by-product waste management. The status of R&D carried out pertaining to postharvest management (PHM) and processing in the country by different ICAR institutes like Central Institute of Post Harvest Technology (CIPHET) (Ludhiana) and State Agricultural University (SAUs) on different aspects of post-harvest management and processing of horticultural crops is given ahead. Depending upon the status report, research scientists can find out the gap/missing links in the available technology to suggest future priorities in the area of R&D.

Maturity is the state where the product is ready for picking. Proper identification of maturity of produce is essential so that the product is less prone to various physiological disorders and diseases. Maturity indices have been developed for various fruits such as mango, pomegranate, apple, grapes, ber, aonla, Nagpur mandarin, etc. Technique to determine the maturity of mango on the tree (CIPHET) and non-destructive method for the maturity of Grand Naine banana (NRCB, Trichur) need to be popularized.

In recent years, rapid industrialization, population growth, and changed lifestyle led to increased demand for processed and packed foods. Currently, ready to eat packed food industry is growing very fast. Packaging is considered as the science, art, and technology of protecting the products during transportation, distribution, storage, sale, and use. Further, the packaging ensures safe and efficient delivery of the commodity to the consumer in good condition. Good packaging attracts the customer to buy the product. It also plays a vital role in reducing the security risks during shipment.

Packaged products are easy in displaying, handling, storing, distributing, opening, reclosing, and reusing. Packaging performs four important functions, such as containment, protection, convenience, and communication. A wide variety of materials, such as cane baskets, wooden boxes, clay vessels, metal cans, China pots, paper bags, and plastics containers are still used for packaging the products in many areas of the world. The packaging material should not cause any environmental pollution. Hence, there is a need to undertake detailed studies to assess the impact of food packaging on the environment.

Postharvest technologies such as chemical and non-chemical treatments are used to maintain fruit quality during storage. For instance, 1-methylcyclopropene (1-MCP) and nitric oxide (NO) have been demonstrated to be effective chemical treatments for preserving mango fruit quality. Postharvest 1-MCP treatments inhibit ethylene biosynthesis, which retards the respiration rate, retain firmness and delay fruit ripening. Nitric oxide, as a postharvest treatment, is known for prolonging the shelf-life by reducing the incidence of postharvest pathogens and chilling injury.

The use of these treatments has shown to be a promising strategy to enhance fruit quality during storage and postharvest handling chain. However, there are growing concerns regarding the postharvest application of chemical treatments mainly because they do not only harm the environment but also pose various risks to human health. As a result, the focus of postharvest research for mangoes has recently shifted toward environmental-friendly and non-chemical treatments.

Edible coatings are biodegradable postharvest treatments applied to fruit and vegetables. They form a thin layer of material over the fruit surface, creating a protective barrier to oxygen, solute movement of food, and moisture. The advantage with edible coatings is that they are natural, contain antioxidants and sometimes vitamins which are beneficial to the consumers. They also possess anti-browning and anti-microbial properties which maintain fruit quality. Natural edible coatings such as chitosan, Gum Arabic, and carboxymethyl cellulose (CMC) can control postharvest disorders and diseases such as anthracnose, stem-end rot, and black spot in mango.

Heat treatment is another non-chemical technique that has proved to be effective in decreasing postharvest diseases, fruit softening and maintaining mango fruit color. Cell wall degrading enzymes, such as β-galactosidase and polygalacturonase (PG) are reportedly inhibited by heat treatments. However, the response of mango fruit to heat treatment significantly depends on various factors including cultivar, temperature, and exposure time. For certain cultivars such as 'Kent' and 'cat Hoa loc', high temperature and increased exposure time can damage the fruit peel, leading to fruit softening and susceptibility to diseases.

Non-chemical postharvest treatments, as well as innovative and environmental-friendly technologies, have received little attention from researchers. Therefore, the current review provides an extensive overview of different postharvest techniques currently used on mango fruit, focusing on non-chemical treatments. Anthracnose and chilling injury are the most commercially important postharvest challenges affecting the mango industry. This review further discusses the potential of emerging postharvest technologies to preserve fruit quality and control postharvest diseases. Research gaps, as well as prospects for future research of the postharvest research in mangoes, are also highlighted.

Based on their growth habit, temperate fruits are categorized into three categories: tree fruits (such as apple, pear, plum, and peach), vine fruits (e.g., grape and kiwifruit), and small fruits, which refers to the plant's size and not necessarily the fruit, for instance, strawberry, raspberry, currant, and blueberry.

TREE FRUITS

Tree fruits belong to the rose family (Rosaceae) and they include the pome and stone fruits. Pome fruits (i.e., apple, pear, quince, and loquat) are characterized by a fleshy outer portion, which is formed by the expansion of floral parts and receptacle. All stone fruits (almond, apricot, cherry, peach, nectarine, and plum) are members of the genus Prunus and they are commonly named as 'stone fruits' because they have a stony pit which encloses a solitary seed.

Apple (Malus × domestica) is the first known fruit of the pome fruits and it has been known since antiquity. It is grown in northern China and Siberia, where temperatures fall to about -40 °C during the winter season, in high elevation locations such as Colombia, and, on the either hand, on the equator, where two crops can be produced.

Pear (*Pyrus species*) is as old a fruit crop as apple and both are consumed in similar ways; however, pear is somewhat less popular in the West. Pears are divided into

European pear (*P. communis*) and Asian pear (*P. pyrifolia* and *P. ussuriensis*). The European pears are usually consumed when they soften after harvest, while the Asian pears are consumed when crisp. The quince (Cydonia oblonga) comes in the third place of importance among the pome fruits and they are usually grown in warm climates and can be consumed raw. Despite this, quinces are not very popular because their tastes are sour and they are astringent before ripe.

Peach (*Prunus persica*), which has become the most popular temperate summer fruit, originated from China and was introduced to Europe via Persia in the first century. There are many types of peaches. The free-stone types, which have melting yellow or white flesh, are usually consumed fresh and the second one, the clingstone types with rubbery flesh, are used in processing; the nectarine, which is characterized by non-fuzzy skin, resulted from a mutation, and Peento, a flat peach with a saucer shape, is another variant.

Almond (*Prunus amygdalus*) originated from the hot regions of Western Asia, but it was prehistorically introduced to Greece and West Africa. Its flesh is leathery and inedible, and its fruits are consumed when quite immature, especially in Arabian countries. Not alike the majority of stone fruits, almonds' seeds have a non-bitter taste and are the consumed part of this fruit, so this species is categorized under nuts. The almond is extensively grown in California and is widely planted in Mediterranean countries.

Cherries, one of the most popular early summer fruits, comprise about 30 species. Among the edible types of cherries, there is sweet cherry (*Prunus avium*), tart (sour) cherry (*P. cerasus*), ground cherry (*Prunus fruticosa*), and duke cherry, which is a hybrid between sweet and tart cherries. Cherries could have yellow, red, or yellow-red bicolor and are consumed fresh, dried, or processed. There are also some semi-processed cherries, such as maraschino cherry, in which cherries are brined, bleached, and then colored, as well as flavored with almond oil.

Apricot (*Prunus armeniaca*), which is native to central Asia and China, is characterized by its soft velvet skin and is consumed fresh, dried, or processed. The apricot blooms very early, is quite sensitive to spring frost, and is difficult to grown on a large scale in such areas. This could be part of the reason why apricot does not have the same degree of importance as peach, cherry, and plum. The beautifully flowered Chinese plum is considered more as an apricot type rather than a plum type.

Plums comprise a diverse group of fruits based on their origins (Europe, Asia, and America). Among the European species, *Prunus domestica*, such as greengage and prune types, and *P. insititia*, which includes bullaces, damsons, mirabelles, and St. Julien types, are characterized by having six sets of chromosomes. Asiatic species contain *P. salicina* and *P. simonii* (apricot plum); the former includes both green- and red-fleshed Japanese plums, while the latter is cultivated in China.

However, there are many American plum species, with none of them being widely planted. Most of the plum industry is mainly based on *P. domestica* in Europe and *P. salicina* in Asia. Plums could be consumed either fresh or dried. The dried plums without fermentation are called plum prunes. Their drying process is run under very low moisture levels, so they can be stored for long periods of time.

VINE FRUITS

Grapes (*Vitis vinifera*), one of the most important temperate fruit species, are usually grown on trellises. Vitis vinifera have been popular for wine production for a long time. However, while any sweet fruit can be used in wine production, the grape is always the preferred species because it has a unique combination of sugars, acids, and astringent substances. Due to this, producers are always reluctant to change the cultivars of grapes to avoid changing the standard characteristics of their produced wine. Grapes can be eaten fresh, especially the seedless varieties. The large-fruited table grape Italia, widely appreciated in Europe, has a Muscat flavor which is sweet and pleasant. Recently, non-alcoholic grape juice, produced from the American species *V. labrusca*, has become common in the US market.

Kiwifruit (*Actinidia deliciosa*) has recently been domesticated. It is derived from a Chinese fruit called Yang Tao, which was collected rather than cultivated. New Zealand growers succeeded in domesticating the crop by selecting suitable male and female clones, as well as techniques for cultivation. After this early adoption, the fruit received a new name, kiwifruit, referring to Kiwi, the popular flightless bird of New Zealand. Kiwifruit has a pleasant but weak flavor and a unique sliced shape, so it has been used in cake decoration or as a component of fruit salad, which has made this fruit globally popular. The fruit is also characterized with the ability of long storage life and this is what makes it possible for New Zealand to export it throughout the four seasons.

SMALL FRUITS

Strawberry is an important small fruit in most temperate climates. Its fruit size has greatly increased, and it has firm flesh with improved flavor and appearance. Grown in all temperate countries, the strawberries industry is focused in southern California in the United States, southern Spain, and Italy.

Blackberries and raspberries belong to the genus *Rubus*, which includes many species and a vast number of hybrids. Rubus is very diverse and its cultivated species are known as brambles, which include black raspberry (*R. occidentalis*), blackberry (*Rubus species*), and various interspecific hybrids between raspberry and blackberry, such as loganberry, boysenberry, and tayberry.

Blueberry (various species) is native to the United States and grows in bushes of various heights. The cranberry (*Vaccinium macrocarpon*) is an unusual berry crop because it is grown submerged in bogs. The fruits are too acid to be eaten raw and are consumed processed as jelly or as a sauce. Lingonberry (*V. vitis*-idaea var. *minus*) is native to northern regions of Europe, Asia, and North America. The loganberry is usually considered to be a hybrid between a blackberry and a raspberry, and its principal use is for juice.

MATURITY AND HARVESTING INDICES

Maturity at harvest is one of the major factors that influence fruit composition, storage life, and final fruit quality. Harvesting at the ideal moment is very important for the development of quality after harvest and for a long postharvest life. When harvested late, fruits may not have a sufficiently long postharvest life. Fruits harvested too early or too late in the season are more subject to physiological disorders and have a shorter storage life than those picked at optimum physiological maturity. Most fruits attain a great eating quality when harvested fully ripe, usually picked at the mature stage, but may also be picked when not ripe to decrease mechanical damage during postharvest handling. Moreover, mechanical harvesting procedures may also damage fruit; thus, the choice of harvest method should maintain fruit quality.

Most of the maturity indices currently used are the result of a combination between the indices which ensure the best food quality to the consumer and those providing the necessary flexibility for transport and marketing. Temperate fruits could be climacteric or non-climacteric. Non-climacteric fruits are not able to continue their ripening process if removed from the tree, while climacteric fruits can be harvested mature and ripened off the plant. Examples of temperate fruit belonging to the two groups are:

- Group 1: Berries (e.g., blackberry, cranberry, raspberry, strawberry), grape, cherry.
- Group 2: Apples, pear, quince, apricot, nectarine, peach, plum.

Fruits of the first group (non-climacteric) produce very small quantities of ethylene and do not respond to ethylene treatment, while those of group 2 (climacteric) produce more significant quantities of ethylene during their ripening. In this group, ethylene treatment will result in faster and more uniform ripening. Many characteristics of fruits have been tested as maturity and harvesting indices, including morphological, anatomical, physiological, and biochemical features and components.

Relevant maturity and harvesting indices are those traits or components that change during the development of the fruit, correlating closely and regularly with the quality of the fruit, and are not strongly affected by factors such as the environment. Maturity indices commonly used for most of the temperate fruits are fruit size and shape, overall and ground color of the skin, flesh color, flesh firmness, soluble solids content, starch content, acidity, etc. Some of these indices are shown in Table 1.

TABLE 1 EXAMPLES OF MATURITY INDICES FOR TEMPERATE FRUIT

Indices	Fruits
Elapsed days from full bloom to harvest	Apple, pear
Size	Most fruits
Shape (fullness of fruit shoulders and suture)	Stone fruits
Dry weight	Kiwifruit
Color, external	All fruits
Firmness	Pome and stone fruits, berries
Compositional factors	
- Starch content	- Apple, pear
- Soluble solids content	- Pome and stone fruits, grapes, kiwifruit
- Acid content	- Most fruits
- Sugar/acid ratio	- Grape

FRUIT QUALITY

'Quality' is a term that refers to a degree of excellence, a high level or value. Quality value and acceptance of fruits depend on the values of a combination of several attributes, such as color, appearance, flavor, texture, nutritional characteristics, and safety. Fruit color is one of the most important appearance attributes, which is due to the presence of several pigments, including carotenoids, betalains, flavonoids (anthocyanins), and chlorophyll.

As consumers, fruit color typically affects our acceptance and perception of sweetness and flavor. Moreover, it can even evoke emotional feelings in humans. Used as a ripening index, fruit color is useful in determining the harvest date and estimating the length of time for which a fruit can be stored and marketed. A large number of temperate fruits contain several pigments which are used to determine the harvesting time and maturity stage.

Fruit texture is another important quality attribute which is defined as the group of mechanical properties determining human sensations when they eat food. This attribute influences considerably the acceptability of the product to the consumer. Cell anatomy, water content, and the composition of cell walls are the most important factors influencing fruit texture. These factors are strongly affected by ripening.

Several fruit texture attributes exist, including crispness, mealiness, juiciness, and firmness. Firmness is the most important attribute of texture for fruit quality, as it decreases during storage and marketing. It also determines the mechanical properties of fruits, their ability to resist the postharvest handling and resistance to mechanical damage.

Sugar content is a crucial attribute for the selection of temperate fruits. This attribute affects fruit flavors and determines the caloric value of fruits. The first stages of fruit development are characterized by a high concentration of starch, which is progressively hydrolyzed during ripening; resulting in an increase in the content of simple sugars, such as glucose, fructose, and sucrose.

Organic acids are also abundant in temperate fruits, mainly citric, malic, and ascorbic. Ascorbic acid is particularly important and it greatly contributes to many aspects of human health. Organic acid content decreases during ripening and is involved in the flavor, texture, pH, and color of fruits, altering their sensorial quality. For some temperate fruits, the ratio of sugar to acids is commonly used as a ripening index.

Temperate fruit quality is also related to the volatile compounds content. Many of them give off a wide range of volatile compounds, some of which are considered extremely important for fruit quality. The flavor of fruits depends on a combination of aroma and taste sensations.

Postharvest procedures considerably affect fruit quality. In developing countries, 40–50% of horticultural crops produced are lost before they can be consumed, due to the high rates of bruising, water loss, and subsequent decay during postharvest handling. Several biochemical and physiological changes occur after the harvesting of temperate fruits.

Most of these changes, such as cell wall degradation, water loss, shriveling, softening, overripening, disease infestation, shelf life decreases, etc., are undesirable. These changes strongly affect fruit quality and they can be slowed down by controlling and improving postharvest handling and development of new postharvest technologies.

CHALLENGES OF POSTHARVEST TECHNOLOGY

To increase fresh tropical fruit production and export volume in Southeast Asia, some problems need to be resolved. There has been rapid progress in the development of postharvest processes in majority of Southeast Asia over the last 20 years. However, Southeast Asian countries still need to reduce postharvest losses.

Majority of the Southeast Asian countries have the world's lowest agricultural research intensity ratios, which affect agricultural production values, and underinvestment in agricultural research affects the development of agricultural technologies. Postharvest losses were 42% for fresh fruits and vegetables in Southeast Asian regions.

The major cause is the short shelf life of the tropical fruits as well as losses during each process such as production, postharvest, and distribution. Poor infrastructure for storage, processing, and market can result in a high proportion of waste (10~40% of the total production). Postharvest losses are caused by improper packaging or handling, lack of cold storage in warm and humid climates, and seasonality that results in surplus.

Major infrastructural limitations also continue to impose severe constraints on domestic distribution as well as export of horticultural produce. Agricultural practices in almost all Southeast Asian countries are still dependent on conventional methods, and a few practices use a mixture of traditional and modern methods on small-scale farms with limited postharvest knowledge.

During the peak season for harvesting tropical fruits, there is a glut in local markets and crops are sold at very low prices because Southeast Asian countries lack storage facilities for tropical fruits. To overcome this issue in Southeast Asian countries, effective elimination in the production of tropical fruits is required through communication and idea exchange among farmers, postharvest engineers, food technologists, and the government. .

Reduction in postharvest losses and extension of the shelf life are possible with advanced research and development of applicable technology for postharvest processes. The direction of research on postharvest technologies should be established in Southeast Asian countries to use postharvest technologies on small-scale farms and improve knowledge on postharvest management.

Developed countries have extensive and effective postharvest systems for extending the shelf life, educated farmers, and good chain systems in the markets. The governments of Southeast Asian countries should take initiatives and allocate resources to improve postharvest handling conditions to catch up to developed countries. Therefore, reduction in postharvest losses is possible with the help of modern techniques and advanced research.

Factors That Affect the Shelf Life of Crops

The short shelf life of fruits is an inherent drawback of the distribution chain. A lot of transportation time for exported perishable agricultural or tropical products is required. Limited shelf life of the fresh fruits is caused by their distinct properties, external conditions, microbial contamination, physiological disorders, mechanical damage, and the level of postharvest treatment. Shelf life is often related to a combination of several factors. Although it is difficult to control all the factors that affect the shelf life of fruits, minimizing the effect of each factor on the shelf life is technically feasible.

Natural Properties

Postharvest losses of tropical fruits are due to a lack of information on their nature (Table 2). Tropical fruits have a shorter shelf life than temperate fruits because of their perishable nature. As a result, these fruits cannot be stored for more than 4~6 days under normal ambient conditions. Good initial quality of fruits encompasses appearance, firmness, aroma, taste, nutritive value, and chemical components. Therefore, it is important to harvest fruits at the proper stage, size, and peak quality. However, Thailand, Laos, and Vietnam have not established any standards or harvest indicators to determine maturit.

TABLE 2. NATURAL PROPERTIES OF TROPICAL FRUITS THAT ARE REQUIRED FOR HANDLING AFTER HARVEST

Natural properties of tropical fruits	
1.	Product with a high moisture content (80~90%)
2.	Some fruits are large and heavy (5 g to 5 kg or more), e.g., litchi, durian, and pineapple
3.	High to very high respiratory rates that lead to heat emission at ambient temperatures
4.	Soft tissues that are highly vulnerable to mechanical injury
5.	Highly perishable with a generally limited postharvest life of days to weeks
6.	Susceptibility to storage at temperatures less than 10°C to 15°C – rotting, failure to ripen, skin, and scalding

	Natural properties of tropical fruits
7.	No kill step during handling; dependence upon layers of protection to minimize microbial contamination
8.	Host to tropical insects, so disinfestation treatment is required prior to export

Harvested fruits are still alive and continue their physiological activities, such as respiration, transpiration, and developmental changes in chemical components that cause loss of color, flavor, texture, and nutritional value. The rate of respiration of the harvested commodities is inversely proportional to the shelf life of the produce: higher respiration rates decrease the shelf life.

Water loss from fresh fruits occurs through transpiration, which affects appearance, salable weight, texture quality, crispness, and juiciness. These physiological behaviors of the fruits are important concerns with respect to postharvest technologies. The quality of the tropical fruit commodities cannot be improved; however, it can be retained until their consumption if the rate of metabolic activities is reduced by using appropriate postharvest handling operations.

External Condition

The external condition has a great influence on the shelf life of fruits. Factors of external condition can be subdivided into temperature, composition of the air surrounding the produce, and humidity of the environment. In general, high tropical temperatures and relative humidity in Southeast Asia regions play a major role in accelerating the rate of chemical and biological degradation as well as providing a conducive atmosphere for the multiplication of tropical insects that contribute to postharvest deterioration.

All fresh produce are subjected to damage when exposed to extreme temperatures that increase the level of respiration. Gas composition surrounding fruits, such as O_2, CO_2, or ethylene, also exerts a great impact on the respiratory and general metabolism of fruits.

Relative humidity is closely associated with water loss, and the transpiration process can be minimized by increasing relative humidity. Most fruit crops retain better quality at high relative humidity (>90%). However, fruits with a high water content are considered more perishable, and nearly 33% of fruits are spoiled during transportation because of bacterial contamination. Humidity chosen for storing produce is frequently a "trade-off" between minimizing water loss and minimizing disease. External condition has a great effect on deterioration in fruit quality because it has a direct relationship with the moisture content in the atmosphere, which determines whether the shelf life will not be exceeded.

Microbial Contamination and Disease

Biodeterioration can be caused by insects, rodents, fungi, or microorganisms. Biological spoilage and disease are the main causes of postharvest loss in developing countries. Two critical requirements for the growth of microorganisms are nutrients, such as organic substrates, and high water activity, and fruits are good substrates for microorganisms.

Microorganisms readily contaminate fresh produce and spread rapidly owing to weak defense mechanisms in the tissues of fresh produce and the abundance of nutrients and moisture.

Contamination not only affects the produce quantitatively but also qualitatively. Since the optimum growth temperatures for most fungi fall between 25°C and 30°C, growth above 40°C is poor and some cases of bacterial mortality may occur.

Tropical fresh fruits can be easily contaminated by microorganisms because the optimum temperature for microbial contamination is the average temperature in tropical countries. Mango is a climacteric fruit and has quite a long shelf life of more than 10 days if not attacked by fungi. The two most serious fungal diseases of tropical and subtropical fruits such as mango, banana, papaya, and avocado are powdery mildew (*Oidium mangiferae*) and anthracnose (*Colletotrichum gloeosporioides*), which are prevalent in regions with high humidity. Approximately 20~25% of the harvested fruits and vegetables decay because of pathogens during postharvest handling even in developed countries. Physiological and storage conditions at the postharvest stage, especially under tropical conditions, accelerate the spread of the abovementioned diseases.

Handling Methods

Poor handling, unsuitable packaging, and improper packing during transportation cause bruising, breaking, wounding, and other forms of injury in fresh fruits and vegetables. Simple processes for picking, transferring, and packaging tropical fruits in Southeast Asian countries make the fruits susceptible to mechanical damage. Handling methods can determine the extent of variability in maturity and physical injuries and consequently influence the composition and quality of fruits.

TECHNOLOGIES TO EXTEND THE SHELF LIFE OF CROPS

Primary factors for maintaining and extending the quality and shelf life of fruits are the determination of optimum harvest maturity, minimization of mechanical injuries, use of proper postharvest treatment, and optimum temperature and relative humidity during all market steps. Postharvest technologies in Southeast Asia are expected to be used for the enhancement of productivity and quality.

Such technologies can extend the shelf life of fruits, consequently enabling the distribution of high-quality produce to distant markets. Recently, many postharvest technologies such as cooling chain, packaging, controlled atmosphere (CA), heat treatments, and coating have been developed in real commercial areas or agricultural laboratories. Postharvest technologies should be applied on the basis of usability, fruit characteristics, and environmental conditions.

Cooling Chain

Delays between harvesting and cooling or processing can result in direct quality losses due to water loss and decay and indirect losses such as those in flavor and nutritional quality. In Southeast Asia, small-scale farmers do not use cooling process after harvesting, and several fruit companies use room storage cooling for extending the shelf life of tropical fruits. Cooling of fruits is necessary at different stages to control degradation: after harvesting (pre-cooling), during storage (cold storage), and during transportation (vehicle with freezer). It extends the shelf life of fresh produce by reducing the rate of physiological changes and by delaying or preventing the growth of microorganisms.

Precooling is a process that removes field heat from freshly harvested produce by using cooling treatments, i.e., room cooling, forced-air cooling, hydro-cooling, and vacuum cooling; each treatment has different advantages. A forced air-cooling system is used for the treatment of exported durian, and hydro-cooling is used for vegetables in Thailand.

Cooling treatment is recommended for most fruits (Intelligent Supply Chain Information, 2010). Refrigeration or cold storage is considered the most efficient method to maintain the quality of most fruits. Durian is a climacteric fruit that is very sensitive to chilling injury; therefore, the storage temperature should not be lower than 15°C. Cold storage is the most effective treatment for extending the shelf life for distribution During the handling stage, proper temperature and relative humidity based on fruit characteristics are important. In general, lower the storage temperature, longer the storage life of the commodity, as long as it is not reached the freezing temperature.

However, the lowest safe storage temperature is not the same for all commodities because a number of tropical fruits are sensitive to low temperature and inappropriate storage temperature may cause chilling injuries such as irregular ripening, failure to ripen, pits on the skin surface, and increased susceptibility to decay.

Sensitivity to chilling injury is commonly associated with subtropical produce and fruits of tropical origin like banana, pineapple, and tomato, which should be stored below 8°C (Singh et al., 2014). Another way to minimize damage due to low temperature storage is the use of a combination of two different treatments (such as good packaging, edible coating, or radiation treatment). For example, preservation of coated papayas at their ambient storage temperature (10~13°C) extends their shelf life. In Malaysia, cooling room and water cooling are used to minimize the damage during the transportation process.

Thermal Treatment

Thermal treatment is a relatively simple method to control deterioration in food quality. Heat treatment between 30°C and 50°C for 10 min to 24 h is commonly used for tropical and subtropical fruits, and it is effective against a wide variety of insects and fungi (Department Agriculture of Malaysia). However, hot water treatment is limited to a narrow range of temperatures and time intervals, and the values are highly dependent on the commodity.

High temperature affects the nutritional quality and sensory properties of fruits. The best combination of suitable temperature and exposure time should be determined to prevent quality loss. Inappropriate set-up of temperature and dipping time during heat treatment causes accelerated fruit ripening, increase in weight loss, reduction in fruit firmness, and increase in the brix of fruit. Hot water, vapor heat, and hot air are commonly used for the thermal treatment of tropical fruits. Hot water is often used in combination with fungicides to control postharvest diseases and physiological and biological activities, extending the shelf life of mango, papaya, and melon. Hot water treatment at 53°C for 9 min is suitable for banana in terms of weight loss, color, firmness, soluble solid, sugar, and â-carotene. However, thermal treatment at 57°C significantly affects the shelf life of banana because it causes burning of skin and damage of tissues.

Most exporters use hot water treatmentfor harvested mangoes; they dip fresh mangoes in hot water (52~55°C) for approximately 10 min. This treatment reduces the incidence of anthracnose and stem-end rot by 10%, resulting in almost a 100% recovery rate and an increase in sales and income. Researchers reported that hot water treatment at 52°C for 5 min was the best treatment for controlling anthracnose in Keitt mangoes during 21 days of storage. Thermal treatment is suitable for tropical fruits because tropical fruits tolerate heat well. Furthermore, thermal treatment is more effective if it is combined with other treatments such as storage in low temperatures.

14

QUARANTINE AND PHYTOSANITORY TREATMENTS FOR FRESH PRODUCE

QUARANTINE TREATMENTS

Until recently, postharvest treatments with the insecticides dimethoate and fenthion were used to provide quarantine security against fruit fly infestation. While preharvest application of these chemicals is still allowed for certain use patterns, they can no longer be applied postharvest to vegetables.

Currently accepted alternatives to chemical controls include cold storage, fumigation with methyl bromide and irradiation.

COLD STORAGE

Cold storage is a commonly used method for ensuring that fruit do not contain live fruit fly larvae. Cold disinfestation is accepted by a number of Australia's trading partners (eg Japan, China, Taiwan, USA) for products such as citrus, stonefruit, grapes and kiwifruit. Cold disinfestation can be done in-transit, with data loggers used to provide proof of treatment.

Protocols may be on a sliding scale according to temperature. So, for example, Australian mandarins must be held at <1°C for 15 days, <2°C for 18 days or <3°C for 20 days in order to access Japan.

Such treatments are very conservative. Research on various fruit crops have all found e"98% mortality after eight days at 3°C. A project on disinfestation of capsicums found that no fruit fly of any life stage survived when held at 3°C for 10 days. However, the shortest currently accepted protocol requires 14 days at <3°C.

There are currently no cold storage protocols for vegetables for a number of reasons:

- Most fruiting vegetables are considered chilling sensitive at the temperatures needed to kill fruit fly larvae (d"3°C).

- The relatively long storage times required to achieve Probit 9 mortality are difficult and expensive to implement within vegetable crop supply chains.
- Cold treatments are both crop and pest specific. Potential export markets are now raising concerns about fruit fly species other than Qfly and Medfly; trials may therefore need to prove effectiveness of a given time/temperature combination against additional species, such as *B. neohumeralis, B. jarvisii and B. cucumis.* This makes the development of disinfestation treatments expensive, logistically difficult and time consuming.

HEAT TREATMENTS

Heat treatments are also primarily used to ensure products are free of live fruit fly larvae. They can be applied through water, air or vapour. Most require exposure times of 30 minutes to several hours at 40–50°C. Heat treatments have generally been focused on chilling sensitive, high-value fruit crops such as mangoes, avocados and papaya. Treatments have also been developed for vegetable crops including zucchini, squash, and tomatoes. However, none have developed into commercial protocols.

Fig. 1 Vapour heat treatment unit

Difficulties with heat treatments include:

- They are expensive and technically difficult to apply.
- The time/temperature combinations necessary can damage the product.
- Like cold treatments, heat treatments are crop and pest specific, making it expensive to develop quarantine protocols. Also, at least one Australian fruit fly species may be highly resistant to heat treatment.

ATMOSPHERE

Although reduced O_2 or increased CO_2 are not effective at killing insect pests when used alone, they can improve the efficacy of other measures. So, for example, high CO_2 combined with cold storage can reduce the time needed to achieve Probit 9 mortality by several days.

However, the gas concentrations needed are extreme; an atmosphere containing <13% O_2 or >%7 CO_2 is deadly to humans, but insects can survive <%0.5 O_2 and >%50 CO_2. The best effects are gained using atmospheres containing >90% CO_2. Such atmospheres raise issues with worker health and safety as well as being expensive to measure and maintain. They can also damage the product.

In addition, regulators may be reluctant to accept treatments that rely on more than one measure, as they are relatively difficult to verify. There are currently no commercial protocols using modified atmospheres to kill pests.

FUMIGATION

Fumigation has a number of advantages in that it is relatively inexpensive, fast, leaves few residues and can be conducted on packed product so long as materials are vented. The fumigant methyl bromide (MB) remains the main generic quarantine treatment allowing international and interstate trade in fresh produce. So far, no other product has been found to be as effective as MB against fruit fly as well as other insect pests.

Fig. 2. Fumigation using a shipping container equipped with MB recapture equipment. This removes MB from the exhaust air after treatment.

Although the Montreal Protocol banned most uses of MB due to its ozone- depleting properties (effective January 2005), quarantine treatments currently have a critical use exemption. Despite this, use is likely to become increasingly restricted; in 2010 the European Union banned quarantine use of MB on both local and imported products.

The emissions of MB responsible for ozone depletion can be greatly reduced using recapture technology. After fumigation, the room or container air is pumped through activated carbon filters that absorb the MB. Once saturated, the carbon filter material can be deep buried, where the MB contained is naturally destroyed by soil bacteria. The carbon can, alternatively, be chemically treated to reduce it to non-toxic compounds or destroyed by pyrolysis. The USA has set a target that all MB used for quarantine purposes should be recaptured by 2020. Similar targets exist in other countries, including New Zealand.

ICA-04 describes a methyl bromide fumigation protocol with sliding scale for dose and temperature. Products need to be warmed to at least 10°C (flesh temperature) before treatment, although at higher temperatures MB dosage can be reduced. An approved fumigation chamber must be used with vegetables occupying 30–50% of the total chamber volume. Dosage is calculated on the total chamber volume (not the chamber volume less the volume of product) and treatment time is standardised at two hours.

Phosphine is another common fumigant widely used against stored product pests (eg weevils in grain). It has the advantage that it can be applied at low temperatures. However, exposure times are much longer than those required for MB, being days rather than hours. Phosphine is also extremely toxic to humans and potentially explosive. A non-explosive formulation of phosphine with CO_2 (ECO_2FUME) has been shown to be effective against Qfly in citrus when applied for 48 hours.

Another alternative to MB is ethyl formate. This fumigant kills surface pests such as thrips and mealybugs yet is very safe for humans. Unfortunately, the high doses needed can damage vegetable skins, causing discolouration and pitting.

A similar problem exists with ethane dinitrile, a relatively new fumigant recently tested on vegetables. A dose of 30g.m EDN for two hours at 15°C resulted in 100% mortality of fruit fly larvae in chillies, zucchini and capsicum. Unfortunately, doses of 5g.m or more caused significant damage, including softening, pitting and rots, to all vegetables tested (tomatoes, capsicum, squash, cucumber and zucchini).

Hydrogen cyanide (HCN) is sometimes used as a fumigant against surface pests such as mealybugs and aphids. It is used in some countries to treat bananas, pineapples and other tropical fruit. However, HCN is extremely toxic, being potentially deadly to humans through skin contact alone. It also causes skin damage to many fresh products.

IRRADIATION

Irradiation involves exposing the product to ionising radiation. This is usually from a radioactive source (cobalt-60), but can also be produced electronically. Irradiation may not kill insects immediately, but stops them from continuing to develop and reproduce.

A dose of 150 Gy is accepted by some countries (including USA) as a generic treatment for fruit fly. This treatment has been approved by FSANZ for capsicums and tomatoes, based on studies showing nutritional qualities were unaffected.

Irradiation has significant advantages, particularly, the acceptance of a generic dose against insect pests. It leaves no residues, is fast, and can be applied to packed product. However, many export markets, including Japan, Taiwan and the European Union, do not permit irradiation. Where it is accepted (eg New Zealand) products must be labelled as such, which limits consumer acceptance.

Other barriers to use of irradiation include cost and logistical difficulties; the only Australian commercial irradiation facility is in Brisbane. The delay in insect mortality also creates issues, although a new ELISA test that can detect changes in irradiated insects may help overcome this barrier.

GENERIC PHYTOSANITARY IRRADIATION TREATMENTS USED COMMERCIALLY

A generic phytosanitary treatment is one specific dose that is used for a group of quarantine pests and/or commodities although not all were tested for efficacy. Although used to a very limited extent with some phytosanitary treatments, the concept has found broad commercial application in PI in that virtually all commercial PI is done using generic doses.

The most used generic dose is a minimum of 400 Gy for all insects except pupae and adults of the order Lepidoptera (moths and butterflies). It is approved for all products exported to the USA. New Zealand has a similar generic dose of 400 Gy for all insects except adult and pupae of Lepidoptera, but it includes mites of the family Tetranychidae (spider mites). Furthermore it is prohibited for use on disease vectoring species, because, although efficacious to prevent further development or reproduction of the insects, 400 Gy has not been shown to prevent disease transmission before the insect vectors die. However, the 400 Gy dose approved by New Zealand is only for mangoes, lychees, tomatoes, and capsicum peppers. A generic dose of 500 Gy is approved in New Zealand for mites other than Tetranychidae.

A generic dose of 250 Gy for import of lychee and mango into New Zealand is approved against regulated pests including a wide variety of species from the insect orders Coleoptera (beetles), Diptera (flies), Hemiptera (scales, mealybugs, whiteflies, among others), Lepidoptera, and Thysanoptera (thrips). A generic 300 Gy dose for Australian mangoes exported to Malaysia includes many of the same regulated pests for New Zealand, but also includes the mango seed weevil. The reason that the dose is 300 Gy for Malaysia instead of 250 Gy is because the mango seed weevil can become established in Malaysia while mangoes are not grown in New Zealand. Hallman discusses why the dose to disinfest mangoes of seed weevil was set at 300 Gy and not lower. A dose of 150 Gy against all weevils has recently been recommended.

Tephritid fruit flies are the most important group of regulated pests requiring a phytosanitary treatment in fresh fruits because of their ubiquity, cryptic nature while infesting fruit, and broad host range. A generic dose of 150 Gy against fruit flies of the family Tephritidae (e.g., Mediterranean fruit fly, oriental fruit fly) is broadly accepted and is commercially used in a few instances where the only regulated pests in certain fruits are these flies.

New Generic Phytosanitary Irradiation Treatments Proposed

A review of literature concluded that 400 Gy would suffice for pupae of Lepidoptera with a measure of efficacy being prevention of hatch of eggs laid by moths emerging from irradiated pupae. This treatment dose was proposed to the International Plant Protection Convention (IPPC) for inclusion in the phytosanitary treatment manual, but was rejected for lack of sufficient large-scale tests with tens of thousands of insects. Nevertheless, although the IPPC also rejected a proposed dose of 400 Gy against all insects except pupae and adults of Lepidoptera, that dose is the one used for ~95% of all commodities irradiated for phytosanitary purposes.

Likewise, a dose of 250 Gy proposed for use against all eggs and larvae of Lepidoptera with the measure of efficacy being prevention of normal-looking adults was rejected by the IPPC for lack of sufficient large-scale testing. However, there is sufficient large-scale testing for the most important family in Lepidoptera, the Tortricidae (e.g., codling moth, oriental fruit moth, light brown apple moth), that would satisfy the requirements of the IPPC.

Although a generic PI treatment for all Tephritidae of 150 Gy is broadly accepted, lower doses for individual species allow for use in specific cases. For example, guavas from Florida may be irradiated with 70 Gy against Caribbean fruit fly and shipped to other parts of the USA. Generic doses for groups of pest species within the family Tephritidae may be justifiable; a generic dose of 70 Gy has been proposed for all fruits infested with species of the tephritid genus *Anastrepha*, which includes all but a few of the species attacking commercial fruit in the American tropics and subtropics. That dose could be used on mangoes exported from Mexico to the USA. Currently almost all mangoes exported to the USA from Mexico are treated with water at 46.1 °C, but irradiation provides for a safer and better quality mango.

After tephritid fruit flies and Lepidoptera, the next most important group of regulated pests on fresh fruits and vegetables is mealybugs, which are general surface pests on many fruits and vegetables. A generic dose of 250 Gy supported by large scale studies with several species has been proposed.

A generic dose of 150 Gy is proposed for weevils of the main weevil family Curculionidae. This dose would allow mangoes from areas that have one of more of the weevils that infest mangoes to be treated with 150 Gy instead of 300 or 400 Gy.

Researchers conclude that the generic doses of 400 and 500 Gy, respectively, for Tetranychidae and all other mites accepted by Australia and New Zealand, although not supported by large-scale confirmatory testing, are probably high enough to be phytosanitarily safe. After large-scale confirmatory testing it might be possible to develop somewhat lower doses.

Researchers argues that the 400 Gy generic dose for insects other than pupa and adult Lepidoptera can be lowered to 300 Gy with minimal increase in phytosanitary risk. This dose would still leave a respectable margin of security because research indicates that these insects can be controlled with ~250 Gy. Data from this study also indicate that doses of ~250 and 200 Gy, respectively, might suffice for armoured scales (family Diaspididae) and leafminers (family Agromyzidae).

EFFECT OF PHYTOSANITARY IRRADIATION ON FRUITS AND VEGETABLES

For phytosanitary treatments to be commercially feasible they must be efficacious and commodities must not be rendered unmarketable by them. At the doses used for PI (70 to 400 Gy), more fresh fruits and vegetables tolerate radiation than any other broadly applicable commercial treatment.

Applying radiation to standard packed pallet loads is economical because commodity can be treated with minimum handling; however, this may result in much of the load receiving a much greater radiation dose than necessary to ensure that all parts of the load receive at least the minimum required dose. For example, fruits on the outside of pallet loads irradiated at a cobalt-60 facility in South Africa can receive almost four times the minimum prescribed dose to ensure that the minimum dose is absorbed by the entire load. This is for two reasons:

(1) the irradiation facility aims to deliver a minimum dose higher than the prescribed PI treatment dose to ensure that the actual minimum dose received is at least the PI treatment dose within a large degree of statistical confidence; and

(2) the dose uniformity ratio for the process (the ratio of maximum to minimum absorbed dose in the production lot) may be as high as three. Electron beams cannot penetrate deeply and therefore these facilities are designed to process boxes of product and not pallets, and can achieve a dose uniformity ratio as low as ~1.2. No matter what source of radiation is used, commodities in their commercial packaging should be tested for radiotolerance at the maximum doses that could be absorbed by any part of the load configuration.

It may also be necessary to adapt harvesting procedures when using PI. Radiation may slow the rate of fruit ripening, and fruit that is commercially harvested before it is ready to be eaten, such as papayas, mangoes, and many other tropical fruits, may not ripen as quickly as non-irradiated fruit. Delaying the physiological stage of harvest may improve ripening quality, and delayed harvest alone should result in improved quality. However, the fact that papaya, mango, and guava have been irradiated commercially from different countries for many years while in the mature hard green stage testifies to the ability of fruit to tolerate PI and still ripen normally.

The International Database on Insect Disinfestation and Sterilization maintained by the Joint FAO/IAEA Programme on Nuclear Techniques in Food and Agriculture is developing an annotated database on tolerance of fresh commodities to ionizing radiation. This database will help facilitate international use of the technology as it can be used by industry to determine the feasibility of using commercial PI. A publication summarizing the literature with a discussion of research methodology and further research that might be warranted is also planned.

PI has been applied commercially to a wide variety of fresh commodities, and the fact that they have been successfully marketed after receiving doses that may considerably exceed the minimum doses required for efficacy attests to the broad tolerance of fresh commodities at doses of radiation required for phytosanitation. The following commodities are irradiated with minimum doses of 400 Gy that results in much of the commodity absorbing approximately twice that dose without rendering them unmarketable: bell peppers, dragonfruit, grape, grapefruit, guava, lime, longan, lychee, mango, manzano pepper, papaya, persimmon, plum, rambutan, sweet potato, and tomato. A few of the commodities suffered initial problems related to bottlenecks in processing and marketing when the treatments were first being done because of delays in moving the product through the novel marketing channels, but these problems were solved with experience. In time all cases resulted in acceptable product reaching consumers. Often the product was in better condition because irradiation resulted in a better quality product than alternative treatments. Other fruits, such as rambutan, could not be successfully marketed until PI became available.

BIBLIOGRAPHY

Adisakwattana, S., Yibchok-Anun, S., Charoenlertkul, P., and Wongsasiripat, N. (2011). Cyanidin-3-rutinoside alleviates postprandial hyperglycemia and its synergism with acarbose by inhibition of intestinal á-glucosidase. *J. Clin. Biochem. Nutr.* 49, 36–41. doi: 10.3164/jcbn.10-116

Aguayo E, Allende A, Artés F. 2003. Keeping quality and safety of minimally fresh processed melon. *Eur. Food Res. Technol.* 216, 494–499.

Alique, R., Martinez, M. A., and Alonso, J. (2003). Influence of the modified atmosphere packaging on shelf life and quality of "Navalinda" sweet cherry. *Eur. Food Res. Technol.* 217, 416–420. doi: 10.1007/s00217-003-0789-x

Alique, R., Zamorano, J. P., Martínez, M. A., and Alonso, J. (2005). Effect of heat and cold treatments on respiratory metabolism and shelf-life of sweet cherry, type picota cv. *Ambrunés Postharvest Biol. Technol.* 35, 153–165. doi: 10.1016/j.postharvbio.2004.07.003

Alkio, M., Jonas, U., Declercq, M., Van Nocker, S., and Knoche, M. (2014). Transcriptional dynamics of the developing sweet cherry (*Prunus avium* L.) fruit: sequencing, annotation and expression profiling of exocarp-associated genes. *Hort. Res.* 1: 11.

Alkio, M., Jonas, U., Sprink, T., Nocker, S., and Knoche, M. (2012). Identification of putative candidate genes involved in cuticle formation in *Prunus avium* (sweet cherry) fruit. *Ann. Bot.* 110, 101–112.

Almenar, E., Siddiq, M., and Merkel, C. (2012). "Packaging for processed food and the environment," in *Handbook of Food Process Design*, ed M.S. Rahman and J. Ahmed (Oxford: Blackwell), 1369–1405.

Andrews, P. K., and Shulin, L. (1995). Cell wall hydrolytic enzyme activity during development of nonclimacteric sweet cherry (*Prunus avium* L.) fruit. *J. Hortic. Sci.* 70, 561–567. doi: 10.1080/14620316.1995.11515327

Arakawa, O. (1993). Effect of ultraviolet light on anthocyanin synthesis in light colored sweet cherry, cv. Sato Nishiki. *J. Jpn. Soc. Hortic. Sci.* 62, 543–546. doi: 10.2503/jjshs.62.543

Arakawa, O., Pungpomin, P., An, S., and Tanaka, N. (2017). "Effect of blue light on red color development and anthocyanin accumulation of sweet cherry," in *8th International Cherry Symposium in Yamagata* (Yamagata), 136.

Artés F, Gómez P, Aguayo E, Escalona V, Artés-Hernández F. 2009. Sustainable sanitation techniques for keeping quality and safety of fresh-cut. *Postharvest Biol. Technol.* 51, 287–296.

Arzani, K. 2019. Asian pear. In: S. Tonetto de Freitas and S. Pareek (eds.) Postharvest Physiological Disorders in Fruits and Vegetables. CRC Press, Boca Raton, FL, pp. 329-349.

Ayadi M, Ghrab M, Gargouri K, Elloumi O, Zribi F, Ben Mimoun M, et al. Kernel characteristics of almond cultivars under rainfed conditions. Acta Horticulturae. 2006; 726: 377-381

Bae YM, Choi NY, Heu S, Kang DH, Lee SY. 2011. Inhibitory effects of organic acids combined with modified atmosphere packaging on foodborne pathogens on cabbage. *J. Korean Soc. Appl. Biol. Chem.* 54, 993–997.

Bai J., R. Prange, and P.M. Toivonen. 2009. Pome fruit. In: E.M. Yahia (ed.) Modified and Controlled Atmospheres for the Storage, Transportation, and Packaging of Horticultural Commodities. CRC Press, Boca Raton, FL, pp. 267-285.

Balbontín, C., Ayala, H., Bastías, R. M., Tapia, G., Ellena, M., Torres, C., et al. (2013). Cracking in sweet cherries: a comprehensive review from a physiological, molecular, and genomic perspective. *Chil. J. Agric. Res.* 73, 66–72.

Ballistreri, G., Continella, A., Gentile, A., Amenta, M., Fabroni, S., and Rapisarda, P. (2013). Fruit quality and bioactive compounds relevant to human health of sweet cherry (*Prunus avium* L.) cultivars grown in Italy. *Food Chem.* 140, 630–638.

Basak, A., Rozpara, E., and Grzyb, Z. (1998). Use of bioregulators to reduce sweet cherry tree growth and to improve fruit quality. *Acta Hortic.* 468, 719–723.

Baskaran SA, Upadhyay A, Kollanoor-Johny A, Upadhyaya I, Mooyottu S, Amalaradjou MAR, Schreiber D, Venkitanarayanan K. 2013. Efficacy of plant-derived antimicrobials as antimicrobial wash treatments for reducing enterohemorrhagic Escherichia coli O157: H7 on apples. *J. Food Sci.* 78, 1399.

Behboudian, M.H. and G.S. Lawes. 1994. Fruit quality in 'Nijisseiki' Asian pear under deficit irrigation: Physical attributes, sugar and mineral content, and development of flesh spot decay. N.Z. J. Crop Hort. Sci. 22: 393-400.

Belge, B., Llovera, M., Comabella, E., Gatius, F., Guillén, P., Graell, J., et al. (2014). Characterization of cuticle composition after cold storage of "Celeste" and "Somerset" sweet cherry fruit. *J. Agric. Food Chem.* 62, 8722–8729.

Bernalte, M. J., Hernandez, M. T., Vidal-Aragon, M. C., and Sabio, E. (1999). Physical, chemical, flavor and sensory characteristics of two sweet cherry varieties grown in Valle del Jerte (Spain). *J. Food Qual.* 22, 403–416.

Bernalte, M. J., Sabio, E., Hernández, M. T., and Gervasini, C. (2003). Influence of storage delay on quality of "Van" sweet cherry. *Postharvest Biol. Technol.* 28, 303–312.

Caleb OJ, Mahajan PV, Al-Said FA, Opara UL. 2013. Modified atmosphere packaging technology of fresh and fresh-cut produce and the microbial consequences: a review. *Food Bioprocess Technol.* 6, 303–329.

Campoy, J. A., Dantec, L. L., Barreneche, T., Dirlewanger, E., and Quero-García, J. (2015). New insights into fruit firmness and weight control in sweet cherry. *Plant Mol. Biol. Rep.* 33, 783–796.

Canli, F. A., and Orhan, H. (2009). Effects of preharvest gibberellic acid applications on fruit quality of "0900 Ziraat" sweet cherry. *HortTecnology* 19, 127–129.

Canli, F. A., Pektas, M., and Ercisli, S. (2015a). Benzyladenine and gibberellin applications improve fruit weight and delay maturity of sweet cherry. *Erwerbs Obstbau* 57, 71–75.

Canli, F. A., Sahin, M., Ercisli, S., Yilmaz, O., Temurtas, N., and Pektas, M. (2015b). Harvest and postharvest quality of sweet cherry are improved by pre-harvest benzyladenine and benzyladenine plus gibberellin applications. *J. Appl. Bot. Food* 88, 255–258.

Carrasco G, Urrestarazu M. 2010. Green chemistry in protected horticulture: the use of peroxyacetic acid as a sustainable strategy. *Int. J. Mol. Sci.* 11, 1999–2009. (10.3390/ijms11051999)

Carrasco, B., Meisel, L., Gebauer, M., Garcia-Gonzales, R., and Silva, H. (2013). Breeding in peach, cherry and plum: from a tissue culture, genetic, transcriptomic and genomic perspective. *Biol. Res.* 46, 219–230.

Caspari, H., M.H. Behboudian, D.J. Chalmers and B.E. Clothier. 1996. Fruit characteristics of 'Hosui' Asian pears after deficit irrigation. HortScience 31(1): 162.

Cengiz MF, Certel M. 2013. Effects of chlorine, hydrogen peroxide, and ozone on the reduction of mancozeb residues on tomatoes. *Turk. J. Agric. For.* 38, 1–6.

Chauvin, M. A., Whiting, M., and Ross, C. F. (2009). The influence of harvest time on sensory properties and consumer acceptance of sweet cherries. *HortTechnology* 19, 748–754.

Chen C, Khir R, Shen Y, Wu X, Zhang R, Cao X, et al. Energy consumption and product quality of off-ground harvested almonds under hot air column drying. LWT Food Science and Technology. 2021; 138: 110768

Chen C, Pan Z. Postharvest processing of tree nuts: Current status and future prospects – A comprehensive review. Critical reviews in food science and food safety. 2021. Accepted for publication

Chen C, Venkitasamy C, Zhang W, Deng L, Meng X, Pan Z. Effect of step-down temperature drying on energy consumption and product quality of walnuts. Journal of Food Engineering. 2020; 285: 110105

Chen C, Weipeng Z, Venkitasamy C, Khir R, McHugh T, Pan Z. Walnut structure and its influence on the hydration and drying characteristics. Drying Technology. 2020; 38(8): 975-986

Choi, C., Wiersma, P. A., Toivonen, P., and Kappel, F. (2002). Fruit growth, firmness and cell wall hydrolytic enzyme activity during development of sweet cherry fruit treated with gibberellic acid (GA_3). *J. Hortic. Sci. Biotechnol.* 77, 615–621.

Clayton, M., Biasi, W. V., Agar, I. T., Southwick, S. M., and Mitcham, E. J. (2003). Postharvest quality of "Bing" cherries following preharvest treatment with hydrogen cyanamide, calcium ammonium nitrate, or gibberellic acid. *HortScience* 38, 407–411.

Conte, A., Scrocco, C., Lecce, L., Mastromatteo, M., and Del Nobile, M. A. (2009). Ready-to-eat cherries: study on different packaging systems. *Innov. Food Sci. Emerg.* 10, 564–571.

Corbo, M. R., Bevilacqua, A., Petruzzi, L., Casanova, F. P., and Sinigaglia, M. (2014). Functional beverages: the emerging side of functional foods commercial trends, research, and health implications. *Compr. Rev. Food Sci. Food Saf.* 13, 1192–1206.

Crisosto, C. H., Crisosto, G. M., and Metheney, P. (2003). Consumer acceptance of "Brooks" and "Bing" cherries is mainly dependent on fruit SSC and visual skin color. *Postharvest Biol. Technol.* 28, 159–167.

Crisosto, C. H., Crisosto, G. M., and Ritenour, M. A. (2002). Testing the reliability of skin colour as an indicator of quality for early season Brooks (*Prunus avium* L.) cherry. *Postharvest Biol. Technol.* 24, 147–154.

Crisosto, C. H., Crisosto, G., and Neri, F. (2006). Understanding tree fruit quality based on consumer acceptance. *Acta Hortic.* 712, 183–190.

Crisosto, C.H. 2016. Asian pear. In: K.C. Gross, C.Y. Wang, and M. Saltveit (eds.) The Commercial Storage of Fruits, Vegetables, and Florist and Nursery Stocks. *USDA Agriculture Handbook* 66, pp. 205-209.

Crupi, P., Pichierri, A., Basile, T., and Antonacci, D. (2013). Postharvest stilbenes and flavonoids enrichment of table grape cv Redglobe (*Vitis vinifera* L.) as affected by interactive UV-C exposure and storage conditions. *Food Chem.* 141, 802–808.

Demir, T. (2013). Determination of carotenoid, organic acid and sugar content in some sweet cherry cultivars grown in Sakarya, Turkey. *J. Food Agric. Environ.* 11, 73–75.

Demirsoy, L., and Bilgener, S. (1998a). The effects of preharvest calcium hydroxide applications on cracking in "0900 Ziraat," "Lambert" and "Van" sweet cherries. *Acta Hortic.* 468, 657–662.

Demirsoy, L., and Bilgener, S. (1998b). The effect of preharvest chemical applications on cracking and fruit quality in 0900 Ziraat, Lambert and Van sweet cherries varieties. *Acta Hortic.* 468, 663–670.

Demirsoy, L., and Bilgener, S. (2000). The effect of chemical applications on cuticular and epidermal properties of some sweet cherry cultivars with respect to fruit cracking susceptibility. *Turk. J. Agric. For.* 24, 541–550.

Demirtas, C., Erturk, U., Yazgan, S., Candogan, B. N., and Soylu, A. (2008). Effects of different irrigation levels on the vegetative growth, flower bud formation and fruit quality of sweet cherry in western part of Turkey. *J. Food Agric. Environ.* 6, 168–172.

Dever, M. C., MacDonald, R. A., Cliff, M. A., and Lane, W. D. (1996). Sensory evaluation of sweet cherry cultivars. *Hortscience* 31, 150–153.

Dhall, R. K. (2013). Advances in edible coatings for fresh fruits and vegetables: a Review. *Crit. Rev. Food Sci. Nutr.* 53, 435–450. doi: 10.1080/10408398.2010.541568

Díaz-Mula, H. M., Castillo, S., Martinez-Romero, D., Valero, D., Zapata, P. J., Guillen, F., et al. (2009). Sensory, nutritive and functional properties of sweet cherry as affected by cultivar and ripening stage. *Food Sci. Technol. Int.* 15, 535–543.

Drake, S. R., and Fellman, J. K. (1987). Indicators of maturity and storage quality of *Rainier* sweet cherry. *Hortscience* 22, 283–285.

Eccher, T., and Granelli, G. (2006). Fruit quality and yield of different apple cultivars as affected by tree density. *Acta Hortic.* 712, 535–540.

ED European Parliament and Council Directive N 98/72/EC (1998). *On Food Additive Other than Colours and Sweeteners.*

ED European Parliament and Council Directive, N 95/2/EC (1995). *On Food Additive Other than Colours and Sweeteners.*

Elansari, A. M. (2009). Design of portable forced-air precooling cooling system. *J. Saudi Soc. Agric. Sci.* 8, 1–15.

Elfving, D. C., Auvil, T. D., Castillo, F., Drake, S. R., Kunzel, H., Kupferman, E. M., et al. (2009). Effects of preharvest applications of ethephon and 1-MCP to "Bing" sweet cherry on fruit removal force and fruit quality. *J. Am. Pomol. Soc.* 63, 84–100.

Engin, H., Gokbayrak, Z., and Sakaldas, M. (2016). Effects of 22S, 23S-homobrassinolide and gibberellic acid on occurrence of physiological disorders and fruit quality of 'Summit' and 'Regina' sweet cherries. *Erwerbs Obstbau* 58, 203–210.

Esti, M., Cinquanta, L., Sinesio, F., Moneta, E., and Di Matteo, M. (2002). Physiochemical and sensory fruit characteristics of two sweet cherry cultivars after cool storage. *Food Chem.* 76, 399–405.

Facteau, T. J., Chestnut, N. E., and Rowe, K. (1996). Tree, fruit size, and yield of *Bing* sweet cherry as influenced by rootstock, replant area, and training system. *Sci. Hortic.* 67, 13–26.

Facteau, T. J., Rowe, K. E., and Chesnut, N. E. (1985). Response patterns of gibberellic acid-treated sweet cherry fruit at different soluble solids levels and leaf/fruit ratios. *Sci. Hortic.* 27, 257–262.

Fallik E. 2004. Pre-storage hot water treatments (immersion, rinsing and brushing). *Postharvest Biol. Technol.* 32, 125–134.

FAO (2017). *Cherry Production.* United Nations: Statistical Database-Agriculture, Food and Agricultural Organisation (FAO).

Farkas J. 2014. Food technologies: food irradiation. *Encycl. Food Safety* 3, 178–186.

Ferretti, G., Bacchetti, T., Belleggia, A., and Neri, D. (2010). Cherry antioxidants: from farm to Ttble. *Molecules* 15, 6993–7005.

Ferrier P. 2010. Irradiation as a quarantine treatment. *Food Policy* 35, 548–555.

Flore, J. A., and Layne, D. R. (1999). Photoassimilate production and distribution in cherry. *Hortscience* 34, 1015–1019.

Food Drug Administration (FDA). (2016). *CFR Title 21: Foods and Drugs, CFR Part 172: Food Additives Permitted for Direct Addition to Food for Human Consumption CFR, Subpart, C. Coatings, Films and Related Substances*. Silver Spring, MD.

Franceschi, P. D., Stegmeir, T., Cabrera, A., Van Der Knaap, E., Rosyara, U. R., Sebolt, A. M., et al. (2013). Cell number regulator genes in *Prunus* provide candidate genes for the control of fruit size in sweet and sour cherry. *Mol. Breed.* 32, 311–326.

Gao, L., and Mazza, G. (1995). Characterization, quantification and distribution of anthocyanins and colorless phenolics in sweet cherry. *J. Agric. Food Chem.* 43, 343–346.

Gecgel U, Gumus T, Tasan M, Daglioglu O, Arici M. Determination of fatty acid composition of γ-irradiated hazelnuts, walnuts, almonds, and pistachios. Radiation Physics and Chemistry. 2011; 80(4): 578-581

Ghasemnezhad M, Zareh S, Rassa M, Sajedi RH. 2013. Effect of chitosan coating on maintenance of aril quality, microbial population and PPO activity of pomegranate (Punica granatum L. cv. Tarom) at cold storage temperature. *J. Sci. Food Agric.* 93, 368–374.

Giacalone, G., and Chiabrando, V. (2013). Modified atmosphere packaging of sweet cherries with biodegradable films. *Int. Food Res. J.* 20, 1263–1268.

Giménez, M. J., Valverde, J. M., Valero, D., Díaz-Mula, H. M., Zapata, P. J., Serrano, S., et al. (2015). Methyl salicylate treatments of sweet cherry trees improve fruit quality at harvest and during storage. *Sci. Hortic.* 197, 665–673.

Giménez, M. J., Valverde, J. M., Valero, D., Guillén, F., Martínez-Romero, D., Serrano, M., et al. (2014). Quality and antioxidant properties on sweet cherries as affected by preharvest salicylic and acetylsalicylic acids treatments. *Food Chem.* 160, 226–232.

Girard, B., and Koop, T. G. (1998). Physicochemical characteristics of selected sweet cherry cultivars. *J. Agric. Food Chem.* 46, 471–476.

Glowacz M, Mogren LM, Reade JPH, Cobb AH, Monaghan JM. 2013. Can hot water treatments enhance or maintain postharvest quality of spinach leaves?. *Postharvest Biol. Technol.* 81, 23–28.

Gol NB, Patel PR, Rao TVR. 2013. Improvement of quality and shelf life of strawberries with edible coatings enriched with chitosan. *Postharvest Biol. Technol.* 85, 185–195.

Gonçalves, B., Landbo, A. K., Knudsen, D., Silva, A. P., Moutinho-Pereira, J., and Rosa, E. (2004a). Effect of ripeness and postharvest storage on the phenolic profiles of cherries (*Prunus avium* L.). *J. Agric. Food Chem.* 52, 523–530.

Gonçalves, B., Landbo, A. K., Let, M., Silva, A. P., Rosa, E., and Meyer, A. S. (2004b). Storage affects the phenolic profiles and antioxidant activities of cherries (*Prunus avium* L.) on human low-density lipoproteins. *J. Sci. Food Agric.* 84, 1013–1020.

Gonçalves, B., Moutinho-Pereira, J., Santos, A., Silva, A. P., Bacelar, E., Correia, C., et al. (2006). Scion-rootstock interaction affects the physiology and fruit quality of sweet cherry. *Tree Physiol.* 26, 93–104.

Gonçalves, B., Silva, A. P., Moutinho-Pereira, J., Bacelar, E., Rosa, E., and Meyer, A. S. (2007). Effect of ripeness and postharvest storage on the evolution of colour and anthocyanins in cherries (*Prunus avium* L.). *Food Chem.* 103, 976–984.

González-Gómez, D., Lozano, M., Fernández-Leóna, M. F., Bernalte, M. J., Ayuso, M. C., and Rodríguez, A. B. (2009). Sweet cherry phytochemicals: identification and characterization by HPLC-DAD/ESI-MS in six sweet-cherry cultivars grown in Valle del Jerte (Spain). *J. Food Compost. Anal.* 23, 533–539.

Goulas, V., Minas, I. S., Kourdoulas, P. M., Lazaridou, A., Molassiotis, A. N., Gerothanassis, I. P., et al. (2015). ^{1}H NMR metabolic fingerprinting to probe temporal postharvest changes on qualitative attributes and phytochemical profile of sweet cherry fruit. *Front. Plant Sci.* 6: 959.

Griggs, W., and B. Iwakiri. 1977. Asian Pear Varieties in California. DANR Pub. No. 4068, University of California, Oakland, CA.

Guajardo, V., Solís, S., Sagredo, B., Gainza, F., Muñoz, C., Gasic, K., et al. (2015). Construction of high density sweet cherry (*Prunus avium* L.) linkage maps using microsatellite markers and SNPs detected by genotyping-by-sequencing (GBS). *PLoS ONE* 10: e0127750.

Guyer, D. E., Sinha, N. K., Tung-Sung, C., and Cash, J. N. (1993). Physicochemical and sensory characteristics of selected Michigan sweet cherry (*Prunus avium* L.) cultivars. *J. Food Qual.* 16, 355–370.

Hadi, M. A. M., Zhang, F., Wu, F., Zhou, C., and Tao, J. (2013). Advances in fruit aroma volatile research. *Molecules* 18, 8200–8229.

Hampson, C. R., Stanich, K., MacKenzie, D. L., Herbert, L., Lu, R., Li, J., et al. (2014). Determining the optimum firmness for sweet cherries using Just-About-Right sensory methodology. *Postharvest Biol. Technol.* 91, 104–111.

Hong P, Hao W, Luo J, Chen S, Hu M, Zhong G. 2014. Combination of hot water, Bacillus amyloliquefaciens HF-01 and sodium bicarbonate treatments to control postharvest decay of mandarin fruit. *Postharvest Biol. Technol.* 88, 96–102.

Horvitz, S., Godoy, C., López Camelo, A. F., Yommi, A., and Godoy, C. (2003). Application of gibberelic acid to "Sweetheart" sweet cherries: effects on fruit quality at harvest and during cold storage. *Acta Hortic.* 628, 311–316.

Intrado GlobeNewswire. Global Edible Nuts Industry [Internet]. 2020.

Ito, Y., Nishizawa-Yokoi, A., Endo, M., Mikami, M., and Toki, S. (2015). CRISPR/Cas9-mediated mutagenesis of the RIN locus that regulates tomato fruit ripening. *Biochem. Biophys. Res. Commun.* 467, 76–82.

Jacob, R. A., Spinozzi, G. M., Simon, V. A., Kelley, D. S., Prior, R. L., and Hess-Pierce, B. (2003). Consumption of cherries lowers plasma urate in healthy women. *J. Nutr.* 133, 1826–1829.

Jakobek, L., Marijan, S., Vóca, S., Šindrak, Z., and Dobrièeviæ, N. (2009). Flavonol and phenolic acid composition of sweet cherries (cv. *Lapins)* produced on six different vegetative rootstocks. *Sci. Hortic.* 123, 23–28.

Kader, A. (1997). "A summary of CA requirements and recommendations for fruits," in *Proceedings of 7th International Controlled Atmosphere Research Conference*, ed A. A. Kader (Davis, CA), 3–10.

Kader, A. (2001). Recent advances and future research needs in postharvest technology of fruits. *Bull. Int. Inst. Refrigeration. LXXXI*, 3–14.

Kader, A. (2002). *Postharvest Technology of Horticultural Crops, 3rd Edn.* University of California, Agriculture and Natural Resources, Publication. 3311, 535.

Kalt, W., Forney, C. F., Martin, A., and Prior, R. L. (1999). Antioxidant capacity, vitamin C, phenolics and anthocyanins after fresh storage of small fruits. *J. Agric. Food Chem.* 47, 4638–4644.

Kang, S. Y., Seeram, N. P., Nair, M. G., and Bourquin, L. D. (2003). Tart cherry anthocyanins inhibit tumor development in ApcMin mice and reduce proliferation of human colon cancer cells. *Cancer Lett.* 194, 13–19.

Kappel, F., Fisher-Fleming, B., and Hogue, E. (1996). Fruit characteristics and sensory attributes of an ideal sweet cherry. *Hortscience* 31, 443–446.

Kappel, F., Toivonen, P., McKenzie, D. L., and Stan, S. (2002). Storage characteristics of new sweet cherry cultivars. *Hortic. Sci.* 37, 139–143.

Kataoka, I., Beppu, K., Sugiyama, A., and Taira, T. (1996). Enhancement of coloration of "Satohnishiki" sweet cherry fruit by postharvest irradiation with ultraviolet radiation. *Environ. Control Biol.* 34, 313–319.

Kelebek, H., and Selli, S. (2011). Evaluation of chemical constituents and antioxidant activity of sweet cherry (*Prunus avium* L.) cultivars. *Int. J. Food Sci. Technol.* 46, 2530–2537.

Khir R, Pan Z, Atungulu GG, Thompson JF. Characterization of physical and aerodynamic properties of walnuts. Transactions of the ASABE. 2014; 57(1): 53-61

Kim H, Ryu J, Beuchat L. 2006. Survival of Enterobacter sakazakii on fresh produce as affected by temperature, and effectiveness of sanitizers for its elimination. *Int. J. Food Microbiol.* 111, 134–143.

Kim, D. O., Heo, H. J., Kim, Y. J., Yang, H. S., and Lee, C. Y. (2005). Sweet and sour cherry phenolics and their protective effects on neuronal cells. *J. Agric. Food Chem.* 53, 9921–9927.

Koc, A., Celik, Z., Akbulut, M., Bilgener, S., Ercisli, S., Gunes, M., et al. (2013). Morphological characterization of cherry rootstock candidates selected from central and east black sea regions in Turkey. *Sci. World J.* 2013: 916520.

Kondo, S., and Danjo, C. (2001). Cell wall polysaccharide metabolism during fruit development in sweet cherry "Satohnishiki" as affected by gibberellic acid. *J. Jpn. Soc. Hortic. Sci.* 70, 178–184.

Koutsimanis, G., Harte, J., and Almenar, E. (2014a). Freshness maintenance of cherries ready for consumption using convenient, microperforated, bio-based packaging. *J. Sci. Food Agric.* 95, 972–982.

Koutsimanis, G., Harte, J., and Almenar, E. (2014b). Development and evaluation of a new packaging system for fresh produce: a case study on fresh cherries under global supply chain conditions. *Food Bioprocess Technol.* 68, 55–669.

Kupferman, E., and Sanderson, P. (2001). *Temperature Management and Modified Atmosphere Packing to Preserve Sweet Cherry Quality*. Tree Fruit Research and Extension Center, Washington State University, 1–9.

Labuza TP, McNally L, Gallagher D, Hawkes J, Hurtado F. Stability of intermediate moisture foods. 1. Lipid oxidation. Journal of Food Science. 1972; 37(1): 154-159

Lallu, N. 1990. Fruit growth, handling and storage. In: A.G. White (ed.) Nashi, Asian Pear in New Zealand, DSIR Publishing, Wellington, New Zealand, pp. 53-74.

Landfeld A, Erban V, Kováªíková E, Houška M, Kýhos K, Pruchová J, Novotná P. 2010. Decontamination of cut carrot by persteril® agent based on the action of peroxyacetic acid. *Czech J. Food Sci.* 28, 564–571.

Lang, G. A. (2017). "The cherry industries in the USA: current trends and future perspectives," in *8th International Cherry Symposium in Yamagata* (Yamagata), 24.

Lang, G. A., Wilkinson, T., and Larson, J. (2017). "Insights for orchard design and management using intensive sweet cherry canopy architectures on dwarfing to semi-vigorous rootstocks," in *8th International Cherry Symposium in Yamagata* (Yamagata), 57.

Lara, I., Belge, B., and Goulao, L. F. (2014). The fruit cuticle as a modulator of postharvest quality. *Postharvest Biol. Technol.* 87, 103–112.

Lara, I., Camats, J. C., Comabella, E., and Ortiz, A. (2015). Eating quality and health-promoting properties of two sweet cherry (*Prunus avium* L.) cultivars stored in passive modified atmosphere. *Food Sci. Technol. Int.* 21, 133–144.

Leong, S. Y., and Oey, I. (2012). Effects of processing on anthocyanins, carotenoids and vitamin C in summer fruits and vegetables. *Food Chem.* 133, 1577–1587.

Li, S. (2014). Transcriptional control of flavonoid biosynthesis: fine-tuning of the MYB-bHLH-WD40(MBW) complex. *Plant Signal. Behav.* 9: e27522.

Lipinski B, Hanson C, Lomax J, Kitinoja L, Waite R, Searchinger T. 2013. *Reducing Food Loss and Waste.* Working Paper, Installment 2 of Creating a Sustainable Food Future. Washington, DC, USA.

Liu, J., Tian, S., Meng, X., and Xu, Y. (2007). Effects of chitosan on control of postharvest diseases and physiological responses of tomato fruit. *Postharvest Biol. Technol.* 44, 300–306.

Liu, Y., Liu, X., Zhong, F., Tian, R., Zhang, K., Zhang, X., et al. (2011). Comparative study of phenolic compounds and antioxidant activity in different species of cherries. *J. Food Sci.* 76, 633–638.

Looney, N. E., and Lidster, P. D. (1980). Some growth regulator effects on fruit quality, mesocarp composition, and susceptibility to postharvest surface marking of sweet cherries. *J. Am. Soc. Hortic. Sci.* 105, 130–134.

Lopez-Galvez F, Ragaert P, Palermo LA, Eriksson M, Devlieghere F. 2013. Effect of new sanitizing formulations on quality of fresh-cut iceberg lettuce. *Postharvest Biol. Technol.* 85, 102–108.

Luo, H., Dai, S., Ren, J., Zhang, C., Ding, Z., Li, Z., et al. (2014). The role of ABA in the maturation and postharvest life of a nonclimacteric sweet cherry fruit. *J. Plant Growth Regul.* 33, 373–383.

Lurie S. 1998. Postharvest heat treatments. *Postharvest Biol. Technol.* 14, 257–269.

Macheix, J. J., Fleuriet, A., and Billot, J. (1990). *Fruit Phenolics.* Boca Raton, FL: CRC Press. 378.

Mahto R, Das M. 2013. Effect of gamma irradiation on the physico-chemical and visual properties of mango (Mangifera indica L.), cv. Dushehri and Fazli stored at 20°C. *Postharvest Biol. Technol.* 86, 447–455.

Manganaris, G. A., Ilias, I. F., Vasilakasis, M., and Mignani, I. (2007). The effect of hydrocooling on ripening related quality attributes and cell wall physicochemical properties of sweet cherry fruit (*Prunus avium* L.). *Int. J. Refrig.* 30, 1386–1392.

Manjunatha G, Gupta KJ, Lokesh V, Mur LAJ, Neelwarne B. 2012. Nitric oxide counters ethylene effects on ripening fruits. *Plant Signal. Behav.* 7, 476–483.

Marsal, J., Lopez, G., del Campo, J., Mata, M., Arbones, A., and Girona, J. (2010). Postharvest regulated deficit irrigation in "Summit" sweet cherry: fruit yield and quality in the following season. *Irrig. Sci.* 28: 181.

Martínez-Esplá, A., Zapata, P. J., Valero, D., García-Viguera, C., Castillo, S., and Serrano, M. (2014). Preharvest application of oxalic acid increased fruit size, bioactive compounds, and antioxidant capacity in sweet cherry cultivars (*Prunus avium* L.). *J. Agric. Food Chem.* 62, 3432–3437.

Martínez-Romero, D., Alburquerque, N., Valverde, J. M., Guillén, F., Castillo, S., Valero, D., et al. (2006). Postharvest sweet cherry quality and safety maintenance by *aloe vera* treatment: a new edible coating. *Postharvest Biol. Technol.* 39, 93–100.

Matias, A. A., Rosado-Ramos, R., Nunes, L. S., Figueira, I., Serra, A. T., Bronze, M. R., et al. (2016). Protective effect of a (poly)phenol-rich extract derived from sweet cherries culls against oxidative cell damage. *Molecules* 21, 1–16.

Matteo, A. D., Russo, R., Graziani, G., Ritieni, A., and Vaio, C. D. (2016). Characterization of autochthonous sweet cherry cultivars (*Prunus avium* L.) of southern Italy for fruit quality, bioactive compounds and antioxidant activity. *J. Sci. Food Agric.* 97, 2782–2794.

Mattheis, J. P., Buchanan, D. A., and Fellman, J. K. (1992). Volatile compounds emitted by sweet cherries (*Prunus avium* cv. Bing) during fruit development and ripening. *J. Agric. Food Chem.* 40, 471–474.

Maxin P, Weber RWS, Pedersen H, Williams M. 2012. Control of a wide range of storage rots in naturally infected apples by hot-water dipping and rinsing. *Postharvest Biol. Technol.* 70, 25–31.

McCune, L. M., Kubota, C., Stendell-Hollis, N. R., and Thomson, C. A. (2010). Cherries and health: a review. *Crit. Rev. Food Sci. Nutr.* 51, 1–12.

Meheriuk, M., Girard, B., Moyls, L., Beveridge, H. J. T. D., McKenzie, L., and Harrison, J. (1995). Modified atmosphere packaging of "Lapins" sweet cherry. *Food Res. Int.* 3, 239–244.

Meheriuk, M., McKenzie, B., Girard, B., Molys, A. L., Weintraub, S., and Hocking, R. (1997). Storage of "Sweetheart" cherries in sealed plastic film. *J. Food Qual.* 20, 189–198.

Milinoviæ, B., Dragoviæ-Uzelac, V., Kazija, D. H., Jelaèiæ, T., Vujeviæ, P., Èièek, D., et al. (2016). Influence of four different dwarfing rootstocks on phenolic acids and anthocyanin composition of sweet cherry (Prunus avium L.) cvs. 'Kordia' and 'Regina'. *J. Appl. Bot. Food Qual.* 89, 29–37.

Misir, J., Brishti, F. H., and Hoque, M. M. (2014). *Aloe vera* gel as a novel edible coating for fresh fruits: a review. *Am. J. Food Sci. Technol.* 2, 93–97.

Mohebbi M, Ansarifar E, Hasanpour N, Amiryousefi MR. 2012. Suitability of Aloe vera and gum tragacanth as edible coatings for extending the shelf life of button mushroom. *Food Bioprocess Technol.* 5, 3193–3202.

Mozetiè, B., Simèiè, M., and Trebše, P. (2006). Anthocyanins and hydroxycinnamic acids of lambert compact cherries (*Prunus avium* L.) after cold storage and 1-methylcyclopropene treatment. *Food Chem.* 97, 302–309.

Mozetiè, B., Trebše, P., and Hribar, J. (2002). Determination and quantification of anthocyanins and hydroxycinnamic acids in different cultivars of sweet cherries (*Prunus avium* L.) from Nova Gorica Region (Slovenia). *Food Tech. Biotechnol.* 40, 207–212.

Mozetiè, B., Trebše, P., Simèiè, M., and Hribar, J. (2004). Changes of anthocyanins and hydroxycinnamic acids affecting the skin colour during maturation of sweet cherries (*Prunus avium* L.). *Swiss Soc. Food Sci. Technol.* 37, 123–128. doi: 10.1016/S0023-6438(03)00143-9

Neilsen, G. H., Neilsen, D., Kappel, F., and Forge, T. (2014). Interaction of irrigation and soil management on sweet cherry productivity and fruit quality at different crop loads that simulate those occurring by environmental extremes. *HortScience* 49, 215–220.

Nora, L., Dalmazo, G. O., Nora, F., and Rombaldi, C. (2012). "Controlled water stress to improve fruit and vegetable postharvest quality," in *Water Stress*, ed Prof. Ismail Md. Mofizur Rahman (InTech.).

Ogundiwin, E. A., Peace, C. P., Gradziel, T. M., Parfitt, D. E., Bliss, F. A., and Crisosto, C. H. (2009). A fruit quality gene map of *Prunus*. *BMC Genomics* 10: 587.

Olaimat AN, Holley RA. 2012. Factors influencing the microbial safety of fresh produce: a review. *Food Microbiol.* 32, 1–19.

Olías, J. M., Sanz, C., Ríos, J. J., and Pérez, A. G. (1995). "Substrates specificity of alcohol acyltransferase from strawberry and banana fruits," in *Fruit Flavors: Biogenesis, Characterization and Authentication* eds R. L. Rouseff and M. M. Leahy (Washington, DC: American Chemical Society), 134–141.

Olmstead, J., Iezzoni, A., and Whiting, M. D. (2007). Genotypic differences in sweet cherry fruit size result from differences in fruit cell number. *J. Am. Soc. Hortic. Sci.* 132, 697–703.

Olmstead, J., Sebolt, A., Cabrera, A., Sooriyapathirana, S., Hammar, S., Iriarte, G., et al. (2008). Construction of an intra-specific sweet cherry (*Prunus avium* L.) genetic linkage map and synteny analysis with the *Prunus* reference map. *Tree Genet. Genomes* 4, 897–910. doi: 10.1007/s11295-008-0161-1

Ozkan, Y., Ucar, M., Yildiz, K., and Ozturkc, B. (2016). Pre-harvest gibberellic acid (GA_3) treatments play an important role on bioactive compounds and fruit quality of sweet cherry cultivars. *Sci. Hortic.* 211, 358–362.

Özkaya, O., Sener, A., Sarida,°, M. A., Ünal, Ü., Valizadeh, A., and Dündar, Ö. (2015). Influence of fast cold chain and modified atmosphere packaging storage on postharvest quality of early season-harvest sweet cherries. *J. Food Process. Preserv.* 39, 2119–2128.

Padilla-Zakour, O. I., Tandon, K. S., and Wargo, J. M. (2004). Quality of modified atmosphere packaged "Hedelfingen" and "Lapins" sweet cherries (*Prunus avium* L.). *HortTechnol.* 14, 331–337.

Pandey N, Joshi SK, Singh CP, Rajput S, Kumar S, Khandal RK. 2013. Enhancing shelf life of litchi (Litchi chinensis) fruit through integrated approach of surface coating and gamma irradiation. *Radiat. Phys. Chem.* 85, 197–203.

Peace, C. P. (2017). DNA-informed breeding of rosaceous crops: promises, progress and prospects. *Hort. Res.* 4: 17006.

Peng, A., Chen, S., Lei, T., Xu, L., He, Y., Wu, L., et al. (2017). Engineering canker resistant plants through CRISPR/Cas9-targeted editing of the susceptibility gene *CsLOB1* promoter in citrus. *Plant Biotechnol. J.* 15, 1509–1519.

Pérez-Sánchez, R., Gómez-Sánchez, M. A., and Morales-Corts, M. R. (2010). Description and quality evaluation of sweet cherries cultured in Spain. *J. Food Qual.* 33, 490–506.

Peschel, S., Franke, R., Schreiber, L., and Knoche, M. (2007). Composition of the cuticle of developing sweet cherry fruit. *Phytochemistry* 68, 1017–1025.

Petriccione, M., De Sanctis, F., Pasquariello, M. S., Mastrobuoni, F., Rega, P., Scortichini, M., et al. (2014). The effect of chitosan coating on the quality and nutraceutical traits of sweet cherry during postharvest life. *Food Bioprocess. Technol.* 8, 394–408.

Pherobase (2007). Available from: http: //www.pherobase.com/database/floral-compounds/floral-taxa-compounds-fg-Carboxylic%20Esters.php

Predieri, S., Dris, R., and Rapparini, F. (2004). Influence of growing conditions on yield and quality of cherry: II. fruit quality. *J. Food Agric. Environ.* 2, 307–309.

Predieri, S., Dris, R., Sekse, L., and Rapparini, F. (2003). Influence of environmental factors and orchard management on yield and quality of sweet cherry. *J. Food Agric. Environ.* 1, 263–266.

Prvuloviæ, D., Popoviæ, M., Malenèiæ, D., Ljubojeviæ, M., Baraæ, G., and Ognjanov, V. (2012). *Phenolic Content and Antioxidant Capacity of Sweet and Sour Cherries. Studia UBB Chemia, LVII* 175–181.

Quirós-Sauceda, A. E., Ayala-Zavala, J. F., Olivas, G. I., and González-Aguilar, G. A. (2014). Edible coatings as encapsulating matrices for bioactive compounds: a review. *J. Food Sci. Technol.* 51, 1674–1685.

Remón, S., Ferrer, A., Marquina, P., Burgos, J., and Oria, R. (2000). Use of modified atmospheres to prolong the postharvest life of "Burlat" cherries at two different degrees of ripeness. *J. Sci. Food Agric.* 80, 1545–1552.

Remón, S., Venturini, M. E., Lopez-Buesa, P., and Oria, R. (2003). Burlat cherry quality after long range transport: optimisation of packaging conditions. *Innov. Food Sci. Emerg.* 4, 425–434.

Ren, C., Liu, X., Zhang, Z., Wang, Y., Duan, W., Li, S., et al. (2016). CRISPR/Cas9-mediated efficient targeted mutagenesis in Chardonnay (*Vitis vinifera* L.). *Sci. Rep.* 6: 32289.

Ren, J., Chen, P., Dai, S. J., Li, P., Li, Q., Ji, K., et al. (2011). Role of abscisic acid and ethylene in sweet cherry fruit maturation: molecular aspects. *N. Z. J. Crop Hortic. Sci.* 3, 161–174.

Rios de Souza, V., Pimenta Pereira, P. A., da Silva, T. L. T., de Oliveira Lima, L. C., Pio, R., and Queiroz, F. (2014). Determination of bioactive compounds, antioxidant activity and chemical composition of Brazilian blackberry, red raspberry, strawberry, blueberry and sweet cherry fruits. *Food Chem.* 156, 362–368.

Ripoll, J., Urban, L., Brunel, B., and Bertin, N. (2016). Water deficit effects on tomato quality depend on fruit developmental stage and genotype. *J. Plant Physiol.* 190, 26–35.

Rivera-Pastrana, D. M., Gardea, A. A., Yahia, E. M., Martìnez-Tèllez, M. A., and Gonzàles-Aguilar, G. A. (2013). Effect of UV-C irradiation and low temperature storage on bioactive compounds, antioxidant enzymes and radical scavenging of papaya fruit. *J. Food Sci. Technol.* 51, 3821–3829.

Rodgers S, Cash J, Siddiq M, Ryser E. 2004. A comparison of different chemical sanitizers for inactivating Escherichia coli O157: H7 and Listeria monocytogenes in solution and on apples, lettuce, strawberries, and cantaloupe. *J. Food Prot.* 67, 721–731.

Rojas-Argudo, C., Pérez-Gago, M. B., and del Río, M. A. (2005). Postharvest quality of coated cherries cv. Burlat as affected by coating composition and solids content. *Food Sci. Technol. Int.* 11, 417–424.

Romanazzi, G., Nigro, F., and Ippolito, A. (2009). Effectiveness of a short hyperbaric treatment to control postharvest decay of sweet cherries and table grapes. *Postharvest Biol. Technol.* 49, 440–442.

Romano, G. S., Cittadini, E. D., Pugh, B., and Schouten, R. (2006). Sweet cherry quality in the horticultural production chain. *Stewart Postharvest Rev.* 6, 1–9. doi: 10.2212/spr.2006.6.2

Rosyara, U. R., Bink, M. C. A. M., Weg, E., Zhang, G., Wang, D., Sebolt, A., et al. (2013). Fruit size QTL identification and the prediction of parental QTL genotypes and breeding values in multiple pedigreed populations of sweet cherry. *Mol. Breed.* 32, 875–887.

Salazar, J. A., Pacheco, I., Shinya, P., Zapata, P., Silva, C., Aradhya, M., et al. (2017). Genotyping by sequencing for SNP-based linkage analysis and identification of QTLs linked to fruit quality traits in Japanese Plum (*Prunus salicina* Lindl.). *Front. Plant Sci.* 8: 476.

Salazar, J. A., Ruiz, D., Campoy, J. A., Sánchez-Pérez, R., Crisosto, C. H., Martínez- García, P. J., et al. (2014). Quantitative trait loci (QTL) and mendelian trait loci (MTL) analysis in *Prunus*: a breeding perspective and beyond. *Plant. Mol. Biol. Rep.* 32, 1–18.

Schirra M, D'Hallewin G, Ben-Yehoshua S, Fallik E. 2000. Host–pathogen interactions modulated by heat treatment. *Postharvest Biol. Technol.* 21, 71–85.

Schmitt, E. R., Duhme, F., and Schmid, P. P. S. (1989). Water relations in sweet cherries (*Prunus avium* L.) on sour cherry rootstocks (*Prunus cerasus* L.) of different compatibility. *Sci. Hortic.* 39, 189–200.

Seeram, N. P., Momin, R. A., Nair, M. G., and Bourquin, L. D. (2001). Cyclooxygenase inhibitory and antioxidant cyanidin glycosides in cherries and berries. *Phytomedicine* 8, 362–369.

Serra, A. T., Duarte, R. O., Bonze, M. R., and Duarte, C. M. M. (2011). Identification of bioactive response in traditional cherries from Portugal. *Food Chem.* 125, 318–325.

Serradilla, M. J., Martín, A., Ruiz-Moyano, S., Hernández, A. M., López-Corrales, M., and Córdoba, M. G. (2012). Physicochemical and sensorial characterisation of four sweet cherry cultivars grown in Jerte Valley (Spain). *Food Chem.* 133, 1551–1559.

Serrano, M., Díaz-Mula, H. M., Zapata, P. J., Castillo, S., Guillén, F., and Martínez-Romero, D. (2009). Maturity stage at harvest determines the fruit quality and antioxidant potential after storage of sweet cherry cultivars. *J. Agric. Food Chem.* 57, 3240–3246.

Serrano, M., Guillen, F., Martínez-Romero, D., Castillo, S., and Valero, D. (2005). Chemical constituents and antioxidant activity of sweet cherry at different ripening stages. *J. Agric. Food Chem.* 53, 2741–2745.

Shackel, K. A., Ahmadi, H., Biasi, W., Buchner, R., Goldhamer, D., Gurusinghe, S., et al. (1997). Plant water status as an index of irrigation need in deciduous fruit trees. *HortTechnology* 7, 23–29.

Shirasawa, K., Isuzugawa, K., Ikenaga, M., Saito, Y., Yamamoto, T., Hirakawa, H., et al. (2017). The genome sequence of sweet cherry (*Prunus avium*) for use in genomics-assisted breeding. *DNA Res.* 24, 499–508.

Singh Z, Khan AS, Zhu S, Payne AD. 2013. Nitric oxide in the regulation of fruit ripening: challenges and thrusts. *Stewart Postharvest Rev.* 4, 3.

Sooriyapathirana, S. S., Khan, A., Sebolt, A. M., Wang, D., Bushakra, J. M., Allan, A. C., et al. (2010). QTL analysis and candidate gene mapping for skin and flesh color in sweet cherry fruit (*Prunus avium* L.). *Tree Genet. Genomes* 6, 821–832.

Spotts, R. A., Cervantes, L. A., and Facteau, T. J. (2002). Integrated control of brown rot of sweet cherry fruit with a pre-harvest fungicide, a postharvest yeast, modified atmosphere packaging, and cold storage temperature. *Postharvest Biol. Technol.* 24, 251–257.

Stern, R. A., Flaishman, M., Applebaum, S., and Ben-Arie, R. (2007). Effect of synthetic auxins on fruit development of "Bing" cherry (*Prunus avium* L.). *Sci. Hortic.* 114, 275–280.

Subedi, P., Walsh, K., and Purdy, P. (2013). Determination of optimum maturity stages of mangoes using fruit spectral signatures. *Acta Hortic.* 992, 521–527. doi: 10.17660/ActaHortic.2013.992.64

Sun-Waterhouse, D. (2011). The development of fruit-based functional foods targeting the health and wellness market: a review. *Int. J. Food Sci. Technol.* 46, 899–920.

Szot, I., and Meland, M. (2001). Influence of rootstocks on size distribution and fruit quality of sweet cherry cultivars. *Int. Agrophys.* 15, 207–214.

Tahir II, Johansson E, Olsson ME. 2009. Improvement of apple quality and storability by a combination of heat treatment and controlled atmosphere storage. *HortScience* 44, 1648–1654.

Tharanathan, R. N. (2003). Biodegradable films and composite coatings: past, present and future. *Trends Food Sci. Technol.* 14, 71–78.

Tian, S. P., Fan, Q., Xu, Y., Wang, Y. S., and Jiang, A. (2001). Evaluation of the use of high carbon dioxide concentrations and cold storage to control *Monilinia fructicola* on sweet cherries. *Postharvest Biol. Technol.* 22, 53–60.

Tian, S. P., Jiang, A. L., Xu, Y., and Wang, Y. S. (2004). Responses of physiology and quality of sweet cherry fruit to different atmosphere in storage. *Food Chem.* 87, 43–49. doi: 10.1016/j.foodchem.2003.10.014

Tomás-Barberán, F. A., and Espín, J. C. (2001). Phenolic compounds and related enzymes as determinants of quality in fruits and vegetables. *J. Sci. Food Agric.* 81, 853–876.

USDA, National Agricultural Statistics Service. 2020 California Walnut Objective Measurement Report [Internet]. 2020. Available from: https: //www.nass.usda.gov/Statistics_by_State/California/Publications/ Specialty_and_Other_Releases/Walnut/Objective-Measurement/202008walom. Pdf [Accessed: 01 December 2021]

USDA-U.S. Department of Agriculture (2017). *Nutrient values and weights – Fruits 09070 Sweet Cherry. National Nutrient Database for Standard Reference.* The National Agricultural Library.

Usenik, V., Fabèiè, J., and Štampar, F. (2008). Sugars, organic acids, phenolic composition and antioxidant activity of sweet cherry (*Prunus avium* L.). *Food Chem.* 107, 185–192. doi: 10.1016/j.foodchem.2007.08.004

Usenik, V., Fajt, N., Mikulic-Petkovsek, M., Slatnar, A., Stampar, F., and Veberic, R. (2010). Sweet cherry pomological and biochemical characteristics influenced by rootstock. *J. Agric. Food Chem.* 58, 4928–4933.

Usenik, V., Kastelec, D., and Stampar, F. (2005). Physicochemical changes of sweet cherry fruits related to application of gibberellic acid. *Food Chem.* 90, 663–671.

Valero, D., Díaz-Mula, H. M., Zapata, P. J., Castillo, S., Guillén, F., Martínez-Romero, D., et al. (2011). Postharvest treatments with salicylic acid, acetylsalicylic acid or oxalic acid delayed ripening and enhanced bioactive compounds and antioxidant capacity in sweet cherry. *J. Agric. Food Chem.* 59, 5483–5489.

Valverde, J. M., Giménez, M. J., Valero, D., Martínez-Romero, D., and Serrano, M. (2015). Methyl salicylate treatments of sweet cherry trees increase antioxidant systems in fruit at harvest and during storage. *Postharvest Biol. Technol.* 109, 106–113.

Vamos-Vigyazó, L. (1981). Polyphenol oxidase and peroxidase in fruits and vegetables. *Crit. Rev. Food Sci. Nutr.* 49, 127–140.

Vangdal, E., Hovland, K. L., Børve, J., Sekse, L., and Slimestad, R. (2008). Foliar application of calcium reduces postharvest decay in sweet cherry by various mechanisms. *Acta Hortic.* 768, 143–148.

Vavoura, M. V., Badeka, A. V., Kontakos, S., and Kontominas, M. G. (2015). Characterization of four popular sweet cherry cultivars grown in Greece by volatile compound and physicochemical data analysis and sensory evaluation. *Molecules* 20, 1922–1940.

Velickova, E., Winkelhausen, E., Kuzmanova, S., Alves, V. D., and Moldao-Martins, M. (2013). Impact of chitosan-beeswax edible coatings on the quality of fresh strawberries (Fragaria ananassa cv. *Camarosa*) under commercial storage conditions. *LWT Food Sci. Technol.* 52, 80–92.

Virtanen JK, Mursu J, Voutilainen S, Uusitupa M, Tuomainen TP. Serum omega-3 polyunsaturated fatty acids and risk of incident type 2 diabetes in men: The Kuopio ischemic heart disease risk factor study. Diabetes Care. 2014; 37(1): 189-196

Vroman, I., and Tighzert, L. (2009). Biodegradable polymers. *Materials* 2, 307–344.

Wang, C., Chen, C., and Wang, S. (2009). Changes of flavonoid content and antioxidant capacity in blueberries after illumination with UV-C. *Food Chem.* 117, 426–431.

Wang, L., Zhang, H., Jina, P., Guo, X., Li, Y., Fan, C., et al. (2016). Enhancement of storage quality and antioxidant capacity of harvested sweet cherry fruit by immersion with b-aminobutyric acid. *Postharvest Biol. Technol.* 118, 71–78. doi: 10.1016/j.postharvbio.2016.03.023

Wang, S. Y. (2006). Effect of pre-harvest conditions on antioxidant capacity in fruits. *Acta Hortic.* 712, 229–305.

Wang, S., Chen, Y., Xu, Y., Wu, J., Xiao, G., and Fu, M. (2014). Super atmospheric O_2 packaging maintains postharvest quality of cherry (*Prunus avium* L.) fruit. *J. Food Process Preserv.* 38, 2037–2046.

Wang, Y., Xie, X., and Long, L. E. (2014). The effect of postharvest calcium application in hydro-cooling water on tissue calcium content, biochemical changes, and quality attributes of sweet cherry fruit. *Food Chem.* 160, 22–30.

Wani, A. A., Singh, P., Gul, K., Wani, M. H., and Langowski, H. C. (2014). Sweet cherry (*Prunus avium*): critical factors affecting the composition and shelf life. *Food Pack. Shelf Life* 1, 86–99.

Warriner K, Huber A, Namvar A, Fan W, Dunfield K. 2009. Recent advances in the microbial safety of fresh fruits and vegetables. *Adv. Food Nutr. Res.* 57, 155–208.

Wen, Y. Q., He, F., Zhu, B. Q., Lan, Y. B., Pan, Q. H., Li, C. Y., et al. (2014). Free and glycosidically bound aroma compounds in cherry (*Prunus avium* L.). *Food Chem.* 152, 29–36.

Wendehenne D, Durner J, Klessig DF. 2004. Nitric oxide: a new player in plant signalling and defence responses. *Curr. Opin. Plant Biol.* 7, 449–455.

Whiting, M. D., and Lang, G. A. (2004). Bing sweet cherry on the dwarfing rootstock Gisela 5: thinning affects fruit quality and vegetative growth but not CO_2 exchange. *J. Am. Soc. Hortic. Sci.* 129, 407–415.

Whiting, M. D., Lang, G., and Ophardt, D. (2005). Rootstock and training system affect sweet cherry growth, yield and fruit quality. *HortScience* 40, 582–586.

Whiting, M., and Das, P. (2017). "A precision pollination system for sweet cherry yield security," in *8th International Cherry Symposium in Yamagata* (Yamagata), 61.3

WorldAtlas. The most popular nuts in the world [Internet]. 2021. Available from: https: //www.worldatlas.com/articles/the-most-popular-nuts-in-the-world.html [Accessed: 01 December 2021]

Xia, M., Hou, M., Zhu, H., Ma, J., Tang, Z., Wang, Q., et al. (2005). Anthocyanins induce cholesterol efflux from mouse peritoneal macrophages. The role of peroxisome proliferators-activated receptor Ó-Liver X receptor á-ABCA1 pathway. *J. Biol. Chem.* 280, 36792–36801.

Yao, H., and Tian, S. (2005). Effects of pre- and post-harvest application of salicylic acid or methyl jasmonate on inducing disease resistance of sweet cherry fruit in storage. *Postharvest Biol. Technol.* 35, 253–262.

Yin, X., Seavert, C. F., and Roux, J. (2011). Responses of irrigation water use and productivity of sweet cherry to single-lateral drip Irrigation and ground covers. *Soil Sci.* 176, 39–47.

Zaharah SS, Singh Z. 2011. Mode of action of nitric oxide in inhibiting ethylene biosynthesis and fruit softening during ripening and cool storage of 'Kensington Pride' mango. *Postharvest Biol. Technol.* 62, 258–266.

Zhang WP, Chen C, Pan Z, Xiao HW, Xie L, Gao ZJ, et al. Design and performance evaluation of a pilot-scale pulsed vacuum infrared drying (PVID) system for drying of berries. Drying Technology. 2020; 38(10): 1340-1355

Zhang, C., and Whiting, M. D. (2011). Improving "Bing" sweet cherry fruit quality with plant growth regulators. *Sci. Hortic.* 127, 341–346.

Zhang, G., Sebolt, A. M., Sooriyapathirana, S. S., Wang, D., Bink, M. C. A. M., Olmstead, J. W., et al. (2010). Fruit size QTL analysis of an F1 population derived from a cross between a domesticated sweet cherry cultivar and a wild forest sweet cherry. *Tree Genet. Genomes* 6, 25–36.

Zhang, X., Jiang, Y. M., Peng, F. T., He, N. B., Li, Y. J., and Zhao, D. C. (2007). Changes of aroma components in "Hongdeng" sweet cherry during fruit development. *Agric. Sci. China* 6, 1376–1382.

Zoratti, L., Karppinen, K., Escobar, A. L., Häggman, H., and Jaakola, L. (2014). Light-controlled flavonoid biosynthesis in fruits. *Front. Plant Sci.* 5: 534.

Zude-Sasse, M., Fountas, S., Gemtos, T. A., and Abu-Khalaf, N. (2016). Applications of precision agriculture in horticultural crops. *Eur. J. Hortic. Sci.* 81, 78–90.

INDEX

A

Abiotic 57-58, 77-78, 80-81
Absorption 228
A-carotene 11
Acrylamide 178
Aflatoxins 170
Alcohols 10
Alkali 159
Almonds 174, 178
Aluminum 122
Antibiotic 86
Antisense 124
Appearance 32, 180
Apricot 238
Aromatic 10
Atmosphere 41, 52-53, 114, 198, 227

B

Bacteria 28, 34, 36, 40, 120, 172
Banana 75, 95, 184
Bataticola 59
B-carotene 11
Bioassays 69
Biotic 57-58
Blanching 131, 176, 185
Breeding 17
Brewing 168
Browning 15, 32, 198, 201-202
Bunchy 76
Burning 208

C

Caffeine 168
Calcium 3-5, 20, 23-24, 197, 273
Canopy 12, 67
Cantaloupe 32-33
Carbon 23, 114, 251
Cardamom 132
Carnation 85, 123-124
Carrots 32, 230
Cashew 169
Cassava 230

Cherry 7-17, 157, 160, 163, 238, 257-274
Chitinase 72-73
Chloride 23
Chlorine 47, 220, 222
Cinerea 36, 72
Cinnamon 137
Climate 58

Climber 148

Climbing 148

Composting 117, 119

Cultivars 2, 8-11, 20-21, 93, 125, 196, 209, 213, 261, 271

Cyanidin- 11

Cystatin 70

Cysteine 70

D

Degreening 229

Dehusking 149

Drought 59-67, 78-79

Drying 91-92, 100, 102, 131-136, 144-145, 149, 155, 160, 163-165, 171, 173-176, 178, 230, 274

E

Endogenous 117

Energy 174-175

Ethanol 28

Ethylene 8, 22, 40, 47-51, 83, 114-115, 117, 123-127, 211, 221, 223-225, 229, 240

F

Farmers 8-9

Flaccid 204

Flavour 103-104, 167-168, 181

Flower 84, 111-113, 116-117, 119, 122-125

Foliage 230

Foliar 207

Freezing 186

Fumigant 252

Fumigation 47, 222, 251

G

Garlic 109, 230, 232

Genetic 71, 78, 84, 90

Genome 88-89

Germicides 121

Ginger 132-133

Glandular 63

Glucanase 73

Glycine 79, 81

Grading 6-7, 91, 95, 131, 216

Grains 92

Greening 233

Grinding 168

H

Harvest 1-7, 9, 19, 24, 50, 92-93, 103, 105, 107-108, 129, 133, 138, 142-143, 155, 172-173, 179, 181, 184, 189, 192, 194, 202, 206-207, 230, 240, 243

Harvester 6, 94, 133

Hazards 172

Health 7, 10-12

Herbal 141

Herbicide 82

Humidity 5, 22-23, 41, 53, 67, 144, 166, 188, 228, 244

Hygiene 173

I

Inhibitors 223

Injuries 24, 41

Inking 206-207

Insects 92, 143, 253

Irradiated 255

Irrigation 2, 13, 21, 190

K

Kiwifruit 239

L

Livestock 143

M

Mangoes 247, 253-254

Manure 172

Marker 87

Market 6, 170, 213, 231-232

Maturity 5, 20, 91, 94, 103-106, 181-184, 236, 240

Microbes 35

Microwave 136

Mortality 69

Mosaic 89

Mucilage 159, 162

Mutants 65

Mutations 89

N

Nematode 70-71

Nitrogen 3-4, 189, 205

O

Olives 206

Onions 232-233

Ornamental 84, 123

P

Papaya 76

Pathogen 59-67, 73, 75, 195

Pathogens 30, 61, 64, 74, 119-120, 194-195

Pectin 36

Peeled 31

Pepper 74, 131-132

Peroxidase 81

Persimmons 211

Pesticides 180

Petals 116, 126

Petunia 82, 85, 125

Phaseolina 64

Polyamines 74, 80

Potassium 3, 19

Potato 72, 78-79, 81, 230-232

Preharvest 13, 19, 112, 260

Proteins 74, 77, 80

Protocol 102

Pruning 20

Q

Quarantine 249, 251

R

Radiation 255

Rhizomes 133

Ripening 3, 8, 10, 22, 40, 46, 49-50, 83, 95, 103, 181, 205, 221, 224-225, 240-241

S

Sanitation 26

Seedless 204

Shelf-life 16, 83

Software 227

Solids 20

Sorbitol 201

Strawberry 72, 76, 81

Sucrose 117

Sunburn 200-201

Superoxide 79

T

Tannins 11

Texture 241

Threshing 92

Tolaasii 37

Toxins 170

Traits 20, 60, 68

Transgenic 67-87, 89, 124-126
Trichomes 62-63
Trypsin 70
Turmeric 133

V

Variation 8
Viruses 76
Volatile 167
Vulgaris 64

W

Walnuts 174-175, 177
Waterberry 205
Watermelon 89
Waxing 109
Weevil 253
Wounding 115

Y

Yeasts 27

www.ingramcontent.com/pod-product-compliance
Lightning Source LLC
LaVergne TN
LVHW062118200926
848484LV00007B/1711